Temperate and Subtropical Fruit Production, Third Edition

Temperate and Subtropical Fruit Production, Third Edition

Edited by

D.I. Jackson

Professorial Fellow
Lincoln University
New Zealand

and

N.E. Looney

Pacific Agri-Food Research Centre
Agriculture and Agri-Food Canada

and

M. Morley-Bunker

Lincoln University
Canterbury
New Zealand

and

Graham F. Thiele

Associate Professor (Retired)
Lincoln University
New Zealand

www.cabi.org

CABI is a trading name of CAB International

CABI Head Office
Nosworthy Way
Wallingford
Oxfordshire OX10 8DE
UK

CABI North American Office
875 Massachusetts Avenue
7th Floor
Cambridge, MA 02139
USA

Tel: +44 (0)1491 832111
Fax: +44 (0)1491 833508
E-mail: cabi@cabi.org
Website: www.cabi.org

Tel: +1 617 395 4056
Fax: +1 617 354 6875
E-mail: cabi-nao@cabi.org

A catalogue record for this book is available from the British Library, London, UK.

Library of Congress Cataloging-in-Publication Data

Temperate and subtropical fruit production / edited by
   D.I. Jackson and N.E. Looney and M. Morley-Bunker. -- 3rd ed.
     p. cm.
   Includes bibliographical references and index.
   ISBN 978-1-84593-501-6 (alk. paper)
1. Fruit-culture. I. Jackson, David, 1935- II. Looney, N. E. (Norman Earl), 1938- III. Morley-Bunker, M. (Michael) IV. Title.

SB355.T33 2011
634–dc22

2010008552

ISBN-13: 978 1 84593 501 6

Commissioning editor: Meredith Carroll
Production editor: Fiona Chippendale

Typeset by SPi, Pondicherry.
Printed and bound in the UK by CPI Group (UK) Ltd, Croydon, CR0 4YY.

Contents

Contributors

William Atkinson

William Atkinson graduated from Massey and Lincoln University as an agricultural engineer specializing in crop mechanization. He has held lecturing posts in New Zealand, Papua New Guinea, Queensland and Samoa. He lives on an avocado and tamarillo orchard that he and his wife developed from pasture. His research interests include tropical root-crop mechanization and the performance of machines in the field under commercial conditions.

Bruce Chapman

Bruce Chapman is a consultant entomologist specializing in integrated pest management of horticultural crops. He was formerly Associate Professor of Entomology at Lincoln University, New Zealand, where he lectured in entomology and pest management and conducted research on pesticide management and biological control. He has published extensively on insect pest management in scientific journals and has written three books on insect pest management.

David Jackson

A graduate of Lincoln College in the late 1950s, he joined the Department of Scientific and Industrial Research, where he investigated aspects of tree growth, flowering, fruiting and storage of apples, pears, peaches, apricots and nectarines. Between 1961 and 1964 he studied for a PhD at the Waite Institute in Adelaide. In 1968 he returned to Lincoln, where he extended his research to include grape production and physiology. He is co-author of *The Production of Grapes and Wine in Cool Climate Viticulture*, *Pruning and Training* and *Edible Tree Nuts in New Zealand*.

Norman Looney

Looney was Head of the Pomology and Viticulture Section at Summerland for more than 15 years. His service to national and international horticultural science includes being Chair of the Fruit Section of the International Society for Horticultural Science (1994–2000), President of the Canadian Society for Horticultural Science (1997–1999; CSHS Life Membership awarded in 2002), President of the 24th International Horticultural Congress (2002) and President of the International Society for Horticultural Science (2002–2010; awarded Fellow of ISHS in 2006). Dr Looney is also a Fellow of the American Society for Horticultural Science (1985). He owns and operates a peach orchard overlooking Okanagan Lake in interior British Columbia.

Peter Lyford

Peter Lyford graduated in Horticulture at Massey University and joined the Ministry of Agriculture as an adviser in subtropical fruit production. He is now a private horticulture consultant in Tauranga, New Zealand, specializing in kiwifruit and other subtropicals.

David McNeil

David McNeil started his career in agricultural science in 1971 with the NSW Department of Agriculture, followed by a PhD at the University of Western Australia. Presently he occupies the Chair of Agricultural Sciences & Director of the Tasmanian Institute of Agricultural Science at the University of Tasmania. He has had a strong involvement in research, development and extension of new crops in general and nut crops in particular, producing over 100 research and extension outputs on nut crops. His career has included 16 years at Lincoln University in New Zealand developing nut production systems, including a personal commercial nut orchard, and ending up as Professor of Horticulture. He has had sabbatical periods in the USA, Europe and Australia, working in the area of nut crops, and was recently the scientific convenor of the 6th International Walnut Symposium in Melbourne.

Michael Morley-Bunker

Mike Morley-Bunker graduated in horticulture from Wye College, University of London. Between 1971 and 1978 he lectured in crop and horticultural sciences at the University of Swaziland and completed an MSc in Applied Plant Sciences with London University. His experience in southern Africa broadened his range beyond temperate fruit production. He emigrated to New Zealand and joined the staff of Lincoln University in 1978. He currently lectures in horticulture and viticulture and supervises postgraduate students. In his research he has worked with subtropical and temperate fruits, including kiwifruit, pepino, persimmon, feijoa, raspberry, strawberry, apples and pears. He has worked on sabbatical leave in Australia and the USA. Recently he visited and lectured in universities in China.

John Palmer

John Palmer graduated from the University of Nottingham School of Agriculture at Sutton Bonington in 1966. After over 20 years at East Malling Research Station, during which time he was awarded a PhD from the University of Nottingham in 1976, he emigrated to New Zealand in 1991 and continued his interests in the crop physiology of apples and pears. He initially joined DSIR, which soon became HortResearch and has recently become Plant and Food Research. His research has included nursery tree quality, rootstocks, planting systems, within-tree factors influencing fruit quality, and leaf and whole canopy gas exchange. He has developed systems for measuring light interception and whole tree canopy gas exchange. Much of his work has been on intensive systems of production of apples and pears, and it is gratifying to him to see the current up-take of intensive systems on dwarfing rootstocks taking place in the New Zealand pipfruit industry, as there was little interest in such systems when he arrived in 1991. He is widely known throughout the international pipfruit research community, and in 2000 was awarded the Distinguished Research Award from the IDFTA.

David Penman

Professor Penman trained as an entomologist with interests in pest management, especially in orchards, and held the position of Professor of Entomology at Lincoln University to 1995. He then moved to research management positions at Landcare Research, focusing on biodiversity. He has played a major role in establishing and chairing the Global Biodiversity Information Facility, and after a period at the University of Canterbury he is now providing independent leadership to a number of science projects.

Graham Thiele

Now retired, Graham Thiele was Associate Professor in the Department of Horticulture at Lincoln University. He is a graduate of Massey and Lincoln Universities and has a wide practical knowledge of fruit production. He joined the Ministry of Agriculture and Fisheries in 1952 as a cadet and later was an adviser in Motueka and Christchurch. He joined Lincoln University in 1962, and in 1967 was given the responsibility for developing new courses in horticulture economics and management. He has lectured and consulted internationally, where his expertise in the systems approach to education, extension and research has created considerable interest.

Preface

This book provides information on the wide variety of fruits that are found throughout the temperate and subtropical regions of the world and their cultivation. It encompasses both large-scale commercial operations and the needs of the enthusiastic amateur gardener. It is in two parts: the first examines the worldwide distribution of fruit, the effect of climate and the principles and general practices of fruit growing. In the second part, details covering all aspects of specific types of fruit are given.

Written by leading experts with many years experience of researching and teaching horticulture, soil science, plant physiology, plant pathology and agricultural engineering, this book is essential reading for commercial fruit growers, students and lecturers in horticulture, and enthusiastic amateurs. It should serve as a basic undergraduate text in pomology.

Fruit production is a highly complicated, continually changing endeavour and no single textbook can provide all the information a reader might require on a given fruit production topic. Neither can it deal with specific local information on, for example, pests and diseases.

This third edition contains specific references for further reading in many chapters. Where additional information is required, the publications of the International Society of Horticultural Science are recommended as a useful starting point. These document the proceedings of international symposia held on a range of horticultural topics and crops.

Our vision is that serious students of pomology will find this book a very good introduction and that producers and practitioners will long value it as an accurate and user-friendly reference.

Acknowledgements

Many people have, in one way or another, helped in the production of this book. Most have been acknowledged in the first and second editions and we hope they will forgive us if we do not mention them again. Sincere thanks to Rachael O'Sullivan, for her extensive and dedicated editing and computer work, and to Christopher Durney, for his skilled artwork in upgrading most of the illustrations in this third edition. The New Zealand Fruitgrowers Federation made a generous financial grant towards the upgrading of the illustrations.

PART I

General Points on Fruit Husbandry

1 The Distribution of Fruits

DAVID JACKSON, NORMAN LOONEY
AND MICHAEL MORLEY-BUNKER

The selection of fruits for any area depends on many factors, although, in the final analysis, the most important will undoubtedly be the climate. All climatic factors will be introduced and discussed in this chapter. The significance of climate will be re-emphasized in a separate chapter (Chapter 2) and in those sections describing specific fruits.

Present-day distribution of fruit production on a world scale is shown in Tables 1.1 and 1.2. Some tropical fruits, e.g. bananas, mangoes and papaya, are included for reference purposes. The increase in world production in fruit crops over the last 30 years is shown in Table 1.3. Many crops show an increase of between 100 and 170% in the last 30 years, although there are some anomalies; for example, production of plums has not doubled (i.e. increased by 100%) in the last 30 years, whereas tangerines, mandarins and clementines have risen by over 200%.

The usefulness of a fruit helps to determine its total production. Thus oranges, which are eaten fresh and made into juice, are much more widely grown than lemons, which are just as easy to produce and grow in similar areas. The difference is that lemons are not eaten fresh and have a more limited use in cooking and drink manufacture. Grapes have four uses – fresh, for juice, wine and as dried fruit – each of which is part of a large or considerable industry. Consequently, they constitute a major world crop. Bananas and apples are popular fruit that are very convenient for fresh eating. They also have important uses as processed products.

A more obvious factor influencing production and distribution is climate. Virtually all crops have specific climate requirements and production is highest where the climate is the most suitable.

The plant kingdom contains somewhere between 250,000 and 400,000 plant species. It contains a wide range of plant forms, but not all plant species are termed flowering plants – those plants that are classed as flowering plants are grouped into plant families and, again depending on the authority, there are somewhere between 150 and 400 flowering plant families. The number of plant families that contain commercially important fruit crop plants is comparatively very small. The number of plant species with known fruit crops may not be more than 500. The small number of fruit crops under cultivation relative to the number of plant species in the world may be a reflection of our lack of detailed knowledge of the world's flora. Temperate fruit crops have tended to be predominant in lists of recognized fruit crops. The awareness of what temperate fruits exist may be more of a reflection of where commercial orcharding has developed. The identification of plant species in the tropics that produce desirable edible fruits and the development of those plants into commercial fruit crops are not complete. The importance of this lack of development of tropical fruit crops is that, as natural plant communities diminish, especially through deforestation, then potential fruit crops may be lost for human use – both humans in the tropics and humans elsewhere in the world.

Although the number of plant species used for fruit cropping may be small, the process of selecting different plants with different desirable characters has expanded the choice of crop plants to use. The entry on Wikipedia suggests that there are more than 7500 known cultivars of apples worldwide. Historically, growers might have called the selected desirable plant a 'variety', but the horticultural convention is that selected cultivated plants should be termed 'cultivars'. Each cultivar is given a unique name. A cultivar should be distinct in at least one character which identifies it from other

Table 1.1. Distribution of world economic fruit crops (production in '000 tonnes).

Crop	Africa	Asia	Europe	North and Central America	Oceania	South America	World
Almonds, with shell	199	422	388	1,043	13	1	2,065
Apples	1,863	38,529	13,950	5,278	601	4,032	64,256
Apricots	422	1,687	800	83	24	52	3,068
Avocados	365	454	103	1,458	54	760	3,363
Bananas	8,008	46,850	406	7,011	1,211	15,907	81,263
Cherries	14	834	811	279	14	44	1,996
Chestnuts	0	1,093	129	0	0	1	1,223
Citrus fruit (nes)	4,007	2,135	48	100	11	828	7,137
Dates	2,339	4,052	13	19	0	1	6,422
Figs	351	509	124	45	0	34	1,062
Grapefruit (inc. pomelos)	681	1,665	63	2,059	14	306	5,061
Grapes	3,678	18,393	29,051	6,441	1,720	6,988	66,272
Hazelnuts, with shell	0	579	165	33	0	0	777
Kiwifruit	0	83	609	23	320	170	1,204
Lemons and limes	822	4,760	1,525	2,839	48	2,968	13,032
Mangoes, mangosteens and guavas	2,863	25,214	0	2,320	58	2,224	33,445
Oranges	5,647	17,419	6,200	12,817	594	20,501	63,906
Papayas	1,398	1,973	0	995	13	2,424	6,937
Peaches and nectarines	808	10,014	4,293	1,280	124	939	17,457
Pears	612	14,517	3,133	843	185	815	20,106
Persimmons	0	3,101	50	0	4	169	3,324
Pineapples	2,673	9,662	3	173	142	3,811	18,874
Pistachios	1	379	11	2,338	0	0	501
Plums and sloes	205	5,751	2,534	678	35	441	9,719
Sour cherries	0	270	756	206	0	2	1,158
Strawberries	218	732	1,449	1,308	31	84	3,823
Tangerines, mandarins, clementines	1,484	18,895	3,000	688	111	2,313	26,514
Walnuts, with shell	35	965	318	359	0	19	1,695

nes, not elsewhere specified or included.
Figures for 2007 (FAO).

similar plants. That unique character must be consistent as new generations are produced through whatever propagation methods might be used. If that propagation method is asexual, meaning that the daughter plants are replicas of the mother plant, then the new plant can also be termed a clone. In theory, at least, a clone should be unvaryingly the same, generation to generation. However, there are small changes over time with asexual propagation. Cellular and tissue mutations do occur, and therefore, over time, some variation may be found in a clonal population. Orchardists may use the term 'strain' to identify progeny lines within a clone. Mutations of tissues can lead to whole organs having a distinct and different character. Plants can often be propagated from organs which are clearly different from the rest of the plant. The new plant will have the genetic composition of the mutated tissue and not the parent tissue. These plants are called 'sports'. There are numerous instances when the 'sport' differs from the parent because of pigmentation change. Many red apple cultivars are sports of less coloured cultivars – this is the case for 'Delicious' and 'Red Delicious' and again for 'Gala' and 'Royal Gala'.

D. Jackson *et al.*

Table 1.2. Distribution of world fruit crops, showing three foremost leading producing countries.

Crops	Annual production ('000 tonnes) in the three leading producing countries in 2007 for some important fruit crops					
Almonds, with shell	USA	1,043	Spain	201	Syria	132
Apples	China	27,507	USA	4,238	Iran	2,660
Apricots	Turkey	528	Iran	280	Italy	212
Avocados	Mexico	1,140	Indonesia	250	USA	250
Bananas	India	21,766	China	7,325	Philippines	7,000
Cherries (sweet)	Turkey	392	USA	270	Iran	225
Chestnut	China	925	South Korea	70	Turkey	63
Currants (black, red, etc.)	Russia	600	Poland	140	Ukraine	33
Dates	Egypt	1,130	Iran	1,000	Saudi Arabia	970
Figs	Turkey	271	Egypt	170	Iran	88
Grapefruit (inc. pomelos)	USA	1,580	China	547	South Africa	430
Grapes	Italy	8,519	France	6,500	China	6,250
Hazelnuts, with shell	Turkey	499	Italy	131	USA	33
Kiwifruit	Italy	455	New Zealand	315	Chile	170
Lemons and limes	India	2,060	Mexico	1,880	Argentina	1,260
Mangoes, mangosteens, guavas	India	13,501	China	3,752	Pakistan	2,250
Olives	Spain	5,788	Italy	3,481	Greece	2,600
Oranges	Brazil	18,279	USA	7,357	Mexico	4,160
Papayas	Brazil	1,898	Mexico	800	Nigeria	765
Peaches and nectarines	China	8,032	Italy	1,719	Spain	1,150
Pears	China	12,625	Italy	841	USA	799
Persimmons	China	2,340	South Korea	345	Japan	240
Pineapples	Brazil	2,666	Thailand	2,320	Philippines	1,900
Pistachios	Iran	230	USA	110	Turkey	78
Plantains	Uganda	9,231	Colombia	3,600	Ghana	2,930
Plums and sloes	China	4,830	Serbia	681	USA	675
Quinces	Turkey	122	China	90	Iran	39
Sour cherries	Poland	195	Turkey	174	Russia	153
Strawberries	USA	1,115	Russia	324	Spain	264
Tangerines, mandarins, clementines	China	14,152	Spain	2,081	Brazil	1,271
Walnuts, with shell	China	503	USA	290	Turkey	184

Figures for 2007 (FAO)

The development of fruit crops has been heavily influenced by successes in selecting and breeding new fruit cultivars. In Europe, in the 18th century, the realization that plants were sexual, and therefore that hybridization was feasible and manageable, led, in some crops, to a proliferation in choice of what to grow. However, more recently the organization of global markets and the management of delivery and retailing have tended to reverse this process. Although there may well be over 7500 cultivars of apple that could be grown, there are probably no more than 30 that are of notable commercial importance. Restricting the range of cultivars for cropping carries risks. A small genetic base

Table 1.3. Increase in world production of selected fruit crops from 1977 to 2007. Data taken from FAO statistics 1977 and 2007.

Fruit crop	'000 tonnes in 1977	'000 tonnes in 2007	% inc from 1977 to 2007
Apples	30,491	64,256	111
Apricots	1,517	3,068	102
Avocados	1,348	3,363	150
Bananas	34,134	81,263	138
Blueberries	45	[a]	395
Carobs	353	187	−47
Cherries	1,142	1,996	75
Chestnuts	461	1,223	165
Cranberries	102	[a]	210
Currants	386	888	130
Dates	2,660	6,422	142
Figs	890	1,062	19
Grapefruit (inc. pomelos)	4,322	5,061	17
Grapes	53,169	66,272	25
Hazelnuts, with shell	447	777	74
Lemons and limes	4,940	13,032	164
Mangoes, mangosteens, guavas	13,134	33,445	155
Olives	8,803	17,456	98
Oranges	33,901	63,906	89
Papayas	2,287	6,937	203
Peaches and nectarines	6,829	17,457	156
Pears	7,552	20,106	166
Persimmons	923	3,324	260
Pineapples	8,082	18,874	134
Pistachios	97	501	416
Plantains	23,631	34,445	46
Plums and sloes	5,485	9,719	77
Raspberries	218	[a]	117
Sour cherries	590	1,158	96
Strawberries	1,535	3,823	149
Tangerines, mandarins, clementines	8,246	26,514	222
Walnuts, with shell	717	1,695	137

[a]no data for 2007, data from 2002 used for computation.

could be overwhelmed by biological threats from pathogen and pest organisms. It may be fortuitous but modern gene analysis offers the opportunity to identify the range of genetically based characters in crop cultivars and what plants carry particular genetic traits. There needs to be a coordinated vision of what to collect to maintain the genetic heritage for future plant breeding of fruit crops.

A number of factors that help to determine fruit crop distribution will now be considered. Many of these factors relate to climate and will be discussed further in Chapter 2, but other factors, such as soils, pests and diseases, and topography, are equally important.

Climatic Factors

Temperature

Temperatures increase as latitude and altitude decrease, and extremes are moderated in sites close to a large volume of water. Temperature is modified by geographic features such as slopes and wind currents that minimize the problem of 'frost pockets'. It is profoundly influenced by mountain ranges that intercept rain clouds and create hot, dry deserts. Major modifications may be induced by ocean currents. The Gulf Stream warms peripheral areas of north-west Europe; the Humboldt (Peruvian) current cools the west coast

D. Jackson *et al.*

of South America. Temperature is the major factor affecting the distribution of fruit plants.

Water

Traditionally, fruit crops were grown in areas where rainfall was adequate to maintain good plant and fruit growth, although irrigation has been known and used for thousands of years. In the 19th and 20th centuries its use increased dramatically, and today it is remarkable in its sophistication and efficiency. In areas with Mediterranean or continental climates, characterized by dry, warm–hot summers, water obtained from underground sources, dams or river diversions provides excellent growth conditions. Low rainfall and humidity, together with adequate irrigation water, provide good growth with low disease pressure. The opposite situation, regions with very high rainfall, is often avoided because of disease problems.

Relative humidity

High rainfall is usually associated with high relative humidity. Some crops, such as subtropicals, blueberries and cranberries, are best grown in humid climates, but most others do better under conditions of low humidity, provided the roots have adequate moisture. However, it must be kept in mind that water loss from evapotranspiration is greater in dry air, and situations can arise where wilting occurs even when soil moisture is considered adequate.

As mentioned above, humid conditions increase disease pressure. High relative humidity at critical stages of fruit development can induce russet and other blemishes on the fruit surface, but there are also positive effects. For example, Cox's Orange Pippin apples develop superior skin texture when grown in a cool, humid climate.

Overall, while warm, humid conditions are conductive to rapid growth and early, high yields, high humidity can add significantly to the cost of production.

Light

Light levels vary with season, latitude, cloud cover and natural and manmade pollution. Generally, southern hemisphere countries have low pollution levels and, since the earth is closer to the sun in southern hemisphere summers than in northern hemisphere summers, light intensity is generally higher. It is usually the case that higher light intensity increases yield, since more leaves in the canopy are adequately illuminated. However, under clear conditions where there is little scattered light, only non-shaded leaves or parts of leaves receive adequate (and often excessive) illumination. In such situations it is not uncommon to see fruit damage from 'sunburn' and leaves performing poorly because of excessive light and heat. This reality must be considered when developing appropriate tree and vine management practices for various regions.

Wind

Regions and districts vary considerably in the amount of wind. Wind (as opposed to air movement) is seldom beneficial; it can inhibit tree growth and often destroy crops. As will be seen later, wind can be controlled to a considerable extent by shelter, and there are many places where, if other things are favourable, orchardists would be wise to erect shelter, artificial or natural, to improve a fruit production site.

Hail

Few areas are free of hail and, for some, hail is a perennial problem. Economic fruit production is very difficult in such regions. A severe hailstorm will annihilate any crop. Even light hail can cause blemishes on fruit which destroy their commercial value. Various methods to reduce hail damage have been tried, but nothing short of hail netting, a very expensive option, has proven totally effective.

Geographic Factors
Soils

Good soils have been favoured by horticulturalists, since they make crop production so much easier. However, successful fruit production can occur on poorer soils if enough attention is given to nutrition, organic matter (water-holding capacity), pH adjustment, drainage and irrigation. Skilful growers may even turn an apparent adversity to an advantage by having a greater degree of positive control over some factor affecting growth and fruiting.

Topography

Preferred sites in cool areas are generally those with a gentle slope facing the sun. The slope allows air drainage in cool conditions and reduces the incidence of spring frost. Sunny slopes are especially important for some crops requiring late summer heat to mature the crop, such as wine grapes growing in a cool climate. On the other hand, flat land is easier to manage and is fully satisfactory where frost is not a problem or where frost protection methods are available. Steep slopes are generally avoided because of mechanization difficulties.

Production Systems

The production of fruit crops involves a wide range of resource and input decisions. Changing circumstances, changing technology and science, and, finally, changing ideas mean that the decisions will also change with time. The long-term view of past developments in fruit production shows changes in approach and decisions about how and what should be done to produce yields of saleable fruit.

High-density planting

One such change in orchard production has been the intensification of tree density together with a rationalization of tree shapes, with the aim of producing fruit-bearing canopies better suited to current management and harvesting practices. HDP, meaning high-density planting, has been used in a wide range of fruit crops, with the intention of delivering early yields in the orchard life cycle and with improved efficiencies. The efficiencies relate to improving the relationships between capital investment and particular inputs with the production system outputs (which can be measured in terms of level of yield, fruit quality or financial return). For example, some orchards have sought efficient returns to labour by developing dwarf trees with less need for pruning and training. The lower labour requirement for training, pruning and accessing the tree at harvesting time aids the competitiveness of the production system.

Organic production systems

Fruit production systems have to be competitive but there is also a realization that fruit production systems should be sustainable. While the term 'sustainable' can be interpreted in many ways, the overall direction for orcharding has been to mini-mize environmentally disruptive actions and to find inputs that are sustainable and, if possible, renewable. Some specific production systems are identifiable by the focus on the choice of production inputs. Some producers choose not to use manufactured fertilizers, crop protection compounds and other chemical inputs. Typically the alternative system inputs chosen by these producers are labelled 'organic', and these producers also follow particular practices, which are also labelled 'organic'. In many countries, government and/or non-governmental bodies have established certification schemes that permit growers to label their produce identifying the produce as 'organic'. The certification schemes set out the criteria (acceptable organic practices and inputs) that must be adhered to. An international organisation, IFOAM (International Federation of Organic Agriculture Movements) acts as an 'umbrella' movement and actively promotes the interests of the organic agricultural movement worldwide.

There is another movement that also promotes organic farming, with an emphasis on principles propounded by Rudolf Steiner. The method of production, termed 'biodynamic agriculture', emphasizes the interrelationship of soil, plants and animals as a unified system and in need of a holistic treatment. There are some methods of production and preparations that are unique to 'biodynamic' principles. Once again, certification of the use and adherence to the principles of biodynamic production is available, offered by approved bodies. Demeter International is the largest certification organization for biodynamic agriculture.

Intergrated fruit production systems

Buyers, consumers and even government agencies have choices about what products are acceptable. Governments have responsibilities for national biosecurity and for the health of their citizens. Fruit consignments must meet set standards for the presence or absence of particular organisms and for the safety of the product – most of the attention is focused on the presence and/or level of particular chemicals deemed to be of concern. Producers have choices about the production methods and inputs they use, and clearly these choices need to be in alignment with the demands of buyers, consumers and governments. The trend towards minimizing pesticides and fungicides, with the benefit of less ecosystem disturbance and lower production costs, has resulted in new systems

D. Jackson *et al.*

of production. One manifestation of this development is called 'integrated fruit production', otherwise known as IFP. IFP emphasizes that the management of the orchard should include less pesticide use and that pesticide use should be targeted for the conditions at hand and for desired outcomes. One such outcome might be the complete absence of any chemical residue on the fruit when the fruit is sold. 'Nil residue' production systems are now feasible for many fruits and in many production areas.

Ecological and Economic Factors

Pests and diseases

Serious problems with pests and diseases can be a major influence on the distribution of a fruit crop. For example, passionfruit is very susceptible to root rots. Soils in a passionfruit-growing district may eventually become so seriously infected that the industry must move to other areas with pathogen-free soils. Chestnut blight is so virulent that it destroyed the chestnut industry when introduced into North America. The phylloxera aphid has profoundly influenced the distribution of grapes at various times in history and continues to influence the rootstocks and cultivars being planted.

As we have seen, climate has a big influence on pest and disease incidence. Later we will see how management can influence these problems.

Labour availability

Traditionally, horticulture was established near centres of population, which, among other things, provided a ready source of labour for harvest and other peak demands. Improvements in product and labour transportation, and the introduction of systems and procedures allowing labour migration have made the need to locate fruit properties near urban centres less important. Additionally, the advent of mechanical harvesting and other technologies to reduce labour requirements has changed crop property location patterns. For example, currants, raspberries and grapes are now commonly machine-harvested and such crops can be grown further away from centres of population. Such moves follow the trend of vegetables like peas, beans and potatoes, which have become predominantly agricultural rather than horticultural crops.

There is, however, a counteracting trend, where the emphasis is on producing high-quality fruit for export or for a domestic niche market. Such products demand high labour inputs, and the size of a production unit must be carefully adjusted to match the available labour supply. These markets are lucrative enough to justify the high labour input required.

Availability and price of land

There is a complex relationship between the potential returns for a specific horticultural crop and the cost and availability of suitable land in the right location or district (due to competition from other agricultural or horticultural crops, industry or residential use). It is sufficient to say that this relationship can markedly affect fruit crop distribution.

Availability of facilities and markets

A grower would be foolish to plant in an area where markets, transport or servicing industries are inadequate. A whole range of factors are important and include:

- Proximity to local or export market facilities, cool stores, packing houses, etc.
- Adequate roads and transportation options
- Possibilities for 'pick your own' and other farm-gate sales
- Availability of advisory services, machinery and chemical retailers, engineering firms, etc.

Availability of research

Although research tends to be seen as the preserve of publicly supported research centres, much is done by individual producers or commercial firms pioneering new techniques which can have dramatic effects on an industry. New cultivars or techniques frequently change growing patterns or induce growers to invest in new districts. Investment in research can pay handsome dividends. It is hard to imagine a successful fruit production district without some level of research support.

Availability of finance

The cost of fruit growing can increase for a number of reasons: land gets more expensive, irrigation is necessary for high yields and quality, and expensive pest management practices are needed to keep fruit completely pest- and disease-free for export. High-density plantings are increasingly popular and require expensive support structures, canopy covers and fertigation systems to be fully practical. Producers wishing to modernize or expand their operation are likely to require financing, and it is not unusual to find that lending agencies are unwilling to risk capital on new technologies, new crops or new districts. This conservatism tends to preserve the status quo and slow down expansion in non-traditional areas. Thus, it is not unusual for governments to provide the financing needed to launch a new horticultural venture. These agencies have learned that once the first commercial venture is successful, expansion can be rapid.

Further Reading

The FAO (United Nations) Production Year Book is an excellent source of information about food production in the world. Data can be accessed through internet sites and in particular http://faostat.fao.org/ (accessed 8th June 2010). The following note was posted on the page at the above web address: From 1 July 2010 the FAOSTAT subscription service changes to a FAOSTAT registered users service offering the same facilities as the current subscription service.

Childers, N.F., Morris, J.R. and Sibbett, G.S. (1995) *Modern Fruit Science. Orchard and Small Fruit Culture.* Horticultural Publications, Gainesville, Florida.

Janick, J. (2005) The origins of fruits, fruit growing, and fruit breeding. *Plant Breeding Reviews* 25, 255–320.

Janick, J. and Paull. R.E. (eds) (2008) *The Encyclopedia of Fruit & Nuts.* CAB International, Wallingford, UK.

Salunkhe, D.K. and Kadam, S.S. (1995) *Handbook of Fruit Science and Technology. Production, Composition, Storage, and Processing.* Marcel Dekker Inc., New York.

Tromp, J., Webster, A.D. and Wertheim, S.J. (2005) *Fundamentals of Temperate Zone Tree Fruit Production.* CAB International, Wallingford, UK.

Westwood, M.N. (1993) *Temperate Zone Pomology: Physiology and Culture,* 3rd edn. Timber Press, Portland, Oregon.

D. Jackson *et al.*

2 Climate and Fruit Plants

DAVID JACKSON

Climate is the most significant environmental variable affecting the production of fruit crops. More than any other factor it contributes to the way fruits are distributed, on both a world and a local scale.

Climatic Zones

Before discussing climatic zones it is necessary to define a very useful measure of heat accumulation – degree days or heat units. These terms are synonymous and mean the accumulated heat above a certain base temperature. For fruit crops, 10°C is a suitable base, since little growth occurs below that temperature. Degree days may be calculated using the formula:

$$\text{degree days} = (M - 10) \times N$$

where M = mean monthly temperature and N = number of days in the month. In berry fruit, <7°C is used as a chill unit.

Annual degree days are calculated for the number of months where the mean temperature is above 10°C and added to give an annual total. Heat units can also be calculated on a daily basis, i.e. the days when temperature is above 10°C (M – 10) are added to give an annual total, which is higher than that calculated on a monthly basis. Authors should always indicate which system is being used (in this book we use monthly). Some calculations are done using the Fahrenheit scale. To convert them to Celsius, multiply by 5/9 (0.555).

It is convenient and illustrative to divide fruit plants into categories based largely on temperature requirements, i.e. tropical, subtropical or temperate.

Tropical fruits

Tropical fruits include banana, durian, plantain, mangosteen, mango, guava, date, litchi, pineapple, papaya and cashew. There are many others but most are of local significance only. These fruit crops grow in areas usually within the latitudes 23° N to 23° S, where there is little temperature difference throughout the year. The mean monthly temperature during the coolest month is 18°C or higher. All of these species need substantial amounts of heat to ripen their fruit. Few tropical fruits will tolerate frosts at any time during the year.

Subtropical fruits

These fruits are adapted to the subtropical zone, which is normally defined as:

- Having a mean temperature in the coldest month between 13 and 18°C.
- Having a seasonal temperature range.
- Having a distinct wet and dry season.
- Having the possibility of some slight winter frosts.
- Having a minimum heat accumulation of about 2000°C days.

There is considerable overlap in plant response to tropical, subtropical and temperate climates. Most citrus, for example, will grow in subtropical climates, but some, such as pomelos and limes, prefer the warm subtropics or even the tropics. Others do quite well in warm temperate areas, e.g. lemons, mandarins. Kiwifruit are happy in both warm temperate and subtropical regions, and passionfruit are grown in all three climatic zones.

The preference of certain subtropical fruit for wet or dry summers is shown below. Those fruits known to be least cold tolerant or more heat demanding are placed higher on each list.

- Tolerant to and may even prefer moist summers: tamarillos, babacos, macadamias, passionfruit, loquats, kiwifruit, feijoas.
- Prefer dry summers: grapefruit, oranges, NZ grapefruit, tangelos, mandarins, lemons.

Temperate fruits

Those fruits classified as temperate have one major feature in common: they need a distinct cold, dormant period to crop satisfactorily. In contrast to subtropicals, this cold period may be quite severe. Sometimes, however, these severe winters can damage even temperate zone fruits, a situation more common in regions with a continental climate and far away from the tempering influence of the sea or large lakes. For many temperate zone plants, winter freeze injury begins when the temperature falls below –15 to –20°C. However, this critical temperature is strongly dependent on the amount of acclimatization (conditioning) that preceded the low-temperature event.

On the other hand, many temperate fruit crops have minimum requirements for summer warmth to achieve good productivity and fruit quality. Table 2.1 lists some important temperate fruits with an indication of their climatic preferences.

Most of those plants tolerant of moist summers will grow satisfactorily in drier areas, provided irrigation is adequate. Likewise, those in the moderately dry list will usually do well in dry climates with irrigation. The table indicates preferences only. The wider tolerances of some of these fruits will be noted under individual crop headings.

Specific Aspects of Climates

Meso- and microclimates

Data for heat units are normally taken from meteorological stations, which may, or may not, be close to local orchards, fields or plantations. Thus, it is important to remember that there can be quite marked differences in heat accumulation owing to local geographic conditions. An orchard or vineyard on a slope facing the sun, for instance, will be warmer than one that is not. If the hill gives shelter from prevailing cold winds it will be warmer still, and if, for example, the soil contains many stones and weeds and ground cover crops are suppressed, further heat will be gained. Such sites can gather up to 200°C days extra in a season and can make an apparently unsuitable district more attractive.

Figure 2.1 shows different temperature zones in a hypothetical geographic area. The differences in temperature and other climatic variables which occur in restricted localities are said to constitute mesoclimates. Differences on an even smaller scale, for example within an orchard area, are termed microclimates.

Trees and shelter belts can be used to create microclimates and will allow warm air to accumulate even on days with a cool wind. Care must be taken to ensure that they do not prevent cool air draining away as in Figs 2.1 and 2.2.

Table 2.1. Some climatic preferences of temperate fruit plants.

	Prefer dry summers and autumns; below 700 mm rain annually	Prefer moderately dry summers; 700–900 mm annually	Tolerant of moist summers; above 900 mm annually
Need heat accumulation above about 850°C days	Grapes Apricots	Peaches and nectarines Plums (Japanese)	Pecans Persimmons (non-astringent)
Wide summer temperature tolerance	Cherries Walnuts	Apples Pears (European and Asian) Chestnuts Strawberries Boysenberries	Blueberries Hazelnuts Strawberries Persimmons (astringent) Raspberries (autumn crop)
Need cold winters			Plums (European) Currants (black, red and white) Gooseberries Cranberries Raspberries (main crop)

D. Jackson

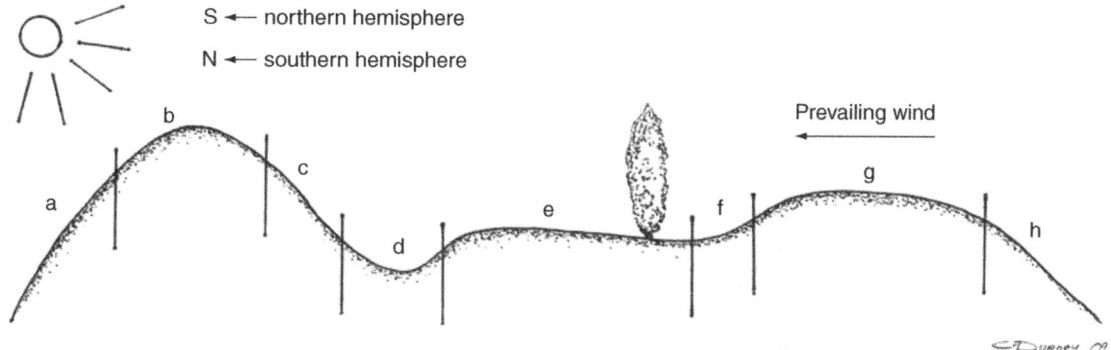

Fig. 2.1. Mesoclimates in a hypothetical area. (a) A warm site catching more sun owing to the lie of the land. It misses late-spring and early-autumn frosts, since the cold air will drain to low-lying areas; there is also shelter from the prevailing wind. (b) The advantages of (a) will be counteracted by the cold which comes with increased altitude. (c) A cold site; although it may miss frosts in spring and autumn, it will accumulate much less heat in summer due to exposure to wind and a poor angle to the sun. (d) A cold site very susceptible to frost; cold air from surrounding districts will drain into this area. (e) Still frosty but less so than (d). Some shelter from wind may be obtained from the windbreak. (f) The windbreak, densely planted at the base of the hill, prevents cold air from draining away and a potentially frost-free site has been lost. This area would be shaded by the windbreak. (g) Less frost than (e), but a prevailing cold wind and altitude may slow the accumulation of heat units in summer. (h) Cold, like (c) above.

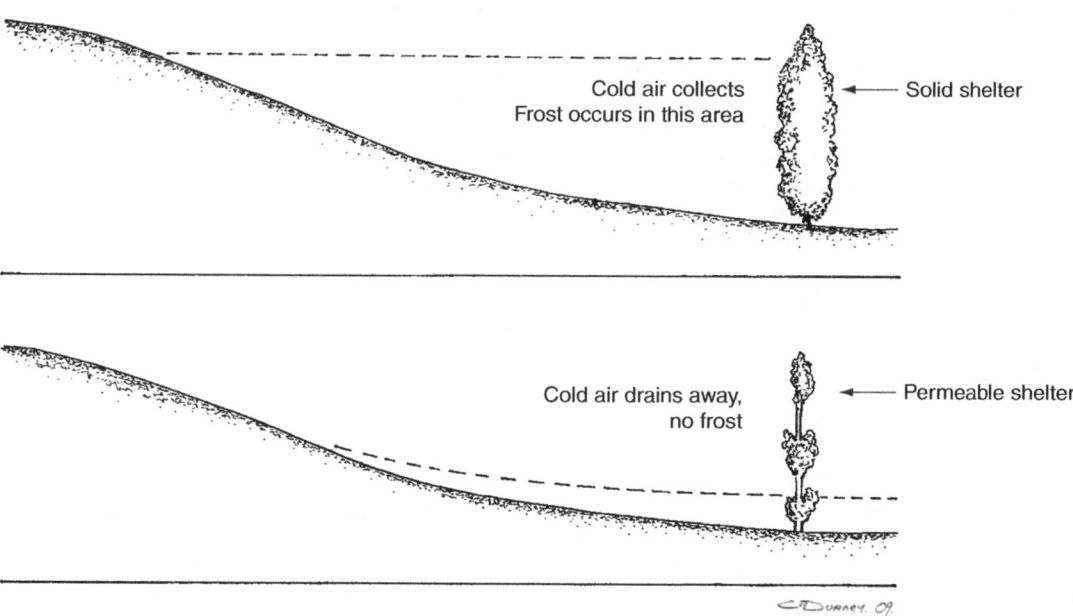

Fig. 2.2. Shelter effects on accumulation of cold air.

Temperature is not the only factor that can vary across mesoclimates. Rainfall amount may vary considerably in hilly sites relatively close to one another. Winds can vary even more; sea breezes or a wind being channelled down valleys may affect not only velocity but also frost incidence, cloud cover and rainfall. Thus, a grower choosing a site for specific attributes will need to consider the obvious factors mentioned above and also seek guidance from locals in the district on specific peculiarities of the mesoclimate.

A survey of plants being grown by local gardeners is often very helpful in this regard. Tamarillos thriving in an area suggest a mild climate with very light winter frosts. Apricots, which produce crops in most years, suggest an area with few damaging spring frosts, especially since home gardeners seldom take measures to control frost. Windswept trees indicate high and persistent winds. People who boast of early potatoes usually inhabit a mesoclimate which is mild in late winter and frost free in spring.

While large-scale producers must learn to live with a range of mesoclimates across a large plantation, smaller producers wishing to provide early fruit or grow a speciality crop must pay particular attention to selecting the best mesoclimate.

Frosts

Spring frosts usually occur on still nights with clear skies and little wind. These are called radiation frosts, because it is the radiation of heat from soil and plants to the cloudless sky which leads to low temperatures at the soil surface. As temperatures drop, cold air accumulates in hollows and, due to lack of wind, is not dispersed. The harder and longer the frost, the deeper the layer of cold becomes. Radiation frosts are a major cause for concern in most temperate zone fruit production regions.

As discussed earlier, winter frosts that can damage tropical and subtropical plants do not normally affect temperate fruits in relatively mild climates. However, once the leaf and flower buds begin to swell in the spring, temperatures below 0°C can be damaging. At bud burst, when the colour of the petals begins to show, temperate fruits will often tolerate –3 to –5°C. However, at full bloom, temperatures below –2 to –3°C can kill the ovary, and as fruits are formed they can be damaged or killed by temperatures below –1 to –2°C. The length of exposure is important: brief periods may do no harm, but longer periods of 30 min or more will usually cause damage.

These damaging temperatures apply to most of the tree fruits. However, walnuts, raspberries and strawberries are susceptible to temperatures between –1 and –2°C from the point of bud swelling onwards. Currants and gooseberries seem to be more tolerant to frost, and flowers can probably survive from –3 to –4°C. Blueberries are even more tolerant (about –5°C), while hazelnut flowers are able to withstand –8°C.

Normally, flowers are more susceptible to frost injury than are leaves. On kiwifruit and grapes, however, the bud which bursts in the spring produces a leafy shoot. Flowers do not appear on that shoot until several weeks later. Thus, a frost occurring after bud burst will kill the whole shoot, and while a new shoot may emerge from the base of the old one, it is inevitably less fruitful. For kiwifruit –1.5 to –2.0°C or below are critical; grapes may tolerate slightly lower temperatures (–2 to –3°C).

The economic consequences of frost damage can be severe. A severe frost may destroy all the fruits, and even a light one may reduce the crop considerably. If frost occurs after the fruit is formed but is not severe enough to penetrate to the seeds, the damaged fruit may not drop but the injury to flesh tissue may cause marking or deformities that persist until maturity. Those fruit trees which blossom early in the spring, like almonds and apricots, are naturally more at risk than those which blossom later, like apples.

Apart from choosing more frost-tolerant fruits, the grower has a number of methods available to counteract the danger. These include the methods listed below. Whatever the frost control method used, it must not be discontinued until the temperature outside the protected area rises several degrees above the danger level.

Choosing a frost-free site

Areas such as (a), (c), (g) and (h), illustrated in Fig. 2.1, can be effectively frost free in the spring. However, it must always be borne in mind that land sloping by more than 10° is difficult and expensive to maintain; it may ultimately cost more than using some of the other frost-control methods.

Choosing late-flowering cultivars

Cultivars of fruit plants vary in their time of flowering, with differences ranging from a few days to 2 weeks. Thus, it is possible to choose late-flowering cultivars purposely to increase the chances of avoiding spring frost injury. However, where cross-pollination is a requirement it is necessary to choose a suitable late-blooming pollinizer. This complication, plus the fact that the fruit of the late-blooming cultivar may not be as valuable as some others, has minimized the use of this strategy.

D. Jackson

Vegetation management

Soil which is weed free, moist (but not water-logged) and moderately compacted will tend to absorb and retain more heat in the day and re-emit it to the cold surface air at night than will a grassed-down site or one with heavy weed growth. On cold nights, this can mean a difference of 1–3°C air temperature in the fruiting zone of the tree or vine and can make the difference between severe damage, light damage or none at all.

Use of burners

These range from metal pots burning diesel oil to large heaters fired with propane placed strategically throughout the orchard or vineyard. They are lit when the temperature falls to the critical level. They can be very effective but they are expensive to run, unpleasant and labour intensive to manage, and some versions can lead to serious air pollution. Thus, they are mainly used in areas or mesoclimates where overhead sprinkling is impossible or wind machines are ineffective. Details relating to their size, fuel, distribution and operation are best obtained from local suppliers, advisory services or other producers.

Use of sprinklers

If a flower bud is continually covered by water in the process of freezing, the tissues of that bud will not fall below 0°C. This is due to the latent heat of freezing. In other words, there may be ice covering the flower but as long as unfrozen water is present, the temperature will not fall below 0°C. Because this temperature is not damaging to temperate fruit crop tissues, over-tree sprinkling during the freezing period can be an effective way to prevent frost damage. Overhead sprinklers are placed in the orchard and nozzles fitted which will deliver 3.8 mm/h. This is sufficient to deal with a frost event where temperatures drop to –6 or –7°C. Keep in mind that the water-delivery system must be capable of maintaining tree wetness across the entire plantation.

To determine when to begin frost fighting with sprinklers, growers should install a wet-bulb thermometer near the bottom of the fruiting zone. Sprinklers should be turned on when the temperature reaches the critical level for that crop, usually –1°C. Alternatively, they can be automated to turn on at pre-set temperatures.

Advisers, consultants or local suppliers of irrigation equipment will be able to advise on installation and operation. Such sprinklers may be able to double for irrigation, although nozzles with a higher delivery rate may need to be installed for that purpose and provisions made to divide the plantation into irrigation zones that match the capacity of the water-delivery system. Furthermore, there are certainly situations where summer disease problems will preclude consideration of over-tree irrigation.

There are some important disadvantages to the use of over-tree sprinklers for frost protection. Since sprinkling cannot stop until air temperatures rise to non-freezing levels, a major hazard to the trees or vines can be limb breakage caused by ice build-up. In poorly drained sites, waterlogging may ensue and root disease problems can increase. Even without waterlogging, diseases such as blast may be serious if excessive water is required.

Use of fans or wind machines

The air temperature on a frosty night rises at increasing heights above the ground. Thus, an air temperature of –2°C at 1 m may be accompanied by a temperature of +2°C at 20 m. The boundary between damaging and non-damaging temperatures is termed the 'inversion layer', and if the two layers can be mixed, non-damaging temperatures may ensue.

Large fans (often referred to as wind machines) strategically positioned in parts of the plantation most likely to suffer frost damage, or smaller fans evenly spaced throughout the planting, can effectively deal with light frost situations. The inversion layer must occur reasonably close to the ground for this technique to be effective. It is difficult to achieve frost control with most fans if the inversion layer is greater than 20 m. Large wind machines are capable of protecting up to about 3 ha each and will raise the temperature by 2–3°C.

Another approach is to use helicopters to achieve the same effect. They are expensive to contract but the grower avoids the high capital investment of a wind machine. Fans and helicopters should begin operating just before the temperature falls to the critical level for the specific crop.

Heat damage

When air temperatures reach 35–40°C, especially on sunny days with little wind, fruit can suffer heat damage. Apples and grapes seem to be the fruit crops most at risk. The surfaces of fruit exposed to direct sunlight will become pale, then brown, then black and, if serious, fungal infection will occur on the damaged area. Fruit showing evidence of sunburn is suitable only for processing.

One successful method of control involves over-tree or over-vine sprinkling. Evaporating water consumes heat and the principle is to maintain a wet fruit surface by intermittent sprinkling. As with frost control, low-volume misting sprinklers are the most appropriate.

Another approach to reducing sun scald is to cover the plantation with a shade cloth or netting that excludes a proportion of the radiant energy. Where hail netting is in place it will serve this dual purpose.

Sunburn can also occur at lower temperatures, but this is usually caused by the movement of branches from one position to another, often due to the weight of fruit. Under such circumstances, fruit surfaces previously shaded by leaves may suddenly be exposed to full sun and surface damage results. For reasons not totally understood, fruits that have been exposed to direct sun for a period of weeks or months develop some resistance to sunburn. Thus, by securing the position of fruiting branches by propping or tying up to a central pole, the latter a feature of the Dutch slender spindle training system, it is possible to reduce sunburn injury. Furthermore, some apple cultivars are more susceptible to sunburn than are others. Experience in British Columbia with Jonagold apple has led some producers to move away from the small trees associated with the use of M.9 rootstocks. Fruit exposure is too high on such trees. Fruit produced on slightly more vigorous trees are less likely to show sunburn.

Wind and shelter

Wind has the following undesirable effects on fruit tree growth and cropping:

- Plant growth is reduced. This is a direct effect of wind on the volume of growth produced in one season.
- Plant growth is uneven. Trees will often lean away from prevailing winds and be difficult to train.
- Physical damage and crop loss may occur. Very strong winds will break branches or even knock down trees. Surface damage may occur to fruit or they may fall prematurely.

- Water loss is increased. By evaporation from the soil and from the surface of the leaves (evapotranspiration), plants lose moisture, and more frequent watering is needed. Sometimes in hot, windy weather, leaves will temporarily wilt, even when the soil is adequately moist.
- Heat accumulation is reduced. Even when the sun is unobscured by clouds, a wind-chill effect may occur and temperatures will fall.
- Winds interfere with pollination. The movement of pollinating insects is greatly reduced on windy days. Furthermore, where there is a wind-chill factor at play, pollen tube growth is slowed. The net result is likely to be reduced fruit set.
- Wind interferes with spraying. Spraying on windy days is at best inefficient and can often be dangerous.
- Conditions for orchard workers may be unpleasant.

Most natural shelter belts are single or double rows of deciduous or evergreen trees, or a mixture of both. Good shelter will alleviate most of the problems mentioned but in itself can cause problems that must be considered. For example, shelter trees will compete with orchard trees for light, nutrients and water; they may harbour pests and diseases; they can increase frost risk; and they always reduce the land available for production.

All crops achieve a degree of self-sheltering, which is greater if the rows are at right angles to the most damaging wind. However, orientating rows with this aim in mind may require some compromises with other considerations. For example, north–south rows achieve the most even illumination and rows parallel to the longest boundary usually waste less space with headlands.

Here are some guides to shelter establishment:

- Establish the shelter first. Shelter should, if possible, precede the planting of fruit trees so the young trees receive the subsequent benefit. A well-sheltered fruit tree planted 1 or 2 years later than an exposed tree may quickly match and possibly overtake its counterpart.
- Properly space and position the shelter belts. Trees provide effective shelter for approximately ten times their height. Thus, 10 m-high shelter trees will be planted approximately 100 m apart. In windy areas, the shelter will be placed on all sides of the orchard; otherwise it may be used only on the sides from whence the

D. Jackson

most-damaging winds originate. Ensure that there will be sufficient headland space (6 m minimum) when the trees are mature.

- Choose trees which are not too dense. A 70% shelter belt is better than a solid, dense one. Solid shelter may prevent air flow on a frosty night (Fig. 2.2) and encourage birds by increasing perching and nesting sites.
- Provide good management of the shelter. To get rapid growth from shelter trees, prepare the soil as thoroughly as for the fruit trees or vines, fertilize and control weeds in the early years, and trickle irrigate. This may double or even treble the rate of growth.
- Choose suitable species. For deciduous fruit crops, deciduous shelter trees are probably better. Air movement in the winter helps to dry out the soil and assists cultivating or mowing and the movement of machinery. However, a mixture of evergreen and deciduous trees may be more suitable for citrus and other evergreen fruit trees. It is important to seek out local advice about the strengths and weaknesses of various shelter species, and combinations of species, before investing heavily in this technology.

Shelter establishment and maintenance

The importance of soil preparation and irrigation has been mentioned. Shelter belts should be planted as close to the boundary as possible. Sometimes the windbreak can be shared with a neighbour. Weeds should be kept under control in a strip about 2 m wide by spraying with appropriate knock-down and pre-emergence herbicides. This is especially important in the establishment years. Deciduous trees can tolerate glyphosate applied directly on the trunk if the bark is mature (not green) and if sprayed prior to bud burst.

Little maintenance is required. Spreading species will need to be trimmed annually or biennially to keep the width to within 1.5 m. Ripping (subsoiling) alongside the shelter belt, say at 2.5 m from the base every 2 years, will help reduce competition with orchard trees.

Artificial shelter

Artificial shelter is often used on high-value crops. Materials are produced by several manufacturers, usually consisting of a plastic or nylon web which can be securely strung between posts. It is expensive but takes up less space, is quickly established and does not rob the soil of water and nutrients. Although it has a long life, it will eventually need to be replaced. There is normally plenty of readily available advice on the construction of artificial shelter from advisory organizations or manufacturers.

Hail

Hail can be one of the most damaging natural phenomena and one of the most frustrating because it is so difficult to deal with. Growers interested in buying or developing a property in a new area should try to establish the hail risk. However, the information obtained may prove unreliable. In some countries where land is exceptionally expensive or wherever the crop is sufficiently valuable, hail netting is worth consideration. It is very expensive but offers the only reliable protection. The other practical way to manage this risk is with the purchase of hail insurance.

Further Reading

Eccel, E., Rea, R., Caffarra, A. and Crisci, A. (2009) Risk of spring frost to apple production under future climate scenarios: the role of phenological acclimation. *International Journal of Biometeorology* 53(3), 273–286.

Faust, M. (1989) *Physiology of Temperate Zone Fruit Trees.* John Wiley & Sons, New York.

Janick, J. (ed.) (2003) XXVI International Horticultural Congress: Genetics and Breeding of Tree Fruits and Nuts. *Acta Horticulturae* 662.

Pérez-González, S., Dennis, F., Mondragón, C. and Byrne, D. (eds) (2001) VI International Symposium on Temperate Fruit Growing in the Tropics and Subtropics. *Acta Horticulturae* 565.

Rodrigo, J. (2000) Spring frosts in deciduous fruit trees – morphological damage and flower hardiness. *Scientia Horticulturae* 85 (3), 155–173.

3 Morphology and Growth of Woody Plants

David Jackson and Roy Edwards

Formation of Bark and Wood

Except for strawberries, all the fruits discussed in this book are members of that group of plants which form woody trunks and branches to support the tree, bush or vine. In some, such as walnuts, apples and cherries, the plants are substantial and the trunk and branches will support the tree for tens or possibly hundreds of years. In others, such as blackcurrants and blueberries, there is no trunk; the plant is a bush and seldom rises above 2 m from the ground. Grapes, passionfruit and kiwifruit produce woody stems and branches, but these are not sufficiently strong to support the plant and so it twines or has tendrils which allow the vine to cling to other plants or trellises.

Woodiness is due to 'lignification', or the laying down of woody tissue within the branch or trunk, as shown in Fig. 3.1.

The part of the branch responsible for growth in thickness is called the cambium. This is a layer of cells in the stem which retains the ability to divide. Such cells are termed meristematic and produce bark (also called phloem) on the outside and wood (xylem) on the inside, such that every season a new layer of bark and wood is produced. On the outside, the bark, as it ages, becomes fissured or may flake or rub off. On the inside, xylem layers can be distinguished and are referred to as annual rings. Trees show these annual rings when cut across, and the age of the tree or branch can be determined by the number of rings present. In such trees the pith in the centre becomes insignificant.

In certain plants, such as raspberries, blackcurrants and gooseberries, the xylem never assumes a significant proportion of the stems. Instead, the soft central pith continues to constitute a large part of the cross section; consequently the stem remains small and fairly weak; these stems are short-lived and regularly replaced by new shoots.

The Shoot

Shoots coming from a seed or from buds on an established tree or bush have a small group of cells at the tip which, like those in the vascular cambium, are meristematic. These are responsible for the production of new cells which will cause the shoot to elongate. Under a microscope, the meristem of a shoot would look somewhat like Fig. 3.2.

As apical cells divide, small bumps appear, called primordia, and these may develop into leaves. The branch elongates to form a typical leafy shoot, as shown in Fig. 3.3.

Roots

Below the ground, roots, while less structured in their branch formation, enlarge in a manner similar to branches and trunks. Phloem and xylem are formed, and old roots can be just as substantial as the branches of the tree. Under the microscope the root tip looks like the drawing in Fig. 3.4.

The apical portion pushes its way through the soil; it is protected by a root cap, which is continually replaced by cells from the meristem, with the outer part of the root cap disintegrating as the apex grows. The meristematic zone at the apex also produces cells which elongate and are responsible for root growth. A little way back from the root tip, fine hair-like structures, 'root hairs', are formed. These are responsible for most of the absorption of water and nutrients from the soil. Thickening of the outer layer occurs behind the root-hair zone; hairs disappear and absorption from the soil is reduced. Branches may form a little way back from the tip and new root hairs appear.

The roots serve to translocate water and nutrients and provide anchorage and support. Root growth and functioning will be discussed later.

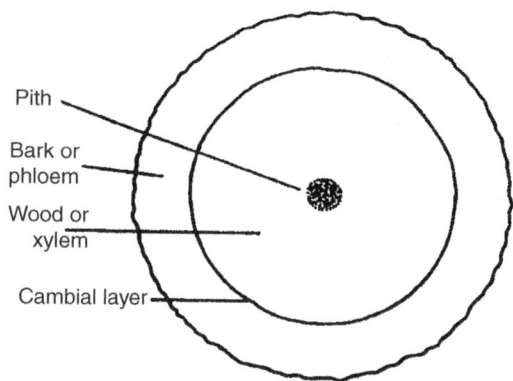

Fig. 3.1. Cross section through branch.

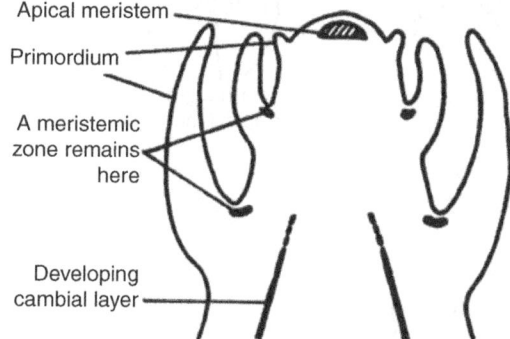

Fig. 3.2. Meristem of growing shoot.

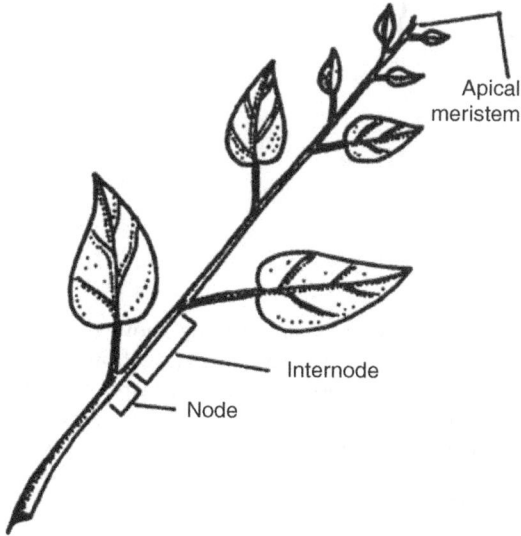

Fig. 3.3. Leafy shoot.

Flowers

Formation of flowers

A flower originates from a meristem in a manner similar to that described for the formation of leaves on a shoot (Fig. 3.2). Instead of leaf primordia, the apical meristem produces the various parts that constitute the flower, i.e. sepals, petals, stamens and carpels. The stamens (male) and carpels (female) contain the reproductive organs of the flower.

The early development of a vegetative and a floral apex is shown in Fig. 3.5. The parts of the flower shown in the process of formation in Fig. 3.5 are arranged in 'whorls' or circles. In a complete flower the whorls are as follows:

- A lower whorl of sepals called the 'calyx'.
- A whorl of petals called the 'corolla'.
- A whorl of stamens called the 'androecium'. This is the male organ, which produces pollen and normally consists of an anther and filament.
- A whorl of carpels called the 'gynoecium' or 'pistil'. A carpel normally consists of a stigma, style and ovary containing ovules. Carpels are sometimes described as floral leaves, since in their primordial form they are similar to leaves. They fold into a cylindrical structure which is the precursor of the fruit. The dorsal suture of the carpel corresponds to the midrib of a leaf, while the ventral suture is formed by the juxtaposition of the edges. The edges contain meristematic tissue giving rise to the ovules, which, when successfully fertilized, will form seeds (Fig. 3.10). As will be shown later, the pistil may develop from one or several carpels. It consists of a stigma, the receptive surface for receiving pollen; a style, or tube to connect the stigma to the ovary; and an ovary containing from one to many ovules.

A simple flower is shown in Fig. 3.6.

Flower types

When analysing the structure of a flower to determine what the potential fruit type (in the botanical sense) may be, the position of the flower parts in relation to the ovary becomes critical. The major situations that can occur, from apical meristem hypogyny, perigyny and epigyny, are shown in Fig. 3.7.

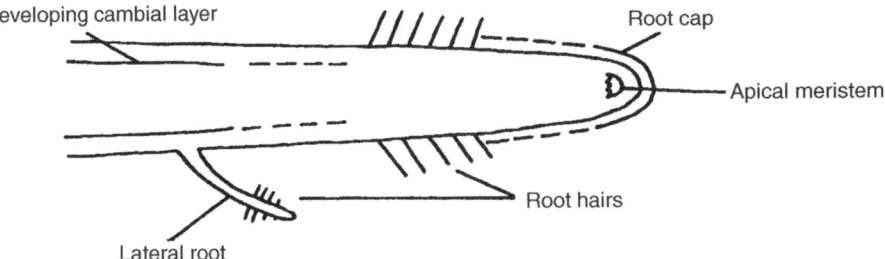

Fig. 3.4. Developing root tip.

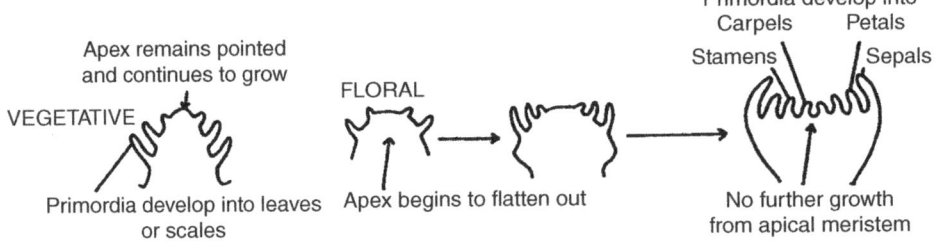

Fig. 3.5. Comparison between vegetative and floral apices.

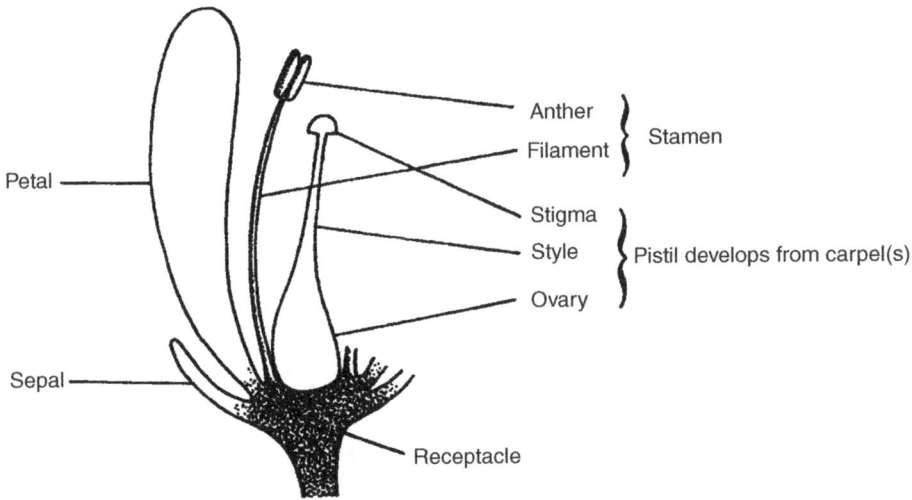

Fig. 3.6. Basic structure of a flower.

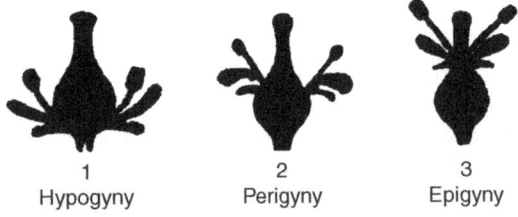

Fig. 3.7. Hypogyny, perigyny and epigyny.

When the ovary is superior, or above the point where the other flower parts arise, the flower parts are termed hypogynous, e.g. citrus, raspberries and strawberries (Fig. 3.6). When the ovary is in a middle position, the flower parts are termed perigynous, e.g. apples, pears, cherries and other stone fruit. When the ovary is inferior, the flower parts arise from above the ovary and are termed epigynous, e.g. blueberries, gooseberries and currants.

D. Jackson and R. Edwards

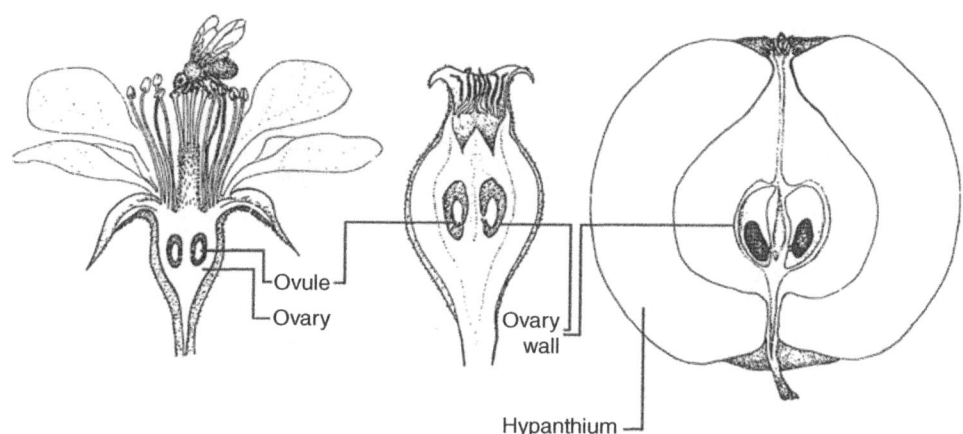

Fig. 3.8. The development of an apple.

Hypanthium tissue

Hypanthium consists of the undifferentiated tissue of sepals, petals and filaments. In the development of the apple, this tissue is attached to and surrounds the ovary. After fertilization of the ovules, the fleshy edible part of the apple develops from the hypanthium tissue (Fig. 3.8). Similarly, stone fruit flowers develop a cup-shaped hypanthium surrounding the ovary, but in this case it is not attached to it and does not develop after fertilization (Fig. 3.9).

Apocarpy and syncarpy and methods of placentation

Flowers are categorized according to whether the ovary consists of free carpels (apocarpy) or fused carpels (syncarpy) (Fig. 3.10). The arrangement of these in the flower and the relative position of the seeds (placentation) form the basis of flower and fruit categorization. This will be considered in more detail when fruits are discussed.

Variation of flower types

Flowers which lack certain parts are called incomplete. Grasses, for example, lack sepals and petals. Kiwifruit, walnuts and hazelnuts have flowers which have either functional stamens or pistils, but not both, i.e. they are single-sexed. Kiwifruit has either male or female flowers on any one plant and is termed dioe-

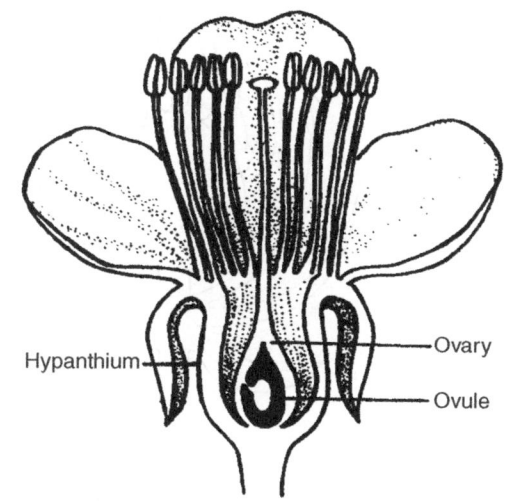

Fig. 3.9. The perigynous flowers of a peach.

cious. Hazelnuts and walnuts have both male and female flowers separately on the same plant and are called monoecious.

Inflorescences

An inflorescence describes the grouping or arrangement of flowers in a branching system on a plant. Flowers which arise singly in the axils of leaves are said to be solitary or axillary, and no conventional inflorescence type is present. Examples of flowers produced in this way include apricots, peaches, passionfruit, kiwifruit, feijoas and walnuts.

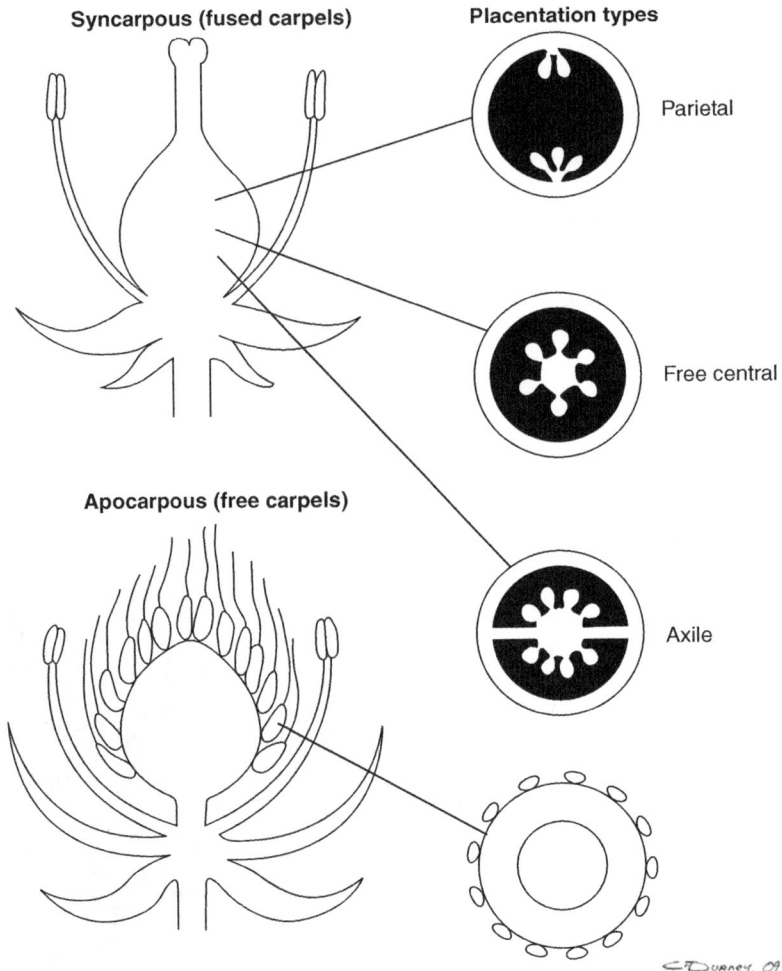

Fig. 3.10. Apocarpous and syncarpous ovaries.

Flowers which are produced in a modified branching system are usually subtended by bracts (much-reduced foliage leaves). Examples include apples, pears, citrus, currants, blueberries, grapes, cherries, raspberries, boysenberries and strawberries.

In a broad sense, inflorescence types are categorized as either racemose (indeterminate) or cymose (determinate). In a determinate inflorescence the terminal or central flower opens first and stops further growth of the axis on which the flowers are borne. In an indeterminate inflorescence the terminal flower is the last to open. Examples of racemose (indeterminate) inflorescence types are found in apples, currants, cherries, blueberries, grapes, strawberries, raspberries and boysenberries. Most pears (and some apples) tend to provide examples of cymose (determinate) inflorescence types.

Seeds

It will be recalled that ovules are produced within the ovary from meristematic tissues at the carpel edges (Fig. 3.10). The male equivalents to the ovules are the pollen sacs, contained in the stamens. Parts of the ovule and stamen are shown in Fig. 3.11.

If the pollen produced in the pollen sacs is successfully transferred to the pistil, a pollen tube will grow via the stigma, style and ovary, to enter between the protective integuments at the micropyle and fertilize the egg. This is further illustrated in Fig. 3.19. The process of pollination and fertilization is described

D. Jackson and R. Edwards

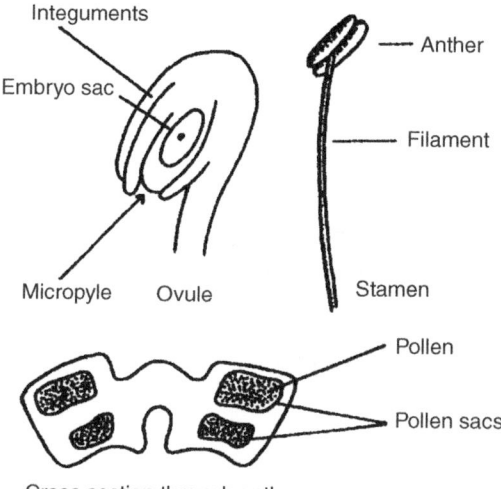

Cross section through anther

Fig. 3.11. Ovules and stamens.

in Chapter 4. Successful fertilization results in the development of the ovule into a seed.

Except in nuts, the seed is of little apparent significance to the orchardist and the consumer. Its presence in the fruit can even be considered an impediment to enjoyment. Nevertheless, the seed is of the utmost importance physiologically and, if it is not present, the fruit may not develop. This will be discussed further in Chapter 4.

Fruits

In the botanical sense, a fruit is the matured ovary of a flower, including its one or more seeds and any part of the flower, such as the hypanthium, which may be closely associated with the matured ovary.

In popular language there can be confusion; for example, botanically all the following are fruits:

- Grain or seeds of corn, oats and wheat.
- Nuts, such as walnuts or chestnuts.
- Tomatoes and pumpkins, normally sold as vegetables.
- Peas and beans with their pods, also sold as vegetables .

After it develops into the wall of the fruit, the wall of the ovary is called the pericarp. Three distinct layers are generally present: the exocarp on the outside, the mesocarp in the middle, and the endocarp on the inside. For stone fruits like peaches or apricots these translate into the peel, the flesh and the pit or stone.

It will be recalled that other parts of the flower may contribute to the tissues that we refer to as the fruit. Fruits are conveniently classified according to their structure.

Types of fruit

Simple fruits

These consist of a single, enlarged ovary, with which some other flower parts may be incorporated. They are the most common types of fruit and can be subdivided as follows.

FLESHY FRUITS (PERICARP FLESHY AT MATURITY)

- Berries. The ovary wall is fleshy and consists of two or more carpels containing seeds, e.g. grapes, tomatoes and blackcurrants. The outer wall (exocarp) may be hard, as in tamarillo (Fig. 3.12), or leathery, as in citrus fruits.
- Drupes. The drupe is a fleshy, one-seeded fruit. It contains a thin outer wall (the exocarp), a fleshy mesocarp and a woody endocarp enclosing a single seed; examples include cherries, peaches (Fig. 3.19), apricots, plums, nectarines and olives. Because of the hard endocarp in *Prunus*, fruits of this genus are often referred to as 'stone fruits'.
- Pomes. These are derived from several fused carpels surrounded by accessory floral tissue (hypanthium). The hypanthium is fleshy and constitutes the flesh of an apple; the inner portion, the pericarp, is papery and constitutes the core. Examples are pears, apples and quince. They are referred to as 'pip fruit' or 'pome fruit' (Fig. 3.8).

DRY FRUITS (PERICARP DRY AT MATURITY)

- Dehiscent fruits (splitting open when ripe). This category includes several subdivisions, but few of these fruits have importance in this book. Examples are peas, beans, poppies and crucifers.
- Indehiscent fruits (not splitting open when ripe).
  - Achenes: these are one-seeded, and the seed is attached to the ovary wall at one point only. Examples are clematis and buttercups, while strawberries produce an aggregate fruit with achenes (Fig. 3.14).

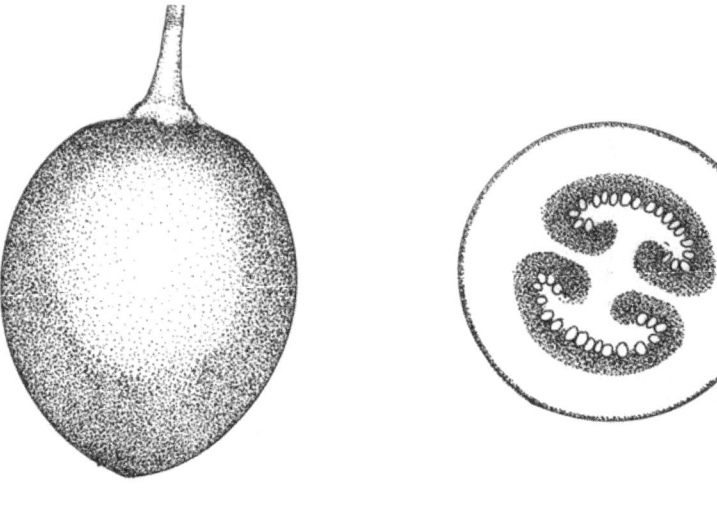

Fig. 3.12. Fruit of tamarillo. This is a berry and the cross section on the right shows a syncarpous ovary with two carpels.

– Caryopsis (grains): these are one-seeded, and the pericarp is firmly united to seed all around. Examples are cereals and grasses.
– Nuts: these are hard, one-seeded fruits, generally produced from a syncarpous ovary. Examples are sweet chestnuts, macadamias, hazelnuts and pistachios. The almond, walnut and pecan, while usually called nuts, are botanically classified as drupes.

Aggregate fruits

An aggregate fruit is one composed of a single receptacle upon which are massed similar fruitlets. It is derived from a single flower having many free carpels (apocarpous) (Fig. 3.13). In blackberries and raspberries, the individual fruits of the aggregate are drupelets (Fig. 3.14a). In strawberries the individual fruitlets are achenes, with the fleshy part being entirely derived from the receptacle (Fig. 3.14b).

Multiple fruits

A multiple fruit is derived from the ovaries of many separate yet closely clustered flowers, e.g. mulberries, pineapples.

Buds

Buds may be considered as unelongated shoots, with or without flower parts, and form either at the tips of shoots or in the axils between the leaves and the stem. Those occurring at the shoot tip form when the primordia, which were developing into leaves, form scales (Fig. 3.15). These scales enclose the apical meristem, which may later produce primordial leaves and flowers (Fig. 3.16). A similar bud may originate from the meristem in the leaf axil (Fig. 3.17).

In deciduous plants the initiation of floral primordia generally occurs towards the end of the growing period of the shoot which carries it. Subsequently, some development occurs during late summer, autumn and early spring, before the bud opens.

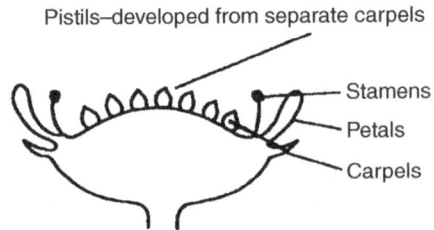

Fig. 3.13. Basic structure of aggregate flower.

D. Jackson and R. Edwards

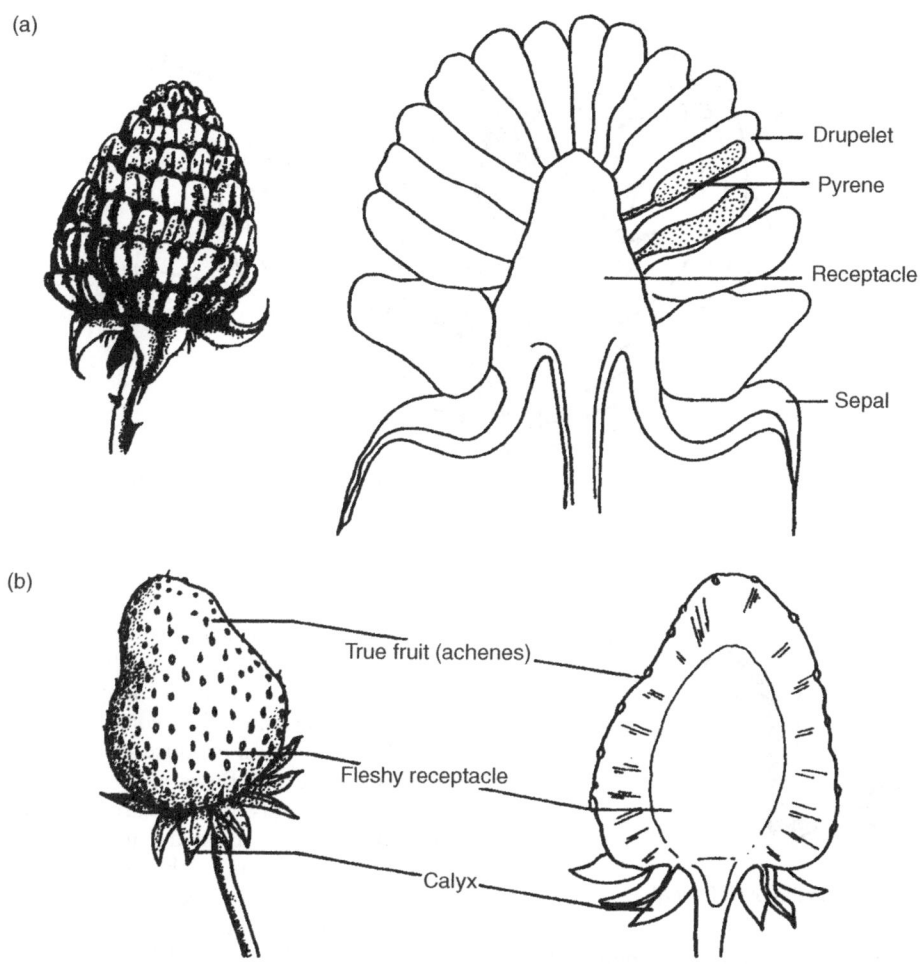

(a)

Drupelet

Pyrene

Receptacle

Sepal

(b)

True fruit (achenes)

Fleshy receptacle

Calyx

Fig. 3.14. (a) Aggregation of drupelets – raspberry. (b) Aggregation of achenes on fleshy receptacle – strawberry.

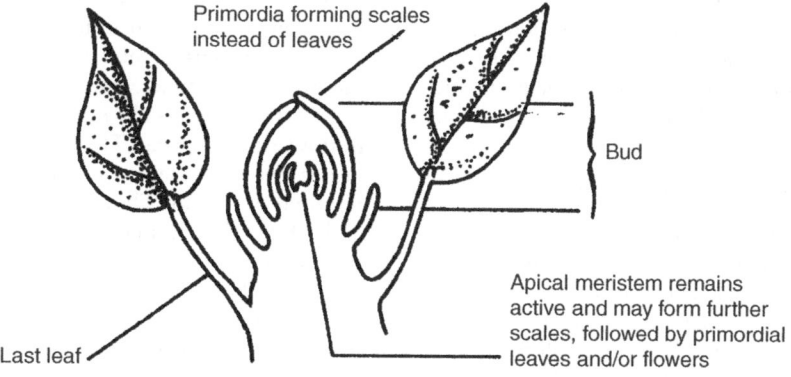

Primordia forming scales
instead of leaves

Bud

Apical meristem remains
active and may form further
scales, followed by primordial
leaves and/or flowers

Last leaf

Fig. 3.15. Formation of apical buds.

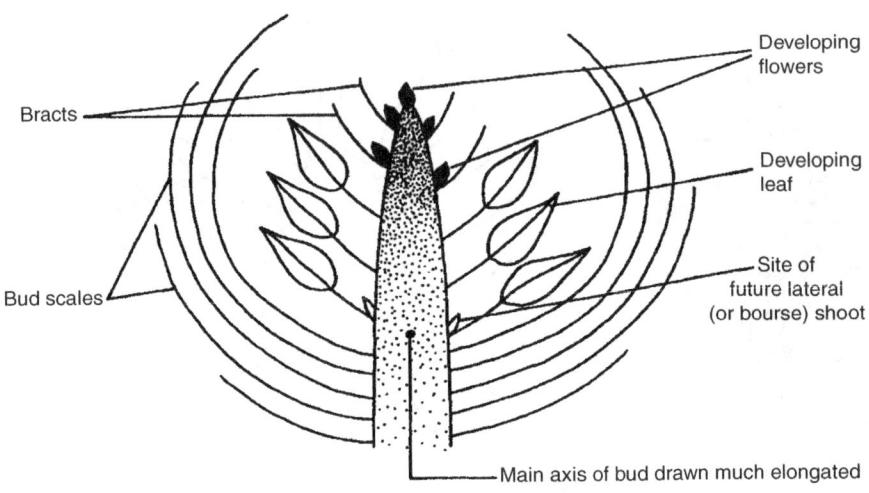

Fig. 3.16. Stylized apple bud.

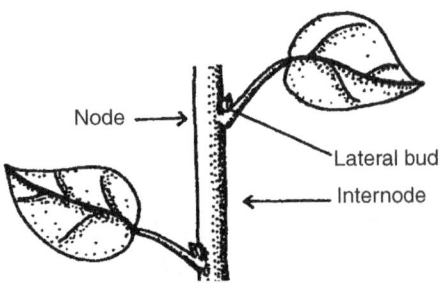

Fig. 3.17. Lateral bud in the axil of a leaf.

Buds are of two main types: simple buds, which contain leaves or flowers, but not both, and mixed buds, containing both leaves and flowers.

Plants may be conveniently grouped according to whether the flower bud is placed terminally or laterally on the shoot; whether the type of bud that occurs is mixed or simple; and whether within the bud the flower or inflorescence is terminal or lateral. Table 3.1 summarizes the types of buds and some examples of fruit plants which contain them.

Examples of bud types

Figure 3.16 shows a terminal bud of an apple examined under a microscope. The flower parts have been drawn in a stylized fashion.

The outer bud scales will begin to form in late spring and the structures shown in Fig. 3.16 will be formed by midsummer. By the end of summer the individual flower parts will be present, in an undeveloped form.

Little development occurs in winter but in spring further and rapid growth occurs in all parts.

Buds of other deciduous tree fruits show similar patterns. Stone fruit, for example, having simple buds in leaf axils, are less complex and develop a little later. They still produce scales and bracts and have one or several flowers.

The flower buds of grapes are produced laterally in axils of leaves. Buds are mixed and, after the formation of about eight primordial leaves, a terminal flower inflorescence is formed. Thus in Table 3.1 one might expect them to be classified alongside raspberries and blackberries. However, in the axil of the eighth leaf, a new primordial shoot is formed. It pushes the inflorescence into a lateral position and assumes itself a terminal position. Tendrils are produced in a very similar fashion to inflorescences and occur when the bud has formed its normal inflorescence number (usually two). In citrus, bud initiation takes place in the previous summer, but, unlike most deciduous trees, citrus flowers do not form until quite close to flowering in spring. The buds of citrus do not fit neatly into any category in Table 3.1. Buds may be terminal or lateral. They may be simple or mixed, and, if the latter, may have terminal or lateral flowers, or both (Fig. 3.18). An understanding of where and how buds are formed is most important for the fruit grower. Such information forms the basis of enlightened pruning and allows the grower to adapt to the plant, rather than prune according to a formula. The development of peach is

D. Jackson and R. Edwards

Table 3.1. Types of flower buds in common fruit trees.

	Simple buds[a]	Mixed buds[b]	
		Inflorescence within bud is terminal	Inflorescence is lateral
Buds produced terminally on (short or long) shoots	Loquat	Apples, pears, female walnut and pecan flowers	Guavas, olives
Buds produced laterally on the shoot	Stone fruit, currants, gooseberries, hazelnuts, male pecan and walnut flowers	Raspberries, boysenberries, blackberries	Figs, passionfruit, grapes, kiwifruit, persimmons

[a]Flower buds contain no leaves; [b]contain both leaves and flowers.

Fig. 3.18. The mixed bud of a citrus tree, which has opened and produced a shoot bearing flowers and leaves.

shown in Fig. 3.19, which summarizes the processes described in this chapter.

Growth of Fruit Plants

Scientific investigations have increased our understanding of plant growth quite dramatically in recent years. Some of this knowledge is complicated and of little relevance to the grower of fruits, but the basic conclusions are important and help us to make wise decisions in the field. This section gives some further information about the structure of plants, both above and below the ground, how they grow and why they respond in certain ways. Later on, in discussing more practical aspects of fruit growing, pruning in particular, these structures and ideas will be further amplified. Understanding them can often mean the difference between an ordinary grower and a good one: a good grower has a feeling for the tree or vine, which is born of this understanding.

Growth of shoots

At the end of winter, the buds of temperate trees and bushes break and start to grow at a rapid rate. As discussed in the previous section on morphology, these buds may be positioned laterally on branches or at their tips. In most cases, they were formed in the previous growing season and remained dormant over the winter

As already described, buds are of two main types: mixed buds, which contain leaves and flowers, and simple buds, which contain leaves or flowers alone (Table 3.1). From leaf or mixed buds, shoot growth begins in spring and continues for a period of time, depending on the position of the shoot on the plant and on the type of plant. In kiwifruit, grapes and other vines, shoots continue to grow throughout the summer, or most of it. It is possible that this is a biological adaptation to enable the vine to remain in a position above the bushes or trees on which it depends for support and in which position it will maintain active photosynthesis. In most other woody plants, rapid growth begins in the spring, continues for a finite period and then slows down and stops, usually for the remainder of the season.

A number of factors will determine the length of this growth period, often called a growth flush, and the following conditions will tend to extend it:

- A warm and humid climate (especially if soil, water and nutrients are plentiful).

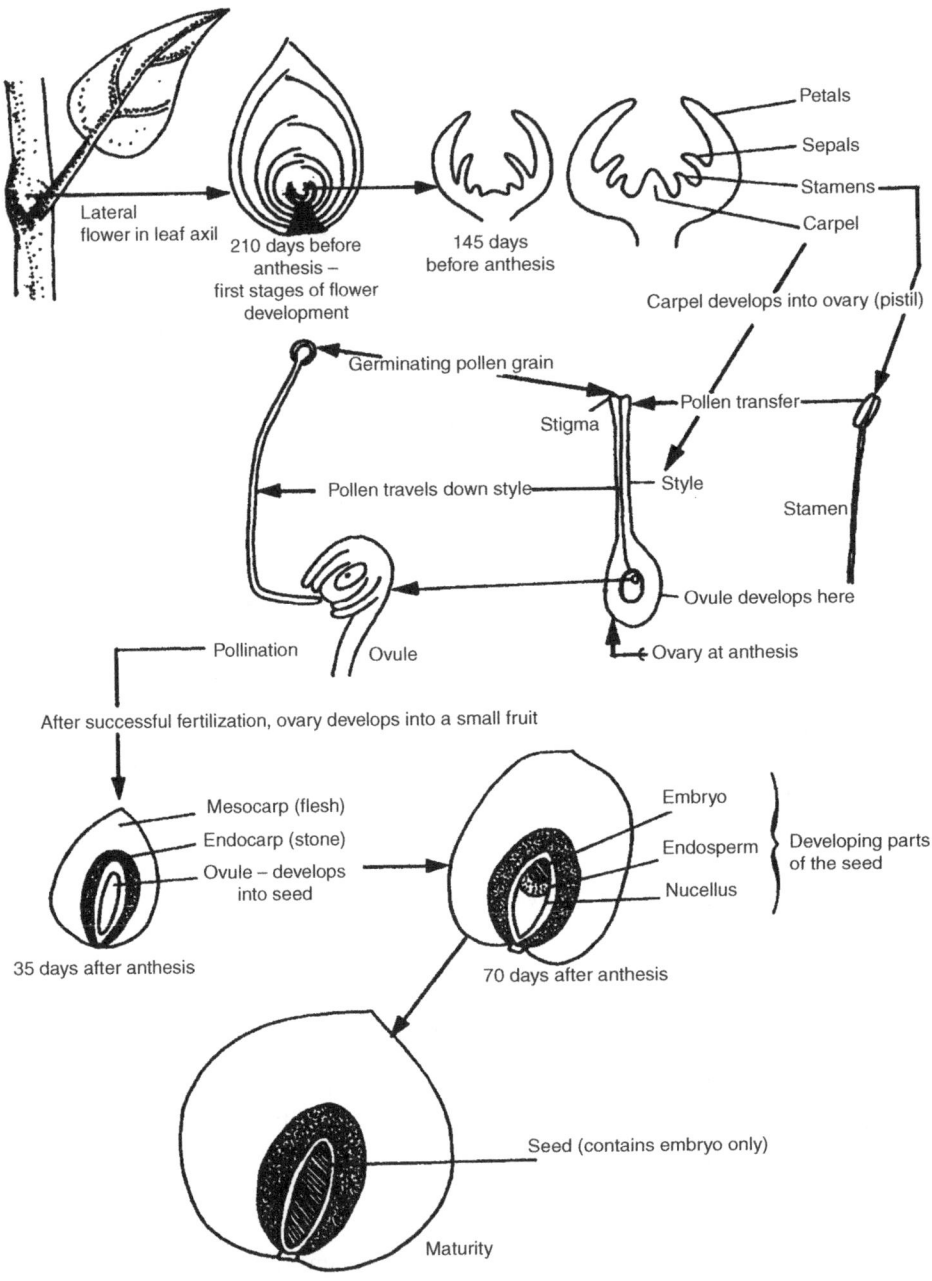

Fig. 3.19. Summary of development of a peach fruit.

- Young plants. Young trees will grow for a long period, often for the whole of the available growing season.
- Favourable bud position. Shoots from buds in dominant positions on the tree, such as buds higher on the tree, buds closer to the main framework and buds on upright shoots, especially if previously pruned hard.

Buds on horizontal branches, far removed from the main trunk and lower in the tree canopy, will

D. Jackson and R. Edwards

tend to be weaker. As will be seen later, however, such buds tend to be more fruitful.

Factors controlling shoot growth

The above-mentioned factors which affect shoot growth probably exert their control by means of hormones, sometimes called plant growth substances. A hormone is a chemical substance, present and active at very low concentrations, which regulates the growth and development of a plant. Such chemicals may be produced in one area and exert their influence elsewhere. Scientists have discovered some artificial compounds which are similar to naturally occurring hormones, although chemically slightly different. The term plant bioregulator is commonly used for these synthetic hormones. The major groups of known hormones are auxins, gibberellins, cytokinins, abscisic acid and related compounds, and ethylene.

Auxins and gibberellins are hormones which promote growth by the induction of cell expansion and elongation. Gibberellins applied to plants have more spectacular effects than auxins, although some synthetic auxins with exceptionally high activity are excellent broadleaf weedkillers (e.g. 2,4-D and other phenoxy herbicides). Important sites of natural auxin and gibberellin production are developing seeds and young expanding leaves. Cytokinins are often referred to as 'cell-division factors', since they are known to have a special role in enabling cells to divide. Root tips and developing seeds are important sites for cytokinin production.

There are several natural growth substances that are not promotive of plant growth and development. Rather, they act to counteract the effects of growth promoters. Abscisic acid is one such substance. It is known to have an important role in the adjustment of plants to stressful conditions such as low moisture. Ethylene, the only gaseous plant hormone, induces fruit ripening in many species and can also slow shoot elongation, where it has an effect opposite to that of the gibberellins.

Thus, the content, distribution and balance of hormones can determine the growth of shoots. In other words, while shoot development is dependent on water, mineral nutrients and the products of photosynthesis, overall growth and development is controlled by a fine balance of hormones arising from tissues that are part of the shoot or that come from other organs, such as roots and fruit.

The purpose of leaves

Leaves serve two main purposes. First, they are small manufacturing units which, using the energy of sunlight, combine water and carbon dioxide to form carbohydrates. This is the process of photosynthesis. Carbohydrates are the fuel of plants. They may be used directly as an energy source or as building materials for other plant organs. They may be stored for later use as, for example, sugar or starch; sugarbeet and potatoes are obvious examples. Generally, growers should encourage maximum leaf exposure to sunlight, which will produce an optimum balance between growth and reproduction. Leaves shaded by other leaves are inefficient and represent wasted energy, while flower buds on shoots with such leaves tend to be unfruitful.

The second main purpose of leaves is to draw up water from the soil and control the water status of the plant. The steady loss of water (transpiration) through small orifices called stomata (plural – we mean more than one when we say 'stomata'; if it was just one, we would say 'stoma') 'pulls' water through the roots, trunk, branches and twigs. These stomata close at night and open or close at other times, depending on the prevailing environmental conditions. To replace the water lost by transpiration, water is drawn up from the soil, bringing up mineral elements essential for plant growth, i.e. phosphorus, potassium, nitrogen, sulphur, calcium, etc. (Chapter 9). Other chemicals, such as cytokinins manufactured in the roots, may be carried in this transpiration stream to the upper parts of the plant.

Loss of water by transpiration helps to keep leaves and plants cool in hot weather, but, of course, depletes the soil moisture. If this is not replaced, it will lead to water stress. Plants reduce water stress either by modifying leaf structure and distributing stomata to reduce evaporation, as do eucalyptus, or by producing an extensive and strong root system, as do grapes. Stomata on all plants close under conditions of water stress.

Apical dominance

The way a tree organizes itself into a typical form or shape is controlled by a phenomenon called apical dominance.

Plant hormones are believed to play important roles in the manifestation of apical dominance.

Auxin produced in the tip of a growing shoot is transported down the stem, where it inhibits the development of lateral buds present in leaf axils. Cutting off the tip of a shoot or the top of the tree will temporarily destroy apical dominance until a new bud or shoot takes over dominance.

Further away from the apex, bud inhibition decreases and some lateral buds will develop. However, cytokinins produced by actively growing roots will stimulate axillary buds to develop into short shoots (spurs) or even into strong lateral branches. Thus, it is the balance of auxin and cytokinins, and perhaps other hormones such as abscisic acid, that influences apical dominance, and this balance can change in response to growing conditions, pruning, nutrition and other environmental factors.

Fruit tree species with strong apical dominance have a different appearance to those with weak dominance. For example, peaches have weak apical dominance, branch freely, commencing 20 cm or less from the shoot tip, and a young tree from the nursery will have many lateral branches. Sweet cherries have stronger apical dominance. Trees of this species branch less freely and are likely to be purchased from the nursery as a single 'whip' with few lateral branches.

The overall form of a mature tree is also governed by a type of apical dominance. Once a lateral bud has burst, its subsequent growth may still be under apical control. A Norfolk Island pine, for example, has a strongly dominant main stem which subjugates lateral branches to a very flat or horizontal orientation. A peach or hazelnut does not, and more than one leading arm may be formed, giving the tree a bushy appearance. Apples and pears are intermediate and are easily adapted to a central-leader method of tree training.

Gravity

A lateral shoot on a horizontal branch will grow for a longer period if it is closer to the centre of the tree. On an upright branch, however, the apical tip grows stronger and suppresses lateral shoot growth. If, therefore, an apical shoot with apical dominance (Fig. 3.20, left) is pulled away from the vertical (right), apical dominance is reduced and the effect of proximity to the main trunk is accentuated.

In apples, pears and European plums, this movement to the horizontal also encourages the formation of short shoots, called spurs, and may increase the fruiting potential of the branch. The reasons why the branch responds in such a manner are not fully understood, but it seems quite likely that gravity, acting on stems oriented at different angles, affects the distribution of hormones, which in turn modifies growth and perhaps flower-bud initiation.

It must be noted that all plants react in their own way to apical dominance and gravity. Of the common fruit plants, those which flower on spurs tend to react most conspicuously. Vines and trees that flower on 1-year-old wood respond less to bending or spreading in terms of growth, and in terms of flower-bud initiation, apparently not at all.

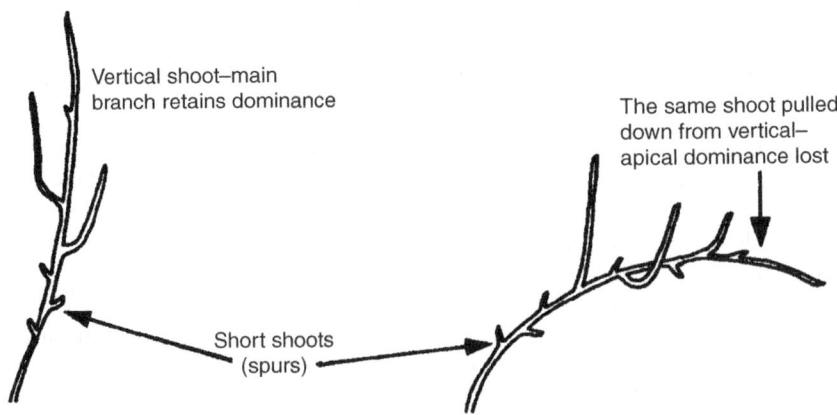

Fig. 3.20. Strong and weak apical dominance due to branch orientation.

D. Jackson and R. Edwards

Rootstocks

In apples and some other fruit plants, the vigour and productivity of a tree can be modified by budding or grafting the selected commercial cultivar on to a specifically selected rootstock seedling or clone. In most cases these rootstocks are clonal in nature and are obtained from stool beds or by rooting cuttings obtained from a mother tree (Chapter 11). However, they may also be derived from growing the seed of a suitable species or cultivar.

Dwarfing rootstocks for apple reduce tree vigour and tend to cause more flower buds to form on young trees. Such rootstocks are said to have induced 'precocity' when this results in higher yields on young trees. At least three ways in which rootstocks could reduce tree vigour have been suggested. First, cross sections through the trunk of dwarfing rootstocks reveal a higher proportion of bark relative to wood than in non-dwarfing rootstocks. This may simply reflect differences in growth rate or it may alter the patterns of transport in such a way that vegetative growth is reduced. The second possibility is that hormones produced by the roots affect the vegetative growth of stems. Thus, dwarfing rootstocks may produce fewer growth promoters. A third suggestion is that dwarfing rootstocks simply produce smaller root systems or are less efficient at absorbing water and nutrients. There is, as yet, insufficient evidence to prove any of these theories. It is certainly known, however, that the most dwarfing apple rootstocks produce trees that must be supported with a post or trellis system, suggesting small and relatively confined root systems.

Dormancy

Dormancy is a state in which viable buds and seeds will not grow or germinate even though conditions of moisture, temperature and oxygen are favourable for growth.

Apical dominance is one manifestation of bud dormancy. In this situation, lateral buds are inhibited from bursting and may remain dormant until the following season, for several seasons or even permanently, when they eventually disappear. If buds do not burst during the season that they are formed, their level of dormancy will increase over the autumn and will no longer be controlled only by the apical bud. They have achieved winter dormancy – an essential feature in deciduous woody plants which have evolved where winters are severe. Dormancy prevents buds bursting in warm spells in late autumn or late winter, times when a sharp frost could kill them.

Dormancy is overcome by exposing buds or seeds to temperatures above freezing but below a threshold level known for that species. The number of chilling hours or days required also varies among cultivars and species. This is equivalent to the time when, in its natural conditions, the plant would be past the danger period and buds could be allowed to burst as spring approaches.

Dormancy is a complex phenomenon but is probably regulated by natural hormones. Bud and seed dormancy can certainly be broken by gibberellin applications. Cytokinins will break summer dormancy of buds inhibited by apical dominance. Physical factors may also play a role: hard seed coats or scales surrounding buds may physically restrict growth or reduce gaseous exchange to the meristem.

In areas with mild winters, lack of winter chilling may mean incomplete breaking of dormancy by spring. As a consequence, not all buds burst at the same time and flowering is extended, sometimes over a 3- or 4-week period. Some buds never break at all.

Theoretically, it should be possible to control dormancy by manipulating natural hormone levels or by applying synthetic plant bioregulators. For example, one might use inhibitors to encourage the onset of winter dormancy or delay flowering as a technique to reduce risk of spring frosts. Unfortunately, this approach has not proved feasible, possibly because the wrong bioregulators or combinations of bioregulators have been used, because the bioregulators do not reach the correct site, or perhaps because hormones may not provide the whole answer to the problem. On the other hand, cyanamide, a plant bioregulator used to break bud dormancy in the spring, is now widely used to facilitate the production of temperate zone fruit crops in the subtropics.

Growth of roots

The function of roots

The roots have the following important functions:

- They support the plant for the duration of its life. This anchorage requirement becomes considerable as the size of the tree increases.

- They provide moisture for the plant. This is not only the water the plant needs to build or repair its structures, but also the water needed for transpiration. Under hot, dry and windy conditions the volume of water needed is much greater.
- They absorb nutrients, such as nitrogen, potassium, phosphorus and a host of minor elements from the soil. Other materials can also be absorbed; for example, poisonous chemicals may kill the plant by absorption through the roots, and it is likely that vitamins, plant growth substances and other complex organic compounds can also gain entry in this way. Roots may also have an association with mycorrhizal fungi which assist the uptake of nutrients that the plant may not otherwise be able to retrieve.
- Roots contribute to plant growth regulation by producing and transporting hormones or other chemical signals, both to the roots and to the above-ground organs of the plant.
- Roots store organic compounds and minerals for use on subsequent occasions, as required. Storage also occurs in the above-ground portions of the plant. Re-mobilization of stored carbohydrates, amino acids and minerals is especially valuable during or after stress periods, when regrowth is demanded, or in spring to develop shoots and flowers.

Growth of roots through the soil

To fulfil the functions described above, roots must grow within the soil. This growth is random and will depend on the type of root and the nature of the soil, especially its structure and texture. Roots do not grow towards areas of high water or nutrition. Nevertheless, if the roots do happen to gain access to water or nutrients they will proliferate and more fully exploit an area.

Figure 3.4 shows the structure of the tip of the root. As it becomes older, the root takes on the appearance shown in Fig. 3.21.

Factors affecting the rate of root growth

The rate of primary root advance into an unexploited area of the soil depends on three factors:

- Photosynthetic products supplied currently or from a stored source. If the plant has no reserves

and if photosynthesis is minimal, there is no chance of rapid root growth.
- Temperature. The warmer the temperature, up to 35–40°C, the more rapid the growth rate. Air temperature has little direct effect on root growth – in other words, if the air is warm and the soil is cold, shoots may grow rapidly and roots hardly at all. The reverse is also true, and sometimes, for propagation purposes, the soil temperature is artificially increased above that of the air to differentially encourage root growth.
- Soil structure and texture. There must be air movement through the soil to carry oxygen to the roots and allow carbon dioxide to escape. In soils with poor structure, especially if particle size is small and waterlogging occurs, roots may die due to lack of oxygen and the build-up of carbon dioxide to toxic levels. Soil structure also affects the amount of water that is held and may become available to the plant, as well as the resistance the root must overcome if it is to continue its advance. Generally, a moist soil is less resistant than a dry one. Soil pore size will also change the degree of branching in a root. Small pores may arrest the advance of the primary root and induce branching, so that the (initially) smaller secondary roots can continue to develop in and exploit the soil. Soils with large pore size, like sands and gravels, have roots which branch less and penetrate further. The roots will range over a wider volume of soil but will not exploit it so thoroughly.

The relationship between the roots and the whole plant

The effectiveness of a root system can, in the final analysis, only be measured in terms of the

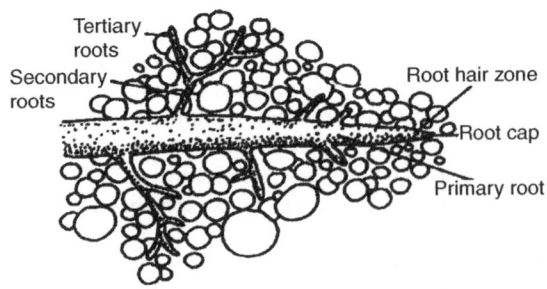

Fig. 3.21. Root structure.

D. Jackson and R. Edwards

whole-plant performance. It is only recently that the subtleties of the relationship between the root and the shoot of the plant have begun to be understood. For example, it has been discovered that there is a fairly stable relationship between the dry weight of the shoots and the roots. This is generally in the ratio, shoots to roots, of 2:1. If part of the top is removed, plants will compensate by growing more top at the expense of the roots. Similarly, trimming roots will cause relatively more regeneration here until the ratio is re-established.

Assuming that this ratio is not unduly upset, it has been found that the top of the plant creates the demand for nutrients and water; it is not the roots by themselves which control overall plant growth.

However, when a tree is under stress the shoot:root ratio decreases.

Further Reading

Heywood, V.H., Brummitt, R.K., Culham, A. and Seberg, O. (2007) *Flowering Plant Families of the World*. Kew Publishing, London.

Kozlowski, T.T. and Pallardy, S.G. (2008) *Physiology of Woody Plants*, 3rd edn, Academic Press, London.

Maib, K.M., Andrews, P.K., Lang G.A and Mullinix, K. (eds) (1996) *Tree Fruit Physiology: Growth and Development: a Comprehensive Manual for Regulating Deciduous Tree Fruit Growth and Development*. Good Fruit Grower, Yakima, Washington.

4 Flowers and Fruit

DAVID JACKSON

Flower Initiation

Chapter 3 discussed types of buds and their formation. The first part of this chapter will consider some of the factors that promote or inhibit the initiation of flowers within the buds. It will be recalled that initiation means the formation of flower parts within the bud, or the change of a bud from a vegetative one to a reproductive one. Sometimes the early stages are termed induction, the later ones development.

Flower-bud initiation in deciduous fruit trees and shrubs occurs some time between leaf formation in the spring and leaf fall in autumn, the exact time depending on the species concerned. Some fruits, for example blackcurrants and many strawberries, initiate flowers during a period of short days, such as occurs in autumn before the leaves fall, but most fruit crops are insensitive to day length. The factors controlling flower-bud initiation are operative 6–12 months prior to the actual emergence of flowers. In evergreen fruit trees, flowers normally form within the buds in the same season, usually just before the flowers appear.

Factors affecting initiation

The following factors or practices have been shown to modify the initiation of flowers, but there is seldom a single limiting or promoting factor.

Crop load

With crops such as apple, where most of the fruit is borne on short shoots (spurs) on 2- to 4-year-old branch sections, the most important factor influencing whether a flower will be initiated on that spur is the presence or absence of a developing fruit. Developing fruits inhibit flower formation either by competing for energy and mineral resources or by producing and exporting hormones that encourage growth rather than flowering. Both mechanisms may be involved. This relationship between fruiting and flower initiation on apple spurs is the reason for early fruit thinning (removing all of the fruit from some of the spurs) to maintain annual cropping.

Light

The products of photosynthesis (carbohydrates) must be in adequate supply if flower initiation and development is to proceed satisfactorily. Thus, low light intensity, such as is found in the lower or interior portions of a tree, vine or bush, will restrict the initiation of flowers. An important reason for pruning and training is to distribute light more evenly through the canopy so that the lower and interior parts of the plant do not decline too much in fruitfulness.

Nutrition

Generally, good nutrition assists in the process of flower-bud initiation. Nitrogen seems to be the important element: it has been shown to promote flower-bud formation in apples and other fruits. There may be conditions, however, when excessive nitrogen, by promoting and prolonging excessively vigorous shoot growth and increasing shading, will have the opposite effect.

Water

Markedly reduced water availability to the roots can limit bud initiation in many fruits. On the other hand, in citrus grown in dry areas, it has been shown that flower-bud initiation can be promoted by withholding irrigation at the time initiation is required. Thus, limited water stress may, in certain plants, be conducive to initiation.

Temperature

High temperatures tend to stimulate vegetative growth and may thus accentuate the depressing effects of poor light penetration on flower-bud initiation. In grapes, however, initiation is known to be better if daytime temperatures at the time of induction are high, irrespective of the light regime.

Gravity

Bending branches of apple, pear, plum and cherry trees to a horizontal or even subhorizontal position reduces vegetative growth and promotes flower-bud initiation.

Rootstocks and tree vigour

Reduced vigour and good internal light conditions imparted by dwarfing rootstocks are often credited for the earlier onset of flowering and fruiting in young apple trees. The specific relationship to dwarfing, however, is not yet clear, since some size-controlling rootstocks for apple do not show this benefit. Furthermore, evidence to date with sweet cherry suggests that dwarfing rootstocks may not promote precocity. None the less, cessation of shoot elongation is a requirement for flower development in all apically flowering plants, and it is generally assumed that even with species such as peach, where flowering occurs in leaf axils, slowing shoot growth allows more energy to be diverted to bud formation. The involvement of root-produced hormones in the regulation of vigour and flower-bud formation is certainly implied but remains to be convincingly demonstrated.

More often than not, high tree vigour is a negative factor in relation to flower initiation. This has already been mentioned within the context of nitrogen nutrition, but there are other ways that this knowledge can be used by producers to improve flowering and cropping. For example, branch positioning is effective. An upright branch will exhibit stronger growth than a horizontal one but tends to form fewer fruit spurs. A young tree tends to grow for a longer period during the season and will initiate few flowers, while an older tree usually finishes

growth sooner and develops more flower buds. Thus, plant bioregulators capable of slowing or even halting shoot elongation are particularly beneficial when used on young apple trees.

Hormones

The plant bioregulators known to slow shoot growth – chemicals like ethephon, daminozide and paclobutrazol – generally do so by inhibiting the natural gibberellins that promote shoot elongation in trees and vines. It is not surprising, then, that gibberellins are often inhibitory of flower formation. This can easily be demonstrated by applying gibberellic acid to peach trees (Fig. 4.1). Trees treated with a high rate of gibberellic acid will develop

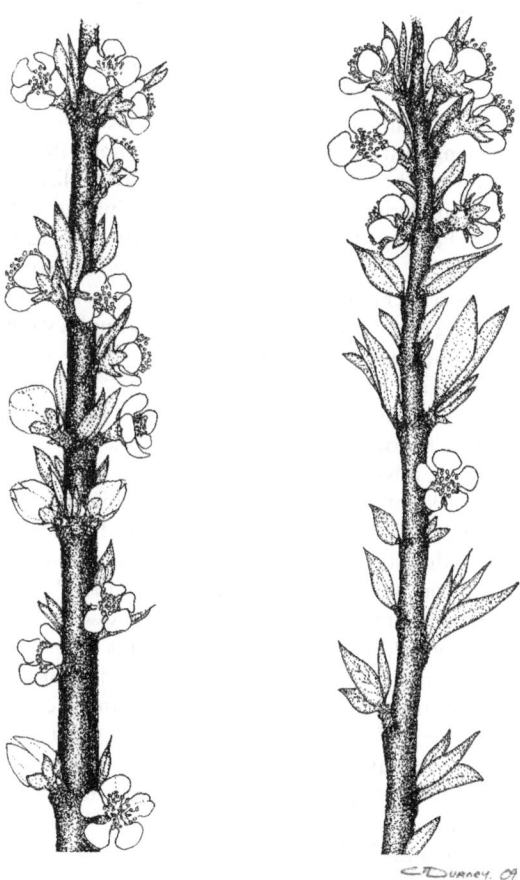

Fig. 4.1. Influence of gibberellic acid treatment on the flowering of peach. Note the high proportion of vegetative buds on the shoot (right) from the tree sprayed with 75 ppm gibberellic acid 10 months earlier.

spindly shoots with long internodes and very few flowers. At lower rates the growth effects are not so evident but the effect on flowering persists.

See Chapter 8 for a more in-depth discussion of plant bioregulator usage in fruit production.

Biennial bearing (also called alternate bearing)

Biennial bearing is chiefly a problem in apples and olives, although it does occur in pears, some stone fruits (e.g. apricots and plums) and tropical fruits. It means that a heavy crop is followed by a light crop. It afflicts whole trees, even whole orchards or districts. Biennial bearing is specifically a problem involving inadequate flowering and should not be confused with poor fruit set due to inadequate pollination.

However, biennial bearing is often triggered by some event, such as frost or disease, causing a severe reduction in crop in one particular year. Flower initiation is excessive in that year and the following-year crop is so large that it limits flower initiation. The biennial bearing cycle is established.

The underlying physiological principle is that spurs bearing fruit in one year are generally unlikely to flower the next, i.e. individual spurs exhibit a biennial tendency. A tree flowering and cropping every year will always have some 'resting' spurs. Biennial bearing occurs when all the spurs on a tree are in phase.

One obvious way to reduce the effect of fruit on biennial bearing is to thin heavily in the 'on' year. However, it has been found necessary to thin within 3–4 weeks of full bloom to affect biennial bearing significantly. The only economical and practical way to do this is to use chemicals that remove flowers or very young fruits. 'Chemical thinning' is an important practice in most apple-producing districts but is less successful for other fruit crops.

One final point about biennial flowering should be mentioned – individual spurs become increasingly biennial as they age. Thus, using renewal pruning to maintain relatively young fruiting wood, coupled with branch positioning to maximize light penetration, will greatly reduce the susceptibility of fruit trees to the biennial-bearing problem.

Fruit Set

Fruit set is a somewhat general and vague term but is widely used in horticulture. It describes the retention of fruit on the tree, bush or vine for a certain imprecise period after full bloom. For example, if an apple tree has a large number of small fruits present one month after blossoming, one would talk about a 'good set'. If, however, many of these fruits drop in June/December, one would talk about a 'heavy drop' or 'poor final fruit set'. The term set implies that the flower has been pollinated and fertilized and, as a consequence, the ovary and accessory tissues will be capable of growing into a fruit. If fertilization does not occur, the fruit normally drops soon after flowering. However, there are other reasons why flowers drop or fruits fail to persist until full maturity. The following periods of flower and fruit drop are often mentioned.

- Early flower drop: flowers that abort before opening or fall during the flowering period are often small and, if examined closely, may be found to have undeveloped pistils.
- Late flower drop: flowers that drop shortly after anthesis commonly do so as a result of inadequate pollination or failure of fertilization.
- Mid-season fruit drop: often called 'June' or 'December' drop. This will be discussed in the section dealing with fruit growth.
- Pre-harvest fruit drop: this phenomenon will be discussed in more detail in Chapter 8.

While pollination and fertilization were previously mentioned, it is necessary to investigate these more closely and consider how they affect fruit set and the control orchardists may exert.

Pollination

When pollen is mature, the anthers open and pollen grains escape – this stage is called anthesis, a term also used to describe 'blossoming' or 'full-bloom'. Pollen is released over a period which varies with the species. Pollen release lasts for 1–2 days in sour cherries, 1–3 days in strawberries, 1–5 days in peaches and apples, 2–7 days in pears and 2–9 days in raspberries.

The stigma is receptive as soon as it becomes covered with the sugary exudates in which the pollen grains can germinate. Stigmas in this condition glisten and remain receptive until they dry and turn brown. The duration of the receptive period depends on the fruit species and climatic factors. In apples and pears, not all the stigmas in a single

D. Jackson

flower become receptive at the same time, so that the duration of receptivity is prolonged. In blackcurrants, the stigma becomes receptive before the petals unfold and remains in this condition for about 5 days.

There are two main methods whereby the pollen reaches the stigma: by air movement and by insects.

Wind pollination

This generally occurs in plants which have inconspicuous flowers. Pollen production is usually profuse and stigmas may be much branched to catch pollen. Wind pollination occurs in walnuts, hazelnuts, sweet chestnuts, mulberries and grapes; only light air movement is required. Peaches and nectarines are most often self-pollinating, in that the pollen transfers to the stigma by contact even before the flower opens.

Insect and bird pollination

In insect-pollinated plants, the pollen does not fall from the opened anthers but sticks to the outside of the shrunken anthers until it is picked up by insects, especially bees. Insects are necessary for the pollination of most fruit crops, with the exception of the crops mentioned above. Flowers which are insect pollinated are often showy. Bees and other insects are also attracted by secretions of nectar by nectaries, which are generally situated on the receptacle at the base of the petals.

Only one common fruit, the feijoa, is pollinated by birds. Birds, like insects, are attracted to the nectar and, in the process of feeding, transfer pollen.

Fertilization

In plants capable of self-fertilization, the pollen fertilizes the ovules in the ovaries of the same flower, other flowers on the same plant and/or flowers on different plants but of the same cultivar. With cross-fertilized plants the pollen is incapable of fertilizing ovules on that plant. The pollen must come from another cultivar of the same species or a plant of a closely related species. For example, the pollen of many crab apple species and hybrids (e.g. *Malus bacata*) will effectively pollinate commercial apple cultivars (*Malus domestica*).

On arrival at the stigma, pollen grains germinate and the pollen tube grows down the style. It penetrates the embryo sac via the micropyle and the nucellus (Figs 3.11 and 4.2). From the pollen grain one male sperm nucleus unites with the egg cell, contained in the embryo sac, to form a zygote. This becomes the embryo of the seed, which is diploid in its genetic content and is the part that can form a new plant. The second sperm nucleus unites with the two polar nuclei of the ovule to form the endosperm. The endosperm is triploid in genetic make-up and forms an important food reserve for the seedling that emerges at the time of seed germination. Unlike the embryo, the endosperm produces hormones. The other elements shown in Fig. 4.2, the antipodals, synergids and tube nuclei, seem to play no further role and disintegrate.

Growers have very little interest in the seed as such, but they should remember that if the seed does not form, due to lack of pollination or fertilization, in most cases fruit set will not occur.

Two abnormalities in the process of fertilization are worthy of mention.

Apomixis

This term describes the production of seed without fertilization. Citrus species are the only common temperate or subtropical fruit plants where this occurs. In species capable of apomixis, the seed

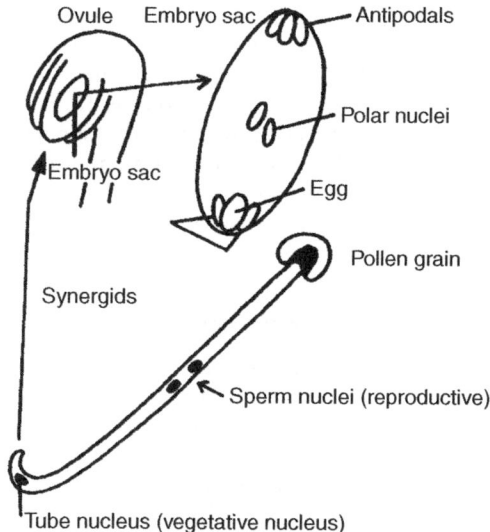

Fig. 4.2. Embryo sac and germinating pollen grain.

may grow to produce several seedlings. One is produced the normal way and will have the characteristics of both parents – it will be sexual. The others originate entirely from part of the seed which has not been involved in fertilization. These will be identical to, and indeed are clones of, the female parent plant.

Parthenocarpy

Sometimes fruit is formed in the absence of pollination, fertilization, or both. This is called parthenocarpy. It is found in the seedless 'currant' grape, Washington navel oranges and bananas, and may be induced in plants which are normally seeded by the application of hormones.

Lack of pollination and fertilization prevents set of all fruit that are not parthenocarpic. Because set is so important, it is necessary to understand the factors which influence it. These can be conveniently discussed under two headings: 'tree factors' and 'external or environmental factors'.

Tree factors causing poor fruit set

Absence of viable pollen

The flowers of some fruits, such as Washington navel oranges or Gravenstein apples, have sterile pollen, which is incapable of germinating or effecting fertilization. Other plants are termed dioecious, which means that the male flowers providing pollen are physically separated from the female flowers and are often on a different 'male' plant; an example is kiwifruit. In walnuts, hazelnuts and, to a lesser extent, a number of common tree fruits, the pollen is released before or after the stigma is receptive. This is known as dichogamy, and careful selection of complementary cultivars may be required to achieve satisfactory overlap (e.g. avocados).

Infertile eggs

Washington navel oranges have not only infertile pollen but also infertile eggs. The same is true of the currant grape and the flowering cherry. The first two species crop because of their inherent ability to set parthenocarpic fruit; the flowering cherry loses its fruit and bears no crop. Even in plants where pollen and eggs are normally viable, a proportion of the ovaries contain infertile eggs. In some stone fruit, up to 20% of all flowers may lack fully developed pistils; these flowers are often smaller than normal and may originate from smaller buds.

Incompatibility

This is a common cause of fertilization failure and occurs despite successful cross-pollination. In these cases the pollen tube is incapable of growing down the style due to genetic, and hence biochemical, factors within these two units. Usually the pollen tube starts to grow, but soon slows after penetrating a short distance into the style. Self-incompatibility is the most common form of incompatibility; it occurs between the style and pollen of the same cultivar. Self-incompatibility is expressed in most cultivars of cherry and almond, many plums and most apples and pears. Cross-pollination is essential, so these cultivars must be inter-planted with other cultivars to serve as pollinators. Sometimes cross-incompatibility occurs, in which case only selected cultivars will fertilize the chosen fruit. The three commonly grown fruits which need specific cross-pollinators are sweet cherries, European plums and pears, but there are specific examples of cross-incompatibility in many other crops. This specific knowledge becomes very important when one is considering a mixture of cultivars and expecting cross-pollination.

There are many degrees of self-fertility. Some apple cultivars are partly self-incompatible and, hence, when cross-pollinated will produce heavier or more reliable crops. Environmental factors may alter the degree of incompatibility; temperature is very important because of its effect on the rate of pollen tube growth. Given optimum temperature conditions the pollen tube of an incompatible pollen source may grow fast enough to reach the ovary before the death of the ovule.

External factors influencing set

The following external factors, some of which are under the grower's control, may affect the setting of fruit.

Mineral nutrition

Autumn applications of nitrogen have been reported to increase fruit set in apples. On the other hand,

D. Jackson

excessive nitrogen application promotes vegetative growth and reduces flower-bud initiation. Boron sprays have been used to improve fruit set of hazelnut, suggesting that minor elements might prove important in certain situations. It is commonly believed that fruit set can be reduced by faulty nutrition in general.

Pruning, thinning and girdling

Pruning may or may not improve set, depending on the extent and type. It might be expected, on theoretical grounds, that if pruning reduces the number of blossoms in relation to the volume of the tree, more food reserves would be available for the remaining blossoms, improving their strength and chance of setting. Thinning blossoms at or near blossoming is now possible using chemicals, and this too should improve nutrition of the remaining blossoms. Evidence is available to indicate that both practices improve set. However, this may not improve yield, especially if too many blossoms are removed or if the treatment is followed by a severe frost which kills many of the remaining flowers.

Girdling or 'cincturing' has been shown to improve set in several grape cultivars. At times it has been found to be successful on other fruits but it is not commonly used.

Age and vigour of the plant

Young trees tend to initiate and set less fruit than older trees.

Locality and season

Variations in set due to locality or season are common. The reasons are not always understood but some of the factors listed below are likely causes.

Temperature

Temperatures below 0°C (particularly −2°C or less) during bloom may kill the pistils. Low temperatures above freezing can interfere with set by reducing bee activity and hence pollination. Honey bees are relatively inactive until air temperature rises above 15°C. As already mentioned, low temperatures following pollination reduce set by preventing pollen germination or slowing pollen tube growth.

Light

The effect of light on fruit set seems to be important in species which flower after leaf formation. For example, low light intensity on individual grape clusters will reduce set. For this reason open canopies are more conducive to good set than dense, thick ones.

Rain

Rain during bloom is a common cause of poor fruit set. It can reduce set by preventing anther dehiscence and by washing off stigmatic secretions. It also reduces set because of associated weather conditions such as low temperatures, low light, reduced bee activity, etc.

Wind

Hot, dry winds may desiccate the stigmatic surface, thus preventing pollen retention and germination. Bee activity is greatest when wind speed is less than 3 km/h and decreases gradually with increasing wind speed, to a maximum of about 35 km/h. These effects have been well demonstrated by increased set in an orchard after the establishment of windbreaks. For wind-pollinated plants, such as walnuts, some light wind is an advantage.

Diseases and insects

Diseases and insects other than bees influence fruit set. Diseases such as brown rot can destroy flowers of apricot and cherry. Many other diseases reduce tree vigour and have a generally depressing effect on fruit set. Control of thrips has been shown to improve the set of apples. These insects consume pollen and destroy anthers and stigmas.

On the other hand, sprays during bloom may affect set indirectly by reducing the population of beneficial insects. It is unwise to use insecticides during blossoming.

Practical methods to improve set

Of all the factors considered above, those of greatest practical significance to the orchardist are incompatibility and pollinating insects.

Where incompatibility is a problem, data are given in the later chapters of this book for individual fruits.

Most orchardists bring bees into the orchards during blossom time. Sometimes these need to be hired from local beekeepers, but the cost will certainly be worthwhile in most crops. Usually, between two and five hives per hectare are required. For kiwifruit, however, it is common to place eight hives on 1 hectare.

The following factors will improve bee activity in the orchard.

- Remove competing nectar sources – kiwifruit, pears, plums and blackcurrants are not very attractive to bees and any other blossom source will encourage them to go elsewhere. On the other hand, if bees are to be kept in the orchard permanently, providing a range of plants which give year-round nectar will help maintain a healthy hive.
- Never spray insecticides at bloom time, and keep spray away from other flowers, such as clover, as much as possible at other times. Mow off flowering weeds. This is especially important if permanent hives are kept.
- A well-sheltered orchard is more attractive to bees, but keep in mind that deciduous shelter trees will not have leaves when crops like plums or apricots are in bloom.

Fruit Growth and Development

Fruit enlargement throughout the growing season is the result of cell division before and after anthesis, cell enlargement after anthesis, or both. The relative importance of these processes varies from one species to another. Generally, cell division predominates in the first few weeks after blossoming but overlaps the cell enlargement phase, which lasts until fruit maturity. Duration of the cell-division phase after anthesis is 2–3 weeks in apricot; 3–4 weeks in apple, peach, plum and grape; 6–8 weeks in pear; and 4–9 weeks in orange. It continues until maturity in avocado and strawberry.

If the cumulative increase in volume, weight or diameter of a fruit is plotted against time after anthesis, the resulting curve may be either sigmoid or double-sigmoid in character (Fig. 4.3). The type of curve bears no relationship to the structure of the fruit. The following fruits exhibit a sigmoid curve growth pattern: avocados, dates, oranges, apples, pears, strawberries and almonds. Apricots, cherries, peaches, plums, olives, rasp-berries, currants, grapes, kiwifruit and figs have double-sigmoid growth curves. The three distinct growth phases in the latter are termed Stages I, II and III. There is some evidence that kiwifruit has a triple-sigmoid growth curve.

Internal factors affecting fruit growth

In discussing the causes of fruit growth, it is useful to consider the possible role of hormones and the consequences of competition between the fruit and other plant parts before looking at external influences such as light, nutrients and water.

As already discussed, if a flower is not pollinated and fertilized, the potential fruit will fail to grow and it will drop at, or shortly after, full bloom. Parthenocarpic fruits are the exception. Since seeds are normally required, it is believed that they contribute a substance or substances necessary for fruit development. These are now felt to be hormonal, since unfertilized fruits can sometimes be induced to grow by the application of hormones. The type or mix varies: auxins, gibberellins and cytokinins will often induce parthenocarpy; sometimes only one or perhaps two will be effective; sometimes a combination is required; and sometimes none seems effective.

In fruits which have several seeds, such as kiwifruit, strawberries, apples and grapes, the size of the fruit usually bears a direct relationship to the number of seeds present. Furthermore, if the distribution of seeds is uneven, fruit growth can be lop-sided, with the side bearing the most seeds growing more. This suggests that seeds influence the growth of adjacent flesh, possibly by the migration of hormones from one tissue to another.

However, although the evidence suggests that there is a growth response of flesh to seed hormones, the flesh does not necessarily grow most rapidly when seed hormone levels are highest. In stone fruit, seed gibberellins are greatest in stage II – a time when overall fruit diameter increase is least.

Apart from affecting growth directly, hormones present in developing fruit function to prevent fruit abscission. Except for physical reasons, such as wind or predator damage, fruits drop as a consequence of changes within a specialized layer of cells between the fruit and the stem or the stem and the branch. This layer, called the abscission zone, is under the control of hormones, and separation generally occurs when the level in the fruit is low.

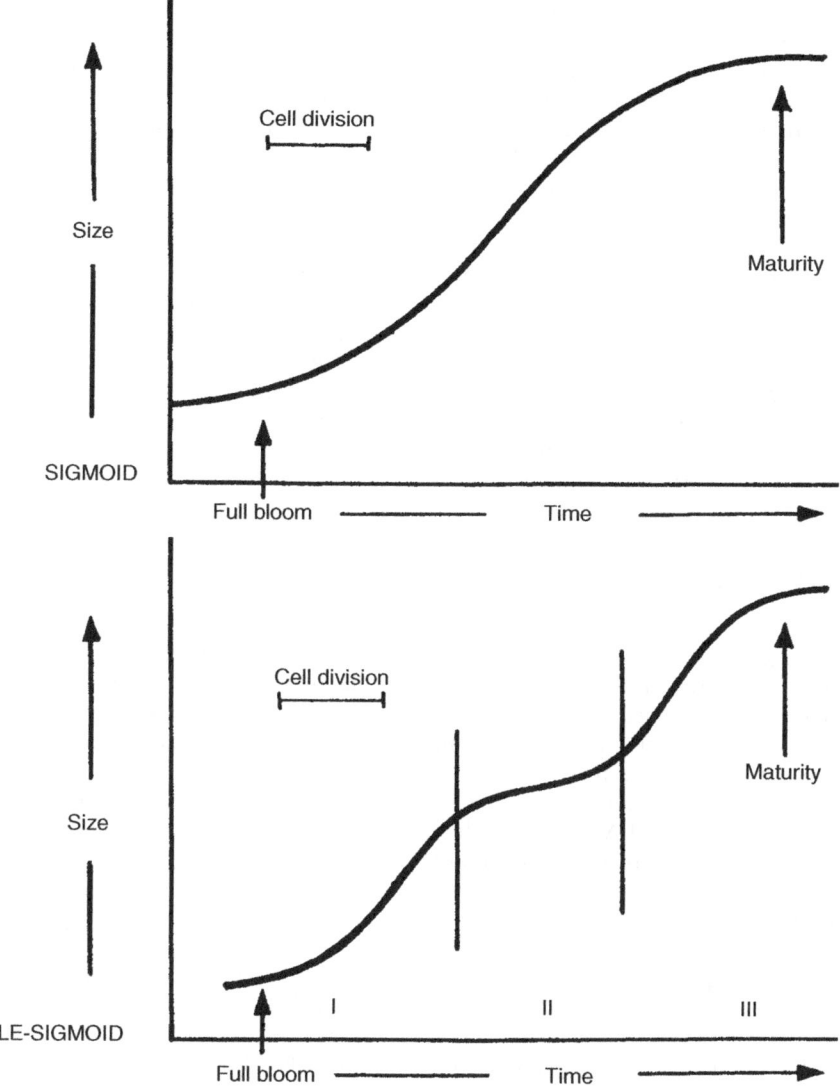

Fig. 4.3. Stages in the growth of fruit.

There are certain times in the development of a fruit when drop is more likely to occur. These are listed and discussed in the section dealing with fruit set. However, it is instructive to comment in more detail about the drop that occurs in mid-season, the June or December drop.

In apples this mid-season drop is found to be associated with the beginning of embryo growth in the seed. Since the embryo cannot develop without fertilization, this mid-season drop can be a delayed response to inadequate fertilization. However, when the initial crop load is very heavy, there is intense fruit-to-fruit competition for resources. This leads to the situation where only the strongest fruits are retained, regardless of fertilization status.

Abscission associated with the June/December drop can be reduced by optimizing pollination. Theoretically, it can also be improved by hormone application to the fruit, but this is seldom commercially feasible since, in most cases, these chemicals must be applied only to the fruit and not to the leaves.

Conversely, synthetic auxins, such as naphthalene acetic acid (NAA), can increase the amount of early to mid-season drop if applied to whole trees shortly after flowering. NAA is one of several such chemicals used to 'chemically thin' apples. However, the same chemicals work in the opposite direction if applied prior to harvest. They are then considered 'stop-drop' chemicals.

With several of the stone fruits, including apricots and some plums, a mid-season application of a synthetic auxin such as Fenoprop (2,4,5-TP) or Dichlorprop (2,4-DP) will reduce mid-season fruit drop as well as advance fruit maturity and increase average fruit weight at harvest. While no longer available to commercial producers in some countries, the dramatic effects of these auxins illustrate the important role that hormones can play in regulating fruit growth and development.

External factors affecting fruit growth

Despite the overall importance of hormones, it must not be forgotten that good-quality flowers and subsequent fruit growth is ultimately dependent on the presence of carbohydrates produced by past or current photosynthesis in leaves, and on water and nutrients transported from the soil via the roots. It is, of course, these factors which are most under the grower's control. While they will be mentioned again elsewhere, it is worthwhile reiterating the importance of the following factors in modifying the growth of the fruit.

Water

If water is in short supply, the fruit may go through its normal cycle of development but will not grow to its optimum size. It may be more highly coloured and possibly keep in storage longer, but generally its small size reduces its value – and the lower yield is not welcomed by the orchardist.

Nutrients

Low levels of mineral nutrients may directly influence flower and fruit development, but there are only a few cases where this has been clearly demonstrated. A deficiency of boron will cause distortions, such as cracking, in the developing fruit. Similarly, low calcium in fruit tissues is a well-known cause of disorders, such as bitter pit of apples, and can be corrected with foliar sprays

during the season. Potassium deficiency may reduce fruit colour. While lack of nitrogen tends to result in smaller fruit, very high nitrogen results in fruit with poor storage life.

While there are many commercial nutrient supplements that claim direct effects on cropping and fruit quality, these claims are seldom backed up with convincing evidence. Thus, it is probably more instructive to consider the effect that most nutrients have on fruiting as being indirect. In other words, a tree growing in a soil adequately supplied with nutrients will have a plentiful supply of healthy leaves, which will photosynthesize sufficient carbohydrates to produce high-quality flowers and support good fruit growth.

Light

Generally, fruit-growing areas are selected because of an abundant supply of bright sunlight. None the less, light remains an important factor for the orchardist to consider in relation to fruit growth. The problem is one of exposing the maximum number of leaves to direct sunlight. This is why dwarfing rootstocks (small trees have less internal shading) and pruning are so important. The well-pruned tree and the well-considered planting and training system will expose the maximum amount of leaves and thus provide sufficient carbohydrates to enable fruit to grow to full size.

Thinning

One of the reasons for thinning fruit is to influence growth by ensuring that the fruits which remain have sufficient water, nutrients and carbohydrates to enable the fruits to grow to full size.

Pests and diseases

It is not often realized that one of the most important reasons for good pest and disease control is to achieve good fruit set and fruit growth. The conspicuous effect of, say, codling moth on apples or brown rot on peaches makes the fruit unsaleable. But the effect of mites, for example, in reducing photosynthetic efficiency, or caterpillars or beetles in consuming a large proportion of the leaf blade can have effects which are almost as damaging. The reduced photosynthesis limits carbohydrate production, which in turn reduces crop load and decreases fruit size.

To summarize: in order to grow good fruit, the producer must ensure the availability of sufficient water, nutrients and carbohydrates. The plant uses these for all processes but regulates their distribution and controls fruit growth by a number of natural plant growth substances or hormones: auxins, gibberellins, cytokinins, inhibitors and ethylene. Humans have been less successful in amplifying or suppressing the effects of these materials, but a few examples, such as thinning and stop-drop hormone sprays, have been used by growers of some crops. An understanding of all aspects of fruit growth will help the orchardist produce better fruit.

Further Reading

Davenport, T.L. (2000) Processes influencing floral initiation and bloom: the role of phytohormones in a conceptual flowering model. *Horticultural Technology* 10, 733–739.

Davies, R.J. (1995) *Plant Hormones – Physiology, Biochemistry and Molecular Biology*. Kluwer Academic Publishers, Boston, Massachusetts.

Glover, B. (2008) *Understanding Flowers and Flowering: an Integrated Approach*. Oxford University Press, Oxford, UK.

Nyeki, J. and Soltesz, M. (eds) (1996) *Floral Biology of Temperate Zone Fruit Trees and Small Fruits*. Akadémiai Kiadó, Budapest, Hungary.

Sedgley, M. (1990) Flowering of deciduous perennial fruit crops. Horticultural Reviews 12, 223–264.

Twyman, R.M. (2003) Flowering and reproduction: flower development: In: Thomas, B. (ed.) (2003) *Encyclopedia of Applied Plant Sciences*. Elsevier, London, pp. 307–315.

Zeevaart J.A. (2008) Leaf-produced floral signals. *Current Opinion in Plant Biology* 11(5), 541–547.

5 Pruning and Training of Deciduous Fruit Trees

DAVID JACKSON, NORMAN LOONEY AND JOHN PALMER

The term 'pruning' implies the cutting away of shoots and branches. 'Training' is a more all-embracing term and incorporates the steps needed to train a tree, vine or bush to the shape desired. It includes pruning but might also involve tying branches to posts, trellises or pergolas. When pruning is referred to in the following section, it must be remembered that this is part of an overall plan for training a tree, bush or vine to a pre-determined shape. Training systems will be considered in detail later.

Pruning – Why is it Required?

Because pruning is one of the most time-consuming and labour-demanding jobs in the orchard, it is very important to ensure that the time is spent wisely and efficiently. First, however, it must be understood why this trouble and expense is required.

Pruning and the appearance of the tree and orchard

It is always impressive to walk into an orchard and see carefully pruned trees and tidy surrounds. The orchard looks efficient and one imagines that the grower is well organized and a good manager. He usually is, in fact, but like everything, this aspect can be overdone. An over-pruned tree will be tidy but may yield little fruit. Nevertheless, appearance can inform us about pruning efficiency. A well-pruned tree will seem balanced in shape: branches should be well spaced and overlapping branches should not be conspicuous, so that light can easily penetrate into the tree. Limbs should not occur in positions where they will interfere with the flow of machinery, either at pruning time or when loaded

with fruit, nor should they be so high as to make picking difficult. Diseased or damaged wood should not be visible.

Over the years a number of attempts have been made to use mechanical pruning. It has generally proven to be not as effective as skilled hand-pruning, with some exceptions, e.g. mechanical summer pruning of apples and pears on a Lincoln canopy. The appearance of mechanically pruned trees, vines and bushes may make the traditionalist's heart sink but a measure of compromise may be needed between what is aesthetic and what is practical. There has, however, been an increasing use of mechanical assistance to the pruning operation, with mobile pruning platforms replacing ladders.

One of the main reasons for the relative lack of success of mechanical pruning is that pruning, particularly of perennial fruit trees, is not an indiscriminate removal of woody tissue. Pruning is done specifically to modify and direct the growth and future cropping of the tree. The skilled pruner is the one who can foresee the response of the tree to his or her intervention.

Pruning to modify fruit yield

It is fairly safe to say that any pruning will reduce yield, at least in the following season. But what should improve is the yield of fruits that are of sound quality, better colour and suitable size, free from disease and easily harvested. These factors are considered below.

Pruning to improve fruit quality

An unpruned tree produces large numbers of small, often poorly coloured fruit. The act of pruning reduces the number of bearing sites on the tree, allowing more assimilates and mineral nutrients to move to those retained and increasing the light falling on the fruit. Within- or between-tree shading has been shown to have a deleterious effect on fruit quality in a wide range of perennial fruit crops, e.g. apple, citrus, peach, cherry, grape, red raspberry. This decline in fruit quality due to shading can be manifested as smaller fruit size, less red skin colour, reduced soluble solids and higher acid concentration, even through to a lower quality of wine, in the case of grapes. In apples, replacement of old wood by new also helps increase fruit size, since fruit growth is generally better on 2- to 3-year-old wood than on older spurs.

Pruning to reduce disease and pests

If a tree is left unpruned and branches are poorly spaced, chemicals applied by spraying may not adequately penetrate the tree, and diseases and/or pests may therefore not be controlled. Pruning can improve air circulation by opening up dense regions of the canopy. Such regions are otherwise prone to fungal infection. Pruning can also help reduce infection by, for example, removing branches infected with powdery mildew, fireblight or canker and thus reduce the inoculum levels within the orchard.

Pruning may, however, help to promote some diseases by creating wounds through which disease organisms may enter. Larger wounds, particularly those on the main trunk, should quickly be painted over with wound-dressing materials. Shears should also be disinfected between cuts when diseased branches are being removed.

Pruning to improve flower-bud initiation

As we have already learned, shading of leaves and buds is one of the major causes of poor flower-bud initiation. Removing or shortening branches will allow more light to fall on lower limbs, inducing more branches to grow and increasing the number of flower buds which will form. Flower initiation and fruiting are therefore more evenly distributed, not restricted to the outermost portions of the canopy. Improving internal light conditions also improves fruit colouring.

Pruning to reduce labour costs

The height and shape of a tree may have a direct bearing on the time it takes to harvest the fruit. If a tree is left unpruned, an increasing proportion of the fruit is produced high in the tree and physical access into the tree is impeded so that picking becomes more difficult and costly. A higher proportion of apples grown on the traditional open-centre tree need to be picked from a ladder in comparison with the more modern centre-leader tree. The aim of pruning should therefore be to have as many of the fruiting branches close to the ground as practicable. High branches should be sufficiently pruned to allow adequate light to penetrate to those below. On larger trees it is often wise to open 'windows' for ladder usage during pruning, thinning and harvesting.

Pruning to accommodate machinery

Pruning should always take into account the machinery and equipment that will be used in the orchard. Trees should not be taller than can be reached by the tallest ladder safe for pickers, and limbs should not extend so far into the row that they interfere with tractors and other equipment. In some cases, pruning may be done to allow mechanical harvesters to move over the rows. This is already being done with blackcurrants, raspberries and grapes, and it will probably be done in future with other berry and tree crops.

Response of Trees to Pruning

Most trees are pruned in the winter (dormant pruning), but on certain occasions, summer pruning is beneficial. Winter and summer pruning will be considered separately. These remarks refer more particularly to deciduous trees and less to vines and evergreen fruit trees.

Winter pruning

New shoots on a heavily pruned older tree will have larger and deeper-green leaves and will continue to grow for a longer period. In many respects, the growth of a heavily pruned older tree has similarities to that of a vigorous young tree. Pruning part of a tree will normally invigorate new branches over the whole tree, although especially long shoots (sometimes called suckers) may be formed near heavy pruning cuts. Thus if one side of the tree is weak, it is not good practice to prune that side hard to invigorate the following year's branches. It would, in fact, be more effective to prune the vigorous side of the tree heavily, tie down branches on the vigorous side and/or strip fruit from the weak side. However, if the tree is somewhat deficient in nitrogen, generally weak or bearing heavy crops, localized pruning may invigorate only the pruned section of the tree.

On large trees, where major pruning is deemed necessary, it is better to use a smaller number of large cuts, removing whole branches back to the main trunk, rather than a larger number of smaller cuts all over the tree. The former will give a pronounced response next to the large cuts and a general increase in growth all over the tree, while the latter will stimulate growth near each cut and probably accentuate an existing shading problem.

How winter pruning invigorates shoot growth

A tree that has been heavily pruned has fewer buds available to develop the following season. Because of this, the stored carbohydrates and minerals are shared by fewer shoots, and they can grow more. Since the roots are undisturbed, the supply of nutrients, water and cytokinins available to the reduced number of shoots is also greater, and this also encourages further growth. By the end of the season, however, the roots and shoots have compensated in their growth for the imbalance caused by heavy pruning. The invigorating effects of pruning usually last for only one season, unless a really old and over-dense tree has been rejuvenated by pruning. Spring growth is dependent on reserve materials stored over the winter in roots and trunks. Because heavy pruning removes a considerable proportion of the active meristems, total shoot growth is reduced. It must be re-emphasized that the invigorating effect applies only to the remaining shoots and not to the total growth of the tree.

Effect of winter pruning on flowering and cropping

Heavy pruning almost always reduces flowering in spring, by removing wood on which flowers might be produced. Although this has a direct effect on reducing the crop, pruning must not be seen as a substitute for fruit thinning. Heavy pruning also results in a longer period of shoot growth, which inhibits flower production. There are three possible reasons for this.

- A longer period of growth means less time for the process of flower initiation.

- The hormone balance is somehow changed to one unfavourable to initiation.
- Nutrients used for the additional growth are unavailable for flower initiation, i.e. the shoots are a stronger sink for nutrients than the developing reproductive meristems.

Summer pruning

In spring, deciduous trees make a flush of growth that, depending on tree vigour, lasts from a few weeks to a few months. This is followed by a period of rest, which, in some trees, is followed by one or more shorter periods of shoot growth (sometimes described as a shoot flush). Summer pruning is, ordinarily, the practice of removing, or partially removing, these current-season shoots.

The timing of summer pruning is important. If trees are pruned before the first growth flush has finished (early-summer pruning), further compensatory growth occurs and the total period of vegetative growth is prolonged. Early summer pruning rarely shows positive effects and is generally discouraged. On plants which flower on 1-year-old wood, the new shoots arising after summer pruning will develop fewer flowers, due to the shorter period available for flower initiation on the new shoots.

If trees are pruned after most shoot growth has finished and terminal buds have formed, a few buds may burst, but usually little further growth is made until the following spring. Because the amount of leaf surface is limited for part of the season, this late-summer pruning reduces total carbon assimilation and is an effective way to reduce vigour of the tree. Consequently, the practice is not advisable on weak-growing trees or those carrying a heavy crop.

A major reason to summer prune apple trees is to improve the colour on red or partially red cultivars. It has also been shown with some cultivars to improve fruit calcium content. It may also increase sunburn on apples in regions of high solar radiation. In general, the earlier summer pruning is performed the greater the deleterious effects on fruit size, soluble solids content and sunburn. Summer pruning of apple is often performed 3–4 weeks before harvest.

Summer pruning is also known to promote flowering of fruit crops that bloom on spurs of 2-year-old or older wood, providing the pruning is restricted to removing the current season's shoots.

This practice exposes these flowering sites and the leaves surrounding them to better illumination. It may also reduce competition from developing shoot tips during a critical period when flower initiation takes place.

Practical Aspects of Pruning

Pruning is mostly done by hand-held secateurs. In the USA secateurs may be called 'pruners'. There are two types of secateurs: roll-cut and parrot-beak. The roll-cut secateurs are typically cheaper, but the parrot-beak type of secateur is generally preferred, since they cut more easily and are less tiring to use. Secateurs must be of a good design and be kept sharp. The blades must not be wrenched sideways. Loppers are used for cutting larger branches, and pruning saws will be needed for large limbs. Pneumatic or hydraulic secateurs, loppers and saws are also available and can considerably speed up the pruning operation and reduce operator fatigue.

Cuts to remove limbs should be made as close as possible to the main trunk or subtending branch. The only occasion when a stub should be left is when further limbs are to be encouraged for replacement purposes, and in this case it is important to angle the cut. The angled cut encourages buds to develop on the underside of the stub. This will produce better-angled branches than buds

from the top side of the stub, which will produce vigorous, upright shoots. Smaller cuts are made close to, though not through, a bud, as shown in Fig. 5.1.

After pruning, large cuts should be covered by a compound that seals the surface and contains a fungicide to help prevent infection. This is particularly important on stone fruit trees, which are susceptible to infection through the cuts by the diseases silver leaf, *Eutypa* (gummosis) or blast, especially if they are pruned in winter. Sealing compounds are seldom used on berry crops.

The importance of branch angle

The angle between the main trunk and the lateral branch, the crotch angle, has a profound effect upon the subsequent growth and cropping of the branch. A narrow angle results in a vigorous, non-precocious shoot. When it finally does bear fruit, such a branch frequently breaks away from the tree, as the branch union is inherently weak. Wider-angled branches, with crotch angles of 60–90°, are better able to bear the weight of a heavy crop. Such branches are also less vigorous and flower and fruit earlier. Shoots drooping below the horizontal tend to become very weak, subtending fruit of poor quality when in shaded regions of the canopy. Pruning therefore preferentially removes shoots that are too strong and upright or too weak and

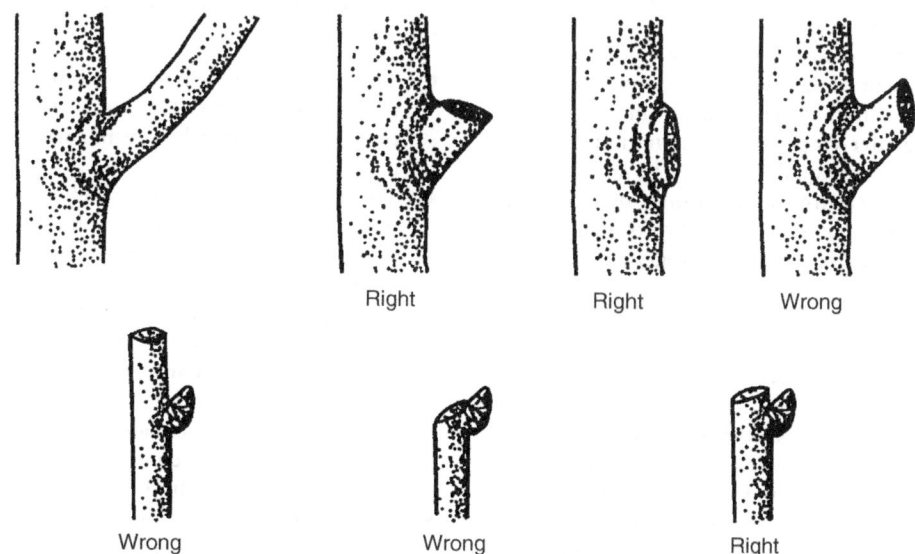

Fig. 5.1. Correct and incorrect ways of pruning.

droopy. Tree training manipulates the branches into more desirable angles.

Systems for training fruit trees

Training systems aim to produce a plant of a certain shape, size and disposition of vegetative and floral tissues. All training systems have to work around various constraints:

- Physical access for spraying and picking.
- Deleterious effects of tree shade on fruit quality and production.
- Limitations of mechanical harvesters where these are used.

At the outset it is important to explain the difference between training systems and plantation systems in fruit growing. A training system aims to produce a plant of a certain shape and disposition of vegetative and floral tissues. It often involves the use of posts and wires to support or manipulate the tree or vine. A plantation system considers the number of trees or vines per unit of land area and the spacing of these units in rows and between rows. It may even involve multiple-row beds separated by access tracks. It is designed to include pollination requirements, irrigation needs, weed control and a host of other considerations. A successful plantation system demands that the natural growth of the tree or vine is accommodated by the spacing between trees and rows. The natural growth of the tree, however, depends upon the vigour of the site and scion/rootstock combination.

In the early years of training the emphasis is on the development of the tree structure and the rapid filling of the allotted space, while in later years there is a greater emphasis on maintaining access and production of quality fruit. The methods of training described are the main ones used for the crops mentioned in this book. Fruits pruned by similar methods will be grouped together.

Deciduous Trees

Deciduous trees are covered in this chapter and other fruits in Chapter 6.

There has been a considerable development of training systems over the last 30 years, particularly in Europe, Australasia and the USA. These developments have not happened in isolation. Through international travel there has been considerable cross-fertilization of ideas and a continuous development of new systems, in response to changes in the economics of production and the limitations of existing systems. It is impossible in this book to describe every system, but we have chosen examples of the more important types, in approximate order of increased tree density and cost of establishment.

In most of the newer training systems there has been a move towards smaller trees at higher tree densities. With smaller trees the emphasis is on branch bending and minimal pruning in the early years of the orchard. Minimal pruning and branch positioning encourages earlier fruit production. This approach contrasts with the severe pruning in the early years of older-style, low-tree-density orchards. Severe pruning was essential to develop the strong branch framework needed for the larger trees in low-tree-density orchards, but it had disadvantages.

The vase or open-centre tree

This method aims to produce a tree shaped like a vase, in which the centre is kept open (Fig. 5.2). The tree has no central axis or leader but is multi-leader. A short trunk is branched at about 30–35 cm to produce four to six main limbs radiating from the centre. Laterals from these limbs carry most of the fruit. This old system was thought to allow light penetration to all parts of the tree and promote good colour and good initiation. It is no longer recommended for apples and pears for the following reasons:

- Trees do not grow naturally into this shape – thus training demands a lot of drastic pruning of young trees, which prevents production at an early age.
- The fruiting zone is too high – as the tree matures, a greater proportion of the fruit is carried high in the tree and cannot be picked without using a ladder.
- Pruning and picking are difficult – the shape of the tree means that ladders must be moved many times to cover the perimeter of the canopy.
- Spray penetration is poor – achieving good penetration to the centre of the vase becomes progressively more difficult as trees get larger.
- Limb junctions are not very strong – unless supported they are liable to break when carrying heavy crops of fruit (Fig. 5.3).

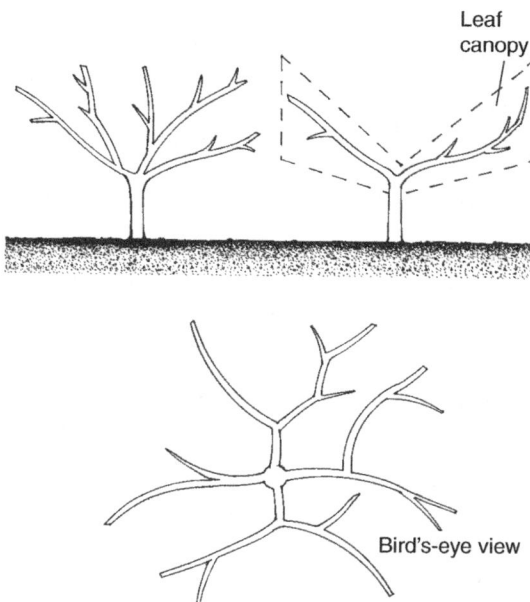

Fig. 5.2. Vase or open-centre training.

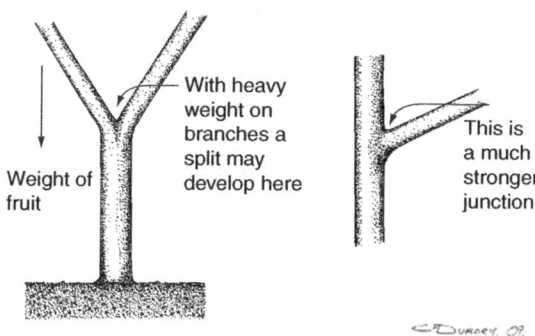

Fig. 5.3. Limb junctions.

Multi-leader trees are still used for pears in parts of western USA, due to an overriding concern with fireblight infection. If one leader is lost due to infection the others remain, whereas an infection on the leader of a centre-leader tree can result in complete loss of the tree.

The centre-leader, semi-intensive system

This was the most popular method of growing apples and pears in New Zealand in the 1970s to 1990s and is still used widely throughout the world, particularly in North America. It overcomes many of the problems associated with the vase method, not only yielding heavier crops but also producing them earlier in the life of the tree. Essential to the success of the system is the use of a semi-dwarfing or lower-vigour rootstock (details are given under specific crops). For strong scion cultivars of apples on good soils, MM.106 and M.7 have been successful; on poorer soils or with weak scion cultivars, MM.111, Northern Spy or M.793 could be used. For European pears, centre-leader trees can be grown on quince or clonal *Pyrus* rootstocks. Tree density would be typically 500–700 trees/ha.

Planting

Apples and pears are frequently bought from a nursery as 1–2 m-high, unbranched trees (sometimes called 'rods' or 'whips'). After planting in the orchard, these are normally cut back to a height of 75–90 cm, to encourage the development of a strong basal tier of branches (Fig. 5.4). If the tree is feathered and suitable laterals are present at the right height, the tree will not be cut back and these feathers will form the first tier. (A feather is a

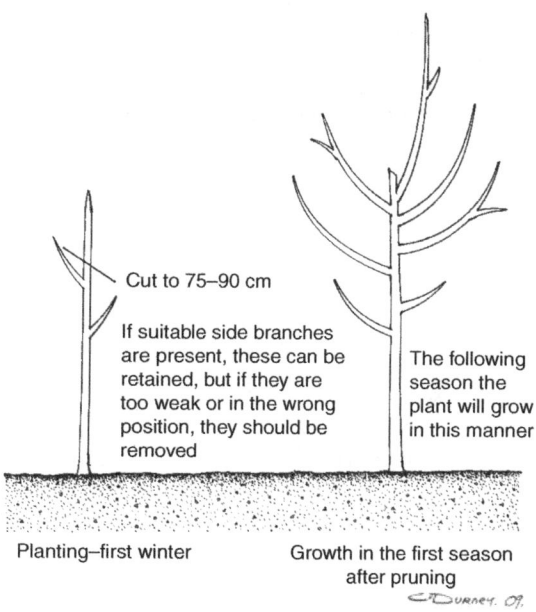

Fig. 5.4. Early stages of growth in the central-leader, semi-intensive system.

sylleptic shoot that develops from the axillary bud in a leaf axil in the current growing season. A proleptic shoot, in contrast, develops in the following season from the same bud position.)

Growth in the first season

It is hoped that the top bud will grow most vigorously in an upright direction and that four other strong subordinate shoots develop to form the basal branches (Fig. 5.5). Two branches should grow along the row and two at right angles to the row, the ideal angle to the horizontal being 30°. If possible, it is better for the branches growing out into the row to be higher than the others. This gives easier tractor movement down rows when the branches are fully extended.

During the first season's growth, vigorous shoots not growing in the correct direction should be removed to encourage vigour in those which are more favourably positioned. Competing strong shoots directly below the central leader can be rubbed out while still short (up to 5 cm long). Small shoots not competing can be retained. When a shoot is about 10 cm long, growers sometimes place a clothes peg or short, sharpened stick between the shoot and the main stem to encourage the branch to grow at a wider angle.

When the branches are larger, further training of this lower tier may be required. Sometimes, a notched piece of wood, a spreader, is placed between the upright and the branch to increase the angle (Fig. 5.6a). Alternatively, branches can be tied down with string, either to the base of the tree or into the ground using a W clip (Fig. 5.6b). This latter method permits good branch positioning, but care must be taken to ensure that the strings do not interfere with machinery or worker access within the alleyway. The W clips are pushed into the ground using a special applicator (Fig. 5.6c). The string should be resistant to rotting, e.g. a polypropylene or nylon-based twine.

Tying down, if required, is best left until midsummer, after extension growth has stopped. Excessive vertical and often unfruitful lateral growth will occur if it is done too early, and growth at the end of a branch may stop too soon or else continue to grow and bend upwards. Once growth is fixed in the new position using these spreading or tying techniques, strings and wires should be removed.

Weaker branches close to the horizontal will make less growth, flower earlier and require little or no tying down. Spreading them further may weaken an already weak shoot, which may never catch up with other branches on the tree. The result can be a lopsided canopy.

The first winter after planting

During the winter, the main leader is treated in a similar manner to the tree at planting. The aim is to begin a second tier of branches 80–100cm above the first. Any strong, upright competing shoots should be removed, otherwise most training is left until the following spring and summer.

Subsequent development

In the second season the second tier will begin to form in a similar manner to the first tier. If vigorous, branches may be tied or trained down in the summer. If the second tier is well established, the

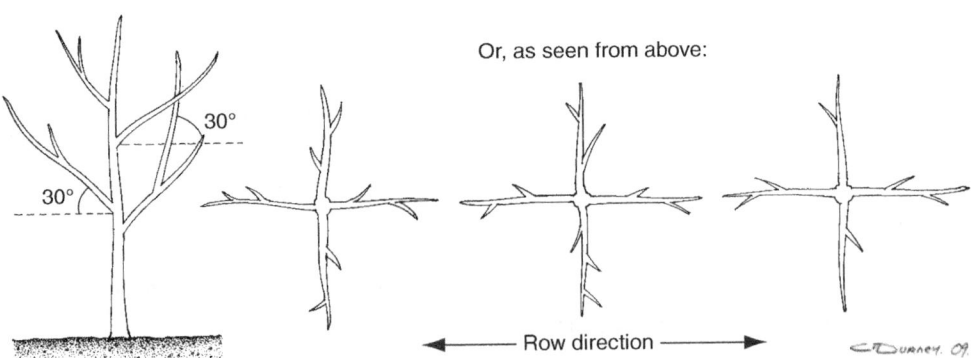

Fig. 5.5. Central-leader, semi-intensive system – side and top views.

D. Jackson *et al.*

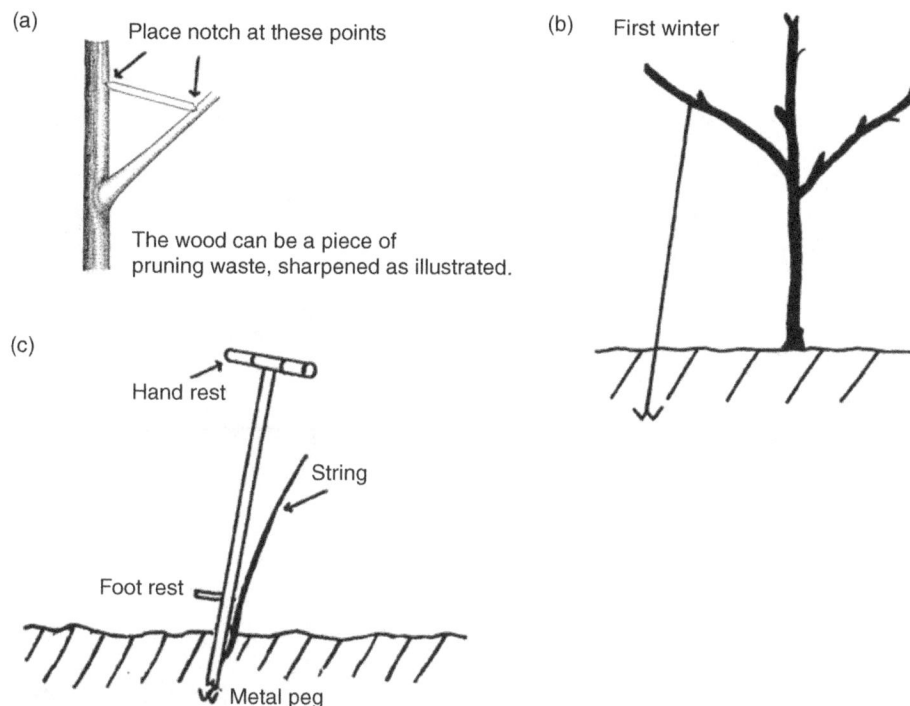

Fig. 5.6. Training the lower tier. (a) Use of a spreader. (b) Tying to the ground with W clip. (c) Applicator used in tying down.

(a) Place notch at these points

The wood can be a piece of pruning waste, sharpened as illustrated.

(b) First winter

(c) Hand rest

String

Foot rest

Metal peg

four main arms should be retained and any competing or vigorous ones removed. Smaller branches are not pruned. Above the second tier, the growing leader is normally left untouched unless it is excessively vigorous, in which case it might be cut back 1 m above the tier to a weaker lateral. If further tying down of the second tier is required, it may be done in the following season (Fig. 5.7a).

It is now common not to allow any more than two tiers to develop. Above the second tier, a form rather like a 2 m spindle-bush is recommended, pruned in a manner similar to that described under the intensive method (see later). This encourages heavy cropping towards the top of the tree, which helps to suppress undesirable excessive vigour.

The geometry of the centre-leader tree is important. For example, the open bays allow all fruit on a mature tree to be picked with only four ladder movements (Fig. 5.7b) and permit easy spray and light access to the centre of the tree. The conical shape brings a greater proportion of the tree close to the ground and up to 70% of the crop may be picked without ladders (Fig. 5.7c)

The disadvantages of the centre-leader system are that tree size is large and precocity, although better than the vase, is poorer than the intensive system. Also, in less-favoured climates, especially those with strong summer winds and dry conditions, the rapid establishment of the tiers is more difficult. Sometimes 2 years may be needed for the production of each tier, in comparison with the single year described above. Persistent winds may bend the centre leader, destroying the geometry of the system. For such conditions, one of the supported systems described later may be an advantage.

Maintenance pruning

Once the framework of the centre-leader tree is established, pruners should ensure that the picking bays are kept open; that strong, vigorous, upward-growing shoots are removed right down to the base; and that the tree is kept to a manageable height – about 4–4.5 m. If growing conditions are very vigorous, rather than cutting the tree back each year to

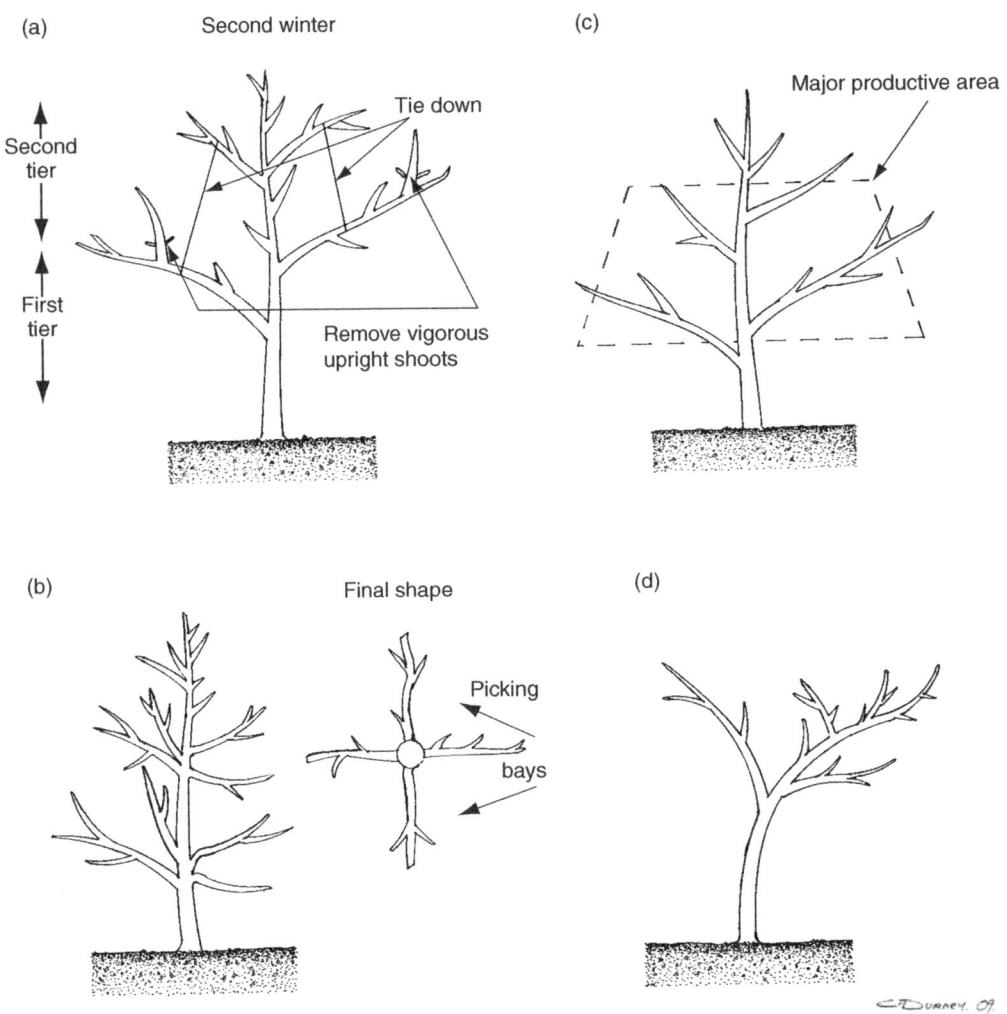

Fig. 5.7. Later stages of the central-leader, semi-intensive system. (a) Tying down the second tier. (b) Final shape of the tree. (c) Major productive area. (d) Response to strong prevailing winds.

4 m, it is better to allow it to grow for 2 years before heading back (Fig. 5.8). This acts as a 'bleeder' to take away excessive vigour and prevents the proliferation of strong shoots at the tip of the tree, which occurs if a vigorous tree is regularly over-pruned.

Laterals on the main arms will gradually fill up with spurs. As a spur system ages, the spurs become less productive and it is wise either to reinvigorate the spurs by reducing their size or to completely replace the laterals at intervals of 4–5 years (renewal pruning). This means that, in the mature tree, a quarter to a fifth of the laterals are removed each year.

This centre-leader system has been modified in New Zealand into the slender pyramid. After planting the central leader is not headed, but during subsequent growth six to eight branches are selected for the lower tier, in the 0.8–1.5 m region of the tree, by removal of other lateral shoots when they are about 10 cm long. Above this lower tier, shoots are selected early for removal during the first growing season if they are competing with the central leader or too vertical. The lower tier of branches is gradually reduced to four or five by year six. As the tree develops, the requirement for good light penetration dictates that the upper branches are gradually

D. Jackson *et al.*

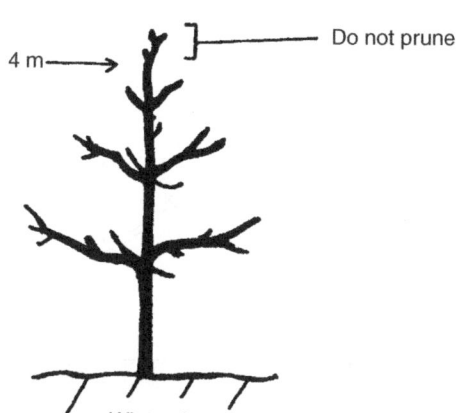
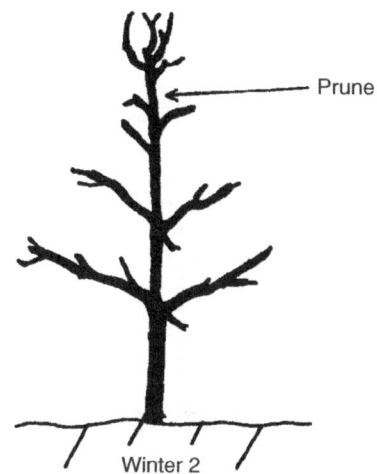

4 m — Do not prune

Prune

Winter 1

Winter 2

Fig. 5.8. Coping with vigour at the top of the tree.

removed completely or renewed to maintain the slender pyramid shape. In less-favourable environments other than New Zealand, the leader might have to be headed to develop the lower tier properly, but better still use a feathered tree.

Centre-leader intensive systems

Slender spindle

This is widely used in Europe and provides a small, precocious, conical-shaped tree which can be picked and pruned from the ground, i.e. 2–2.2 m tall. The original spindle-bush trees were developed in Germany in the 1930s, with a tree density of 1600 trees/ha, but were developed further in the Netherlands in the 1950s as the free spindle-bush and later as the slender spindle, with tree densities of 1900–3300 trees/ha.

It is essential in this high-density system to use dwarfing rootstocks such as M.9 or, where growth is weak, M.26. The trees are spaced from 1.2 to 2.0 m apart, in rows 3–3.7 m apart. The tree spacing depends on the vigour of the apple cultivar/rootstock/site combination (Table 14.2). The roots of many of the dwarfing stocks are weak, and this, coupled with the high cropping, means the trees must be supported throughout their life. They can be either staked individually with a 4–5 cm diameter stake, staked with a thinner wooden or metal post with a top wire support, or the trees supported by wires on a two- to four-wire trellis.

The advantages of this system are:

- Very early cropping – there is a crop in the second season and full production in 4–6 years.
- No ladder work – this means picking, pruning and fruit thinning are more efficient.
- Trees are more resistant to strong winds (when supported).
- Trees are easier to spray.
- Fruit quality is improved.

Some disadvantages include:

- Higher cost of establishment – this is due to the greater number of trees (1900, compared with 650–700/ha in the semi-intensive system) and the extra cost of supports, although many growers are now using posts and two wires to support semi-intensive, centre-leader trees while they are young.
- Greater attention to detail required by the grower.

METHOD OF PRUNING This is summarized in Fig. 5.9. As early cropping is important, to offset the high cost of establishment, feathered trees offer the two advantages of earlier cropping and easier tree management, as the basal tier of branches is already in place. Feathers below 50 cm in height and any that are too vertical are removed. The leader is headed to 20–30 cm above the highest useful feather. Unfeathered trees are headed to 80–90 cm

---

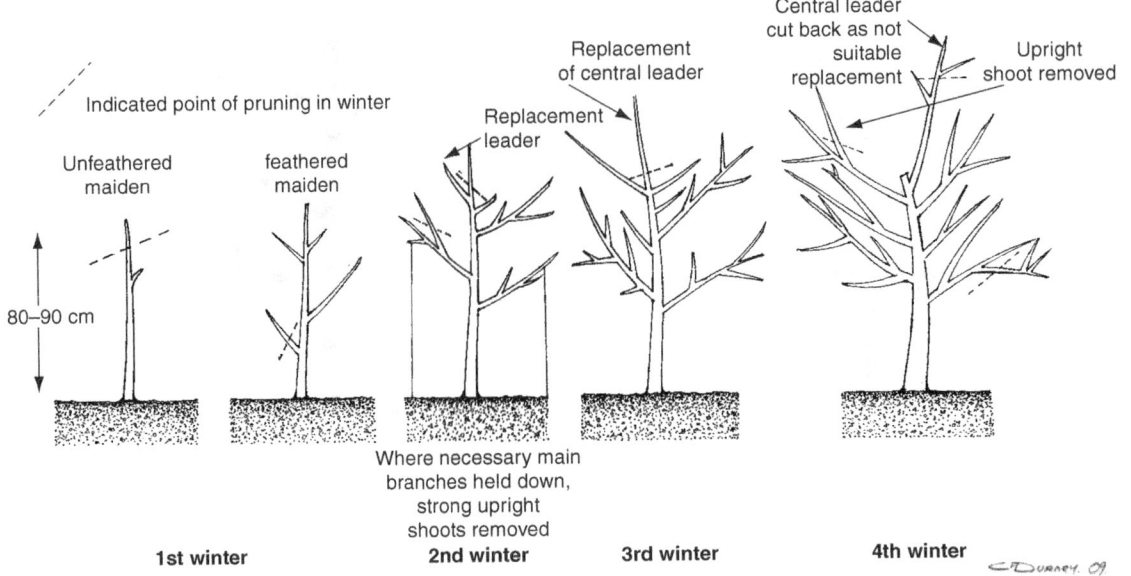

Fig. 5.9. Training the intensive spindle bush.

to encourage the development of a strong lower tier of branches. Again to hasten early cropping and the development of the tree, tying down of branches is important (using techniques described before). Where the trees have individual wooden posts, galvanized nails can be driven into the posts near ground level as an anchor for the string. With these smaller trees, weighted pegs or rubber bands are also used to bend the branches. Early cropping means that sometimes the basal branches have to be tied up in the summertime to prevent the branches breaking or becoming permanently set into a too pendant position. In contrast, side branches in the upper part of the tree can be tied below the horizontal, if necessary, to reduce their vigour.

The central leader is generally replaced each year by tying up a lower, weaker shoot; this produces a somewhat twisted leader shape, which is designed to reduce excessive vigour in the top of the tree. Where growth is vigorous the leader can be cut out and replaced by tying up one of the lower branches, including some 2-year-old wood, which is normally well furnished with flower buds. When mature, fruiting laterals in the top of the tree are shortened or cut back to the leader for replacement and to maintain the conical shape of the tree. The basal branches either receive similar renewal pruning or their laterals are renewal pruned. The lowest branches on the

bottom tier often become weak and shaded by the rest of the tree and these are gradually removed.

The French axe

The system is based on an approach to training developed in France and called *L'axe centrale* and known elsewhere as the axe, central axe or vertical axis (Fig. 5.10). It sets out to produce a tree which is about 30% below the natural size expected from the specific combination of rootstock and scion. It is designed for precocity, high yields and ease of pruning, and has a greater emphasis of working with the natural growth habit of the tree than the more regimented slender spindle system. Rootstocks are typically M.9, M.26 or even M.7 or MM.106 for spur-type cultivars on weak growing sites.

The basic difference between the axe and the slender spindle is the treatment of the central leader. The spindle bush is designed to be 2m tall and the repeated replacement of the central leader reduces its vigour and prevents the development of excessive vigour in the top of the tree, with its attendant problems of heavy shading of the lower branches. The axe training system, by contrast, allows the central leader to grow unhindered and eventually slow up with cropping after 5–6 years, consequently forming a taller tree 3–4 m tall. Since the tree is much taller than

D. Jackson *et al.*

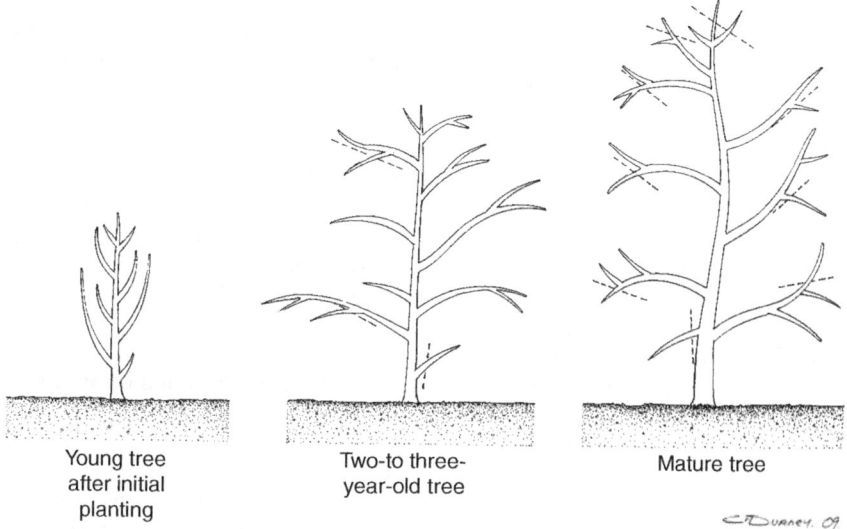

Young tree after initial planting

Two-to three-year-old tree

Mature tree

C.Durney, 09

Fig. 5.10. *L'axe centrale.*

the slender spindle tree, tree densities range from 1000 to 2500 trees/ha.

As with the slender spindle, feathered trees are preferred at planting. Lower feathers below 50 cm and feathers that are too upright are removed. The leader is left untouched. During the first growing season competitive shoots directly below the leader are removed or pinched back. Early pruning is minimal. A well-grown axe tree does not have a pronounced tier structure but has many breaks up the leader, with a general conical shape maintained. Lower branches are renewed as they gradually bend under the weight of crop by cutting back to a suitable lateral, and branches higher up the tree are renewed by cutting back to a suitable leader. As with the slender spindle, the very lowest branches are gradually removed as they become too low and too shaded.

Solaxe and centrifugal training

The vertical axis was further modified by the French in the late 1980s into the solaxe. The leader is bent over at the desired maximum height (2.5–3.5 m) to avoid excessive height development. The fruiting branches do not receive renewal pruning, but all vigorous, upright, unwanted shoots are removed as soon as they begin to develop from both the main trunk and the lateral branches. This produces a tree with many pendulous fruiting branches. One of the drawbacks of this approach is

that the number of flowering sites increases rapidly, which has led to the development of 'centrifugal training', a key part of which is spur extinction.

Spur extinction entails the deliberate removal of all lateral shoots or spurs on the underside of the fruiting arms – these often produce inferior fruit. Further spurs will be removed on the fruiting arms close to the central leader to produce a 'light well' down through the tree. All of these operations are designed to produce higher-quality fruit (size and colour) on the remaining fruiting sites. This type of detailed pruning can only be achieved on trees on dwarfing rootstocks grown intensively.

Further changes in centrifugal training over the solaxe include some removal of whole branches to avoid overcrowding in the later years and a greater reliance on natural bending under the weight of crop, rather than relying on artificial bending with string or weights. Thus the French have continued their development of training systems on the basis of experience gained with their earlier systems and a continual desire to work with the natural growth habit of the tree rather than trying to manipulate the tree excessively, which often results in more work to overcome the growth reaction of the tree.

Hybrid forms

There are numerous other types of intensive centre-leader tree forms. For example, the hybrid tree cone or HYTEC, developed by Bruce Barritt in Washington State, was somewhat of a hybrid between the slender spindle and the axe, and was designed to give a tree 3 m tall without excessive vigour at the top of the tree.

Over the last 10 years or so, there has been a growth in the use of other hybrid systems, e.g. 'tall spindles' in a number of parts of the world (Fig. 5.11). These have really taken ideas from the slender spindle, the vertical axis and the solaxe, as growers seek for higher yields, good fruit quality and more economic systems of growing apples.

In a tall spindle, the central leader is not headed at planting, and the ideal tree from the nursery has 10–15 feathers, starting at a height of 80 cm from the ground. Such high feathers mean the resulting branches can droop without the need for tying up, as was common with the slender spindle, when lower feathers were used. In the case of the tall spindle, the feathers are deliberately tied below the horizontal at planting to induce early fruiting and to decrease their vigour. If growth is moderate, then further development of new laterals from the central leader results in short, relatively weak shoots, which with terminal flowering and fruiting will naturally bend down. The final tree is tall and narrow with weak, pendant branches. Where growth is vigorous, the new laterals may need tying down each year. As the tree ages, some removal of whole branches becomes necessary.

Trees trained to any of these systems can be planted either in single rows with tractor alleyways between each row or in double- or multiple-row beds separated by tractor alleyways. The slender spindle tree system (the system providing the least tree to tree shading) has been planted and satisfactorily managed in multi-row beds of two to five rows. Bays for pedestrian access are maintained on two- and three-row beds, and walking paths are provided in wider beds. Although yields have been high with such systems, establishment costs are also high and problems of poor light and spray penetration are more common in the multi-row plantings. Weed control also becomes a major problem if modern chemical herbicides cannot be used.

Divided canopies on trellis supports

The Lincoln canopy

The training systems for apples and pears described so far rely on hand harvesting the crop. Some efforts have been made in various parts of the world to mechanize harvest on these or similar systems, but the machinery has been expensive and, except for cider or processed fruit, not very effective. However, advances in robotics and rapid pattern recognition software offer exciting future possibilities for mechanical picking.

The New Zealand Agricultural Engineering Institute at Lincoln University proposed a canopy system of growing apples which adapted the apple tree to fit the most simple method of mechanical harvesting – shaking followed by a very short fall on to a padded catching tray. The horizontal canopy was 3.0 m wide and 1.6 m above the ground, with the trees grown on MM.106 rootstock at 2.0–2.4 m within the row and 4.25 m between rows.

(a)

(b)

Fig. 5.11. Tall spindles. (a) Royal Gala on M.9. (b) Fuji on M.9 (also showing hail netting and Extenday reflective mulch).

D. Jackson *et al.*

After planting, unfeathered trees (rods) were topped at 1 m and, from subsequent re-growth, four main leaders were selected. These were grown in an H shape on the first two wires of the trellis, which were 40 cm apart. Laterals from these leaders were selected at 20–25 cm intervals and taped down horizontally to fill the single canopy. (Fig 5.12). Considerable labour was required to tie down laterals but, once in position, summer pruning could be done with a mechanical cutter bar. Lateral replacement was instituted on a rotational basis. Spraying could be achieved with a twin-boom applicator. The system, however, was not widely adopted. Some reasons included problems with the mechanical harvesting system (both technical and economic) and also poor colour development on red or striped apple cultivars. Heavy shading from vigorous upright shoots on the horizontal canopy was responsible for poor fruit colour development. On the other hand, skin finish was excellent. The lack of any central leader on the tree was a factor in the shoot growth response. In our opinion, this system has more merit for pears (Asian and European) than for apples.

V- and Y-shaped canopies

The Lincoln canopy is basically two planar canopies side by side. Other arrangements have consisted of dividing the canopy into two inclined planes. The Australian Tatura trellis, which will be described under the stone fruit section, was, like the Lincoln canopy, designed for use with mechanical picking and has been used with some success with apples, particularly what has become known as a 'mini-Tatura' (Fig. 5.13). Various other configurations of the inclined canopy form have been described so that in cross section the canopy resembles a V or Y shape, supported on a wire trellis. In some cases this entails training alternate trees to the right or to the left on to the trellis; in other cases the trees are headed and the branches trained out to the left or the right.

Many other training methods have been used, especially for home gardens, where they can be very labour intensive. Cordons and espaliers of many shapes and sizes have been described and adorn the pages of many gardening books, but they are rarely used commercially nowadays.

Stone Fruit with Flowers on 1-year-old Lateral Shoots

The vase or open-centre tree

Most stone fruit trees naturally have a bushy and open habit, and adapt easily to the vase system of pruning. As a consequence, until recently, most have been grown in this way. Now, however, the intensive centre-leader method, described later, is becoming the more popular for peaches and nectarines.

To establish a vase tree, growers use one of two methods – the first is more detailed and the second employs a minimum-pruning technique.

Detailed pruning

Trees are normally bought from the nurseryman, who has headed them back to about 50 cm to form three or four branches. The following season, vigorous growth will ensue, as shown in Fig 5.14. Only two of the leaders are shown in lateral views.

Some growers head back the main leaders by about two-thirds and cut out more of the other shoots, believing that a good framework built up at this stage will stand the tree in good stead for its

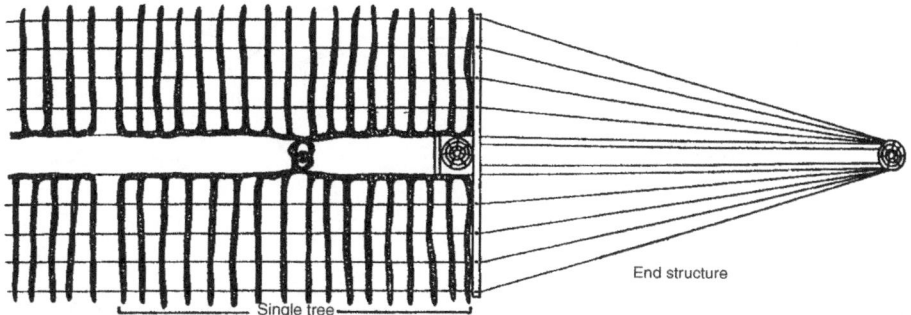

Fig. 5.12. H canopy training system – plan view.

Fig. 5.13. Comice pear on a mini-Tatura system.

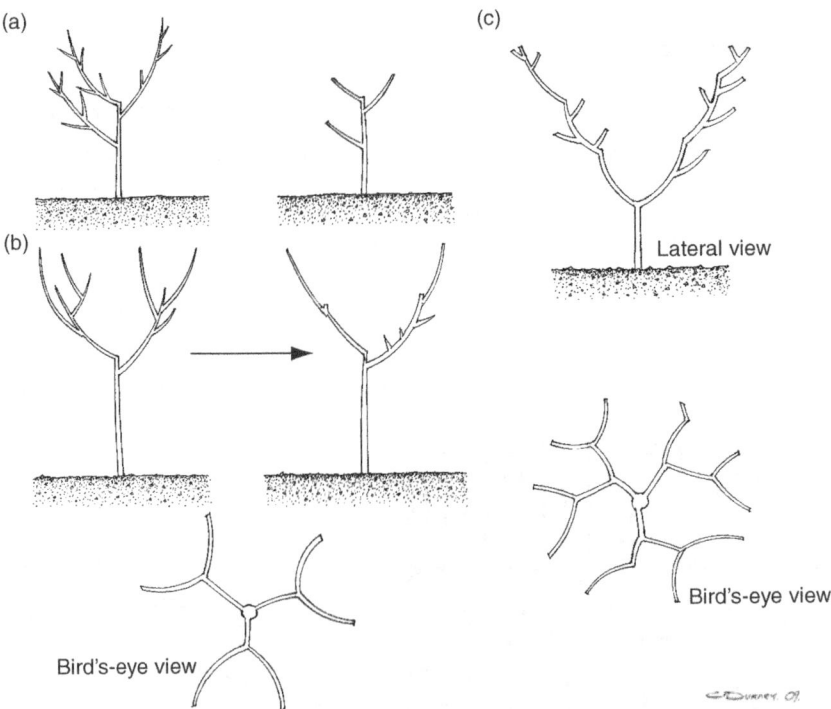

Fig. 5.14. Detailed pruning to establish a vase tree. (a) On planting, the orchardist cuts the shoots to 20–30 cm at outward-pointing buds. (b) The shoots growing outwards are retained and the more vigorous inward-growing ones are removed. The number of main leaders will normally be increased by retaining two leaders not one. (c) Final shape of tree.

D. Jackson *et al.*

subsequent life. However, the more a stone fruit tree is cut, the later it will come into bearing and the more likely it will be to succumb to such disorders as silver leaf, blast or gummosis.

Ultimately, there should be about eight leaders placed in a vase shape, with an open centre in the middle, as shown.

Minimum pruning

Many growers now prefer to do an absolute minimum of pruning in the first 3–4 years and rely on the natural tendency for stone fruit trees to adopt a vase habit. This has several advantages. First, it encourages early cropping and quickly gives a financial return to the grower. Second, the very much lower pruning reduces silver leaf and other disorders, as has already been mentioned. Third, it has now been found that marketplace-preferred stone fruit cultivars, particularly peach and nectarine, change very quickly. By cropping early, one can more easily afford to replant with another cultivar. There is less financial loss in doing so, since there is a lower non-productive period between changeovers.

The technique is very simple. On receiving the tree, which has been headed back by the nurseryman, it is left unpruned, except for any very weak or broken shoots, or those at an inappropriate angle. From then on, virtually no further pruning is done for the next 3 years, although branches obviously growing in the wrong direction will be removed. After 4–5 years, the tree will be nearly full height and approaching full production. At this time corrective pruning to tidy up the vase shape will be applied.

Maintenance pruning

Stone fruit trees are best pruned, if possible, after cropping and before leaf fall. This is difficult to do, both because the leaves make appraisal of the overall framework difficult and because this is a very busy time of the year, especially for growers with other fruit crops requiring attention, such as pip fruit. Detailed attention to small branches is not recommended and the orchardist should seldom cut a branch less than 1 cm in diameter. The aim should be to produce a tree that is not overcrowded and on which a proportion of laterals, say a third to a quarter, are removed each year. Even those stone fruit which have spurs (apricots, plums and cherries) also crop on 1-year-old shoots, and such a

replacement programme ensures a supply of new productive growth. Having said this, there are occasions where detailed pruning is still warranted: when pruning is also used to assist with fruit thinning and with some early cultivars, where thinning is often impractical. In such cases, most laterals are shortened to one-half or one-third, depending on the degree of thinning required.

Intensive stone fruit production

This system is now becoming very popular for dessert peaches and nectarines. Trees are planted at 1.0–2.5 m in rows 4–5 m apart. Sometimes a 'dormant bud' is used; this is a seedling rootstock budded the previous season. The bud has not grown but will make vigorous growth the following year. A tree is normally preferred, however, and the nurseryman should be instructed not to head it to cause branching, as for open-centre trees.

Pruning in the first season (spacing between 1.5 and 2.5 m)

In the first season of growth, some summer or winter pruning is done: any branches closer than 50 cm to the ground are removed and any shoots competing with the main leader are also pruned away. Both of these operations are done as soon as possible, i.e. when the undesirable shoots first appear.

If the tree grows vigorously in the first season, some branches will be removed to begin forming a tier. After a series of lateral branches have formed at 50–100 cm, a gap of 40–50 cm may be left before more laterals are allowed to grow. This allows light into the lower branches and helps flower initiation and spray penetration.

On good nursery-grown trees, a small crop of six to eight fruits may be retained, but this crop should immediately be removed if any signs of loss of tree vigour occur.

Pruning in the second season (spacing between 1.5 and 2.5 m)

In the second growing season, a reasonable crop should be obtained. This has a vigour-regulating effect on the tree and is part of the plan to contain the vigour in closely planted trees without the assistance of dwarfing rootstocks. Nevertheless, some thinning should be done as soon as fruits are of suitable size (see Chapter 8).

At the same time as thinning, vigorous water-shoot growths with sharp crotch angles should be removed. In late summer further light summer pruning is recommended to remove any further water shoots and to avoid congestion of branches.

The main pruning of spent fruiting wood and thinning of those laterals which will fruit next year is done in autumn. (or winter – depending on disease pressure). The informal tier structure, which may have begun in the first year, is kept open and a general thinning of branches elsewhere on the tree is continued (Fig 5.15).

Pruning in the third and subsequent seasons (spacing between 1.5–2.5 m)

By the third growing season, a full canopy will be developed and cropping will also be approaching mature orchard levels. Training will consist of maintaining a pyramid- or spindle-bush-shaped tree and will require removal of potential water shoots at fruit-thinning time. This will be followed by a further thinning out of vigorous shoots in mid- to late summer to allow good light penetration into the tree, give good fruit colour and promote flower initiation.

If leader growth has slowed up by this time, it may be cut back into the previous season's extension growth to a fruiting side lateral. If leader extension growth is still too vigorous, it should be left for a further growing season before cutting back. As in the second year, the main detailed pruning of fruiting laterals and spent fruiting wood should be done in the autumn after vigorous growth ceases.

Closely spaced trees

Trees at close in-row spacing of between 1.0 and 1.5 m should take the form of a narrow spindle-bush or axis-type tree, with fruiting laterals from the main axis rather than defined tiers. Pruning is more or less a recycling process, with those laterals that have fruited being removed each year and new and weaker laterals being retained.

Mature, intensive stone fruit trees are expected to reach 3–3.5 m height and, providing a good vigour and cropping balance can be maintained, it should be possible to stabilize them at this height. Ultimate tree height is dependent on cultivar, soil type, fertilizer, irrigation and crop load factors. Weak-growing cultivars or trees on poorer soils will produce trees less than 3 m in height, while with strong-growing cultivars on fertile soils, it may not be possible to hold trees at the 3.5 m level without upsetting their crop and growth balance. The majority of the crop should be borne between 0.5 and 2.5 m from the ground.

The life of the trees in this system will depend on the district, the presence of disease, the ability to contain vigour and the usefulness of the cultivar planted. Experience around the world suggests that productive life under correct tree management would be similar to that of a standard stone fruit orchard, but with the advantage of the pyramid shape maintaining the crop lower in the tree due to better light penetration to the lower branches.

For vigorous peach cultivars, such as Golden Queen, Armking, and O'Henry, growth in close-planting systems is difficult to control. Such cultivars may be planted at 3.5 m × 5.0–6.0 m but are treated in the early years in a similar way to the 1.5–2.5 m-spaced intensive trees. The tree will be allowed to grow larger and more attention is given to fruiting arm development. Less summer pruning is required. This approach might be suitable for the more vigorous plum cultivars and for apricots and cherries, which have not adapted so well to closer-spaced, intensive planting.

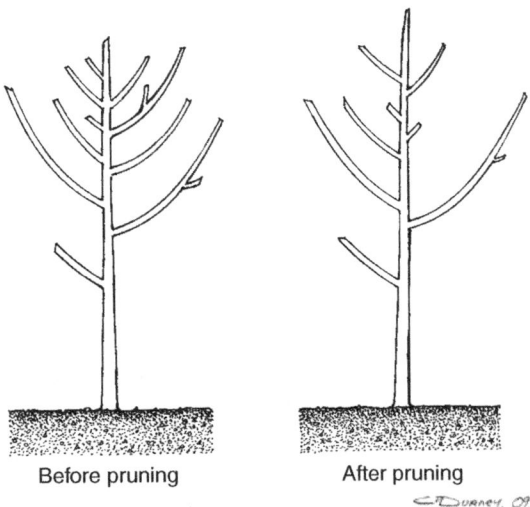

Before pruning   After pruning

Fig. 5.15. Pruning in the second season in intensive stone fruit production. Only cross-sectional branches are shown.

D. Jackson *et al.*

The Tatura trellis

This method of growing peaches was developed at the Tatura Research Station in Victoria, Australia. Its genesis came from theoretical calculations of maximum light interception at the latitudes of Southern Australia. It was argued that if the ground was covered by ridges of leaves running north–south, each at an angle of 60° to the horizontal, then maximum useful light interception (and therefore yield) would be obtained.

The first practical attempts were very successful and high yields were obtained. The method now uses trees planted at 1 m, in rows 5.5–6.0 m apart; the total height is 4 m with the arms 5 m long (above the ground). There is a gap of 1–1.2 m between the tops of adjacent arms. Each tree is allowed to produce two main leaders, which grow up either side of the ridge. Growth is encouraged to spread over the trellis but cannot be allowed to become too thick – 60 cm maximum from upper surface to lower. If any more than this is left, light interception by lower leaves is too low and leaf area is wasted. To maintain this thickness, summer cutting of foliage is needed, and specially mounted cutter bars mow the inner parts of the V, where the strongest, most upright shoots develop.

Experience in California with the Tatura system resulted in alternative V-type trellis systems. The perpendicular V used trees planted at 2 m within the row, with one main arm extending one side of the row and another to the other side. Further refinement resulted in the quad V or even the hex V, where two or three branches were trained out each side of the row, with increased spacing between the trees within the row. In the absence of dwarfing rootstocks, this was an attempt to reduce vigour by spreading growth among a larger number of branches.

The Italians have designed a wider V system, the Sibari Y, with the angles of the V being extended so that the adjacent rows are joined together. This means that the overall height of the canopy is reduced and the cost of the trellis reduced. Tree density in this case is 900–1100 trees/ha. Early yields are high in this system compared to conventional delayed vase systems, largely as a function of increased light interception. Establishment costs, however, remain higher than for more conventional systems.

The New Zealand experience with Tatura trellis has not been very encouraging, especially with peaches and nectarines. Much of the high yields claimed probably result from the high density and, for these fruits, similar results can be obtained with the free-standing, high-density systems just described, which are cheaper to establish and easier to manage. It has also been found difficult to maintain even cropping on all sections of the Tatura trellis as trees age, and mechanical cutting of shoots, although theoretically possible, is, in practice, rather difficult to manage without very skilled operators.

It may be more promising for other fruit such as apricots, cherries, plums, kiwifruit and dessert grapes. Apricots have done well on Tatura trellis in New Zealand's Central Otago region, and it is likely that plums might respond similarly, since they need some form of support to produce high-quality fruit without windfall problems close to harvest. Neither apricots nor plums behave as well as peaches and nectarines on the free-standing, high-density systems described earlier. Some growers are attempting to grow kiwifruit on Tatura trellis and experiments are being done to investigate the potential of the trellis for outdoor table grapes. It is still too early to make recommendations as to the merit of this trellis for many of these crops.

Further Reading

Barritt, B.H. (1992) *Intensive Orchard Management*. Good Fruit Grower, Yakima, Washington.

Barritt, B.H. and Kappel, F. (eds) (1997) VI International Symposium on Integrated Canopy, Rootstock, Environmental Physiology in Orchard Systems. *Acta Horticulturae* 451.

Brunner, T. (1990) *Physiological Fruit Tree Training for Intensive Growing*. Akademiai Kiado, Budapest

Faust, M. and Miller, S. S. (eds) (1992) First International Symposium on Training and Pruning of Fruit Trees. *Acta Horticulturae* 322.

Ferree, D.C. and Warrington, I.J. (eds) (2003) *Apples: Botany, Production, and Uses*. CABI Publishing, Wallingford, UK.

Forshey, C.G., Elfing, D.C. and Stebbins, R.L. (1992) *Training and Pruning Apple and Pear Trees*. American Society for Horticultural Science, Alexandria, Virginia.

Hrotkó, K. (ed.) (2007) VIII International Symposium on Canopy, Rootstocks and Environmental Physiology in Orchard Systems. *Acta Horticulturae* 732.

Oberhofer, H. (1990) *Pruning the Slender Spindle*. Province of British Columbia, Victoria, B.C., Canada.

Palmer, J.W. and Wünsche, J.N. (eds) (2001) VII International Symposium on Orchard and Plantation Systems. *Acta Horticulturae* 557.

6 Pruning and Training of Vine, Bush and Cane Fruit

GRAHAM THIELE, DAVID JACKSON
AND MICHAEL MORLEY-BUNKER

Berry Fruits

Pruning of berry fruits is very simple in comparison with pruning of tree fruits. Indeed, it is becoming simpler as new and easier ways of managing the bush are being established. Both conventional and new methods of pruning berry fruits will be considered.

Blackcurrants

The most fruitful branches of blackcurrants are those which arise at the base of the bush and grow vigorously for one season; such a 1-year-old shoot bears a high proportion of fruitful lateral buds, which flower and crop the following season. One season later, if unpruned, much of the fruit is produced on the previous season's laterals, on shoots which are now 2 years old, thus creating a situation where most fruit is on new shoots near the top of the plant. Because of the high proportion of non-fruitful 2-year-old and older wood, past practice was to remove most of these and select as many as possible 1- year-old replacements coming from the base of the plant. This served to thin out the bush, give good light and spray access, and encouraged further growth on strong shoots.

The advent of machine harvesting dramatically changed planting patterns for blackcurrants. Instead of bushes being 1–2 m apart in the rows, they are now planted 2–3 plants per metre of row 30–40 cm apart. To facilitate the passage of machinery, the rows might be 3–4 m apart.

Mechanical pruning followed the advent of mechanical harvesting. This involves cutting wood, usually from the centre of the bush, to promote vigorous, new, annual growth. Spreading and older wood can be cut from the outside of the bushes along the rows as well, using a mechanical, vertical saw.

There may be some hand-pruning in the first 2 or 3 years after planting. The aim is to promote vigorous, 1-year-old fruiting wood to shoot from the base of the plant. Every 8–10 years the bushes can be cut back hard to near ground level for renewal. In this case they will not fruit in the next season as they regrow fruiting wood. To avoid complete loss of crop for 1 year, growers may alternate the renewal of growth by adopting a one row in eight to ten cycle.

Some growers still use hand-pruning if labour is available but this is generally on smaller areas. The pruning in this case can be more accurate than mechanical pruning, although it is doubtful if the extra cost can be compensated by additional yields. Hand-held, hydraulically operated loppers can speed up this hand process.

Care should be taken in areas where the silver leaf fungus disease is a problem. Growers may prune earlier in the autumn so that the open wounds have time to seal before winter sets in.

Redcurrants, whitecurrants and gooseberries

In these fruit crops, unlike blackcurrants, most fruit is carried on spurs a few centimetres long on older wood or on short shoots 10–15 cm long. Since the plants do not produce the same number of long shoots from the base as do blackcurrants, they have tended to be grown on bushes with a permanent framework, more like a miniature, vase-shaped apple tree. Pruning consists of a thinning operation aimed at replacing a proportion of laterals each year and keeping the bush open to light and sprays.

There is no reason why redcurrants and whitecurrants cannot be grown in a similar manner to blackcurrants for mechanical harvesting, but they are mainly grown in small areas in Eastern European countries such as Poland, the Czech Republic and Slovakia, where they are still hand harvested. Gooseberries pose more problems, being of a more spreading, often drooping, habit. Plants trained on a 30–40 cm standard, close planted, could be adapted for mechanical harvesting.

Blueberries

Most blueberries are planted 1.2–2 m apart in rows and picked by hand. The high price for quality fruit makes hand harvesting commercially feasible. The bush grows rather like a redcurrant, and training tends to restrict it to five to eight leaders arranged in a vase pattern. Once again, pruning involves the thinning of laterals to open up the bush for light and sprays. The occasional replacement leader may also be required to help regenerate the plant and provide a new source of younger wood.

Some tipping back of shoots in summer may be practised on mature bushes to promote fruit bearing within the bush. When required, bushes can be totally regenerated by cutting back to 10 cm above the ground.

When blueberries are grown for processing, mechanical methods of removal will be required and the plant can easily be adapted to mechanical harvesting.

Raspberries and brambles

Raspberries and brambles are now planted in rows 2–3.5 m apart – commonly 2.5–2.8 m – the distance being primarily determined by the machinery to be used. Within the row, raspberries are placed at about 30 cm and brambles at 2 m.

Pruning of these plants consists of removing all the floricanes as soon after cropping as possible. This gives the new developing canes – primocanes– space to grow and better access for sprays and light penetration. It can also be important in removing pests and diseases, so preventing their spreading to primocanes. This is particularly significant in brambles, where old canes should be severed at the base immediately after cropping, even if they are not removed from the wires.

Raspberries

Supporting trellises are used for both raspberries and brambles. The basic trellis for raspberries is shown in Fig. 6.1. Double wires on the upright posts are kept loose to allow primocanes to grow up between them. They are clamped together when training is required (Fig. 6.2).

Several methods to separate primocanes and floricanes have been adopted. The hoop system, whereby floricanes from plants 1.5–2 m apart in the row are tied together, has largely been discontinued. It has been found that close tying of floricanes in both rasperries and brambles inhibits the production of good flower initials.

Similarly, the Lincoln canopy system (Fig. 6.3), which showed promise for mechanical harvesting of good-quality fruit, has not been adopted to any extent either, perhaps because of capital costs of the trellis. It provided an excellent method of separating the floricanes and primocanes.

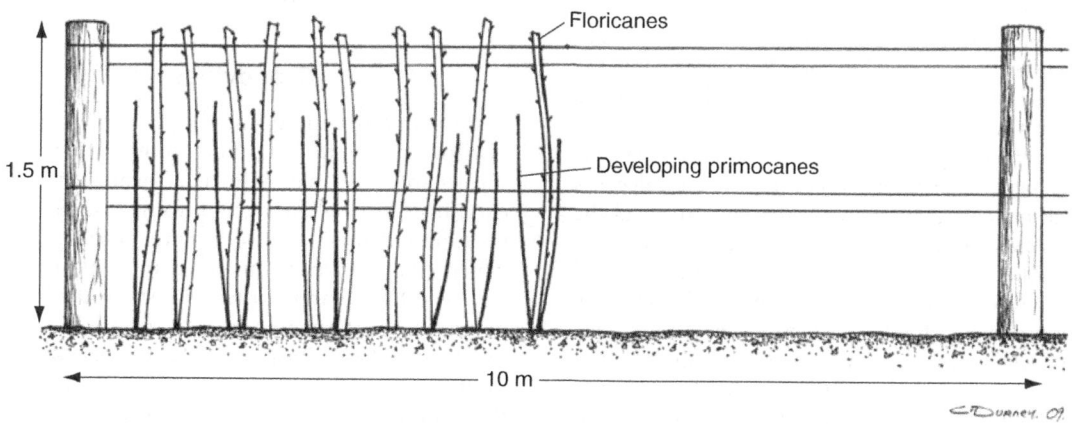

Fig. 6.1. Basic upright trellis for raspberries – the hedgerow system.

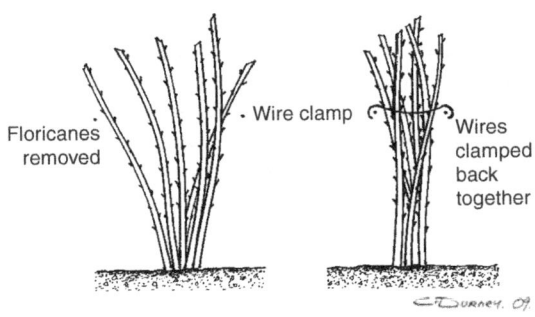

Fig. 6.2. Clamps for bringing wires together.

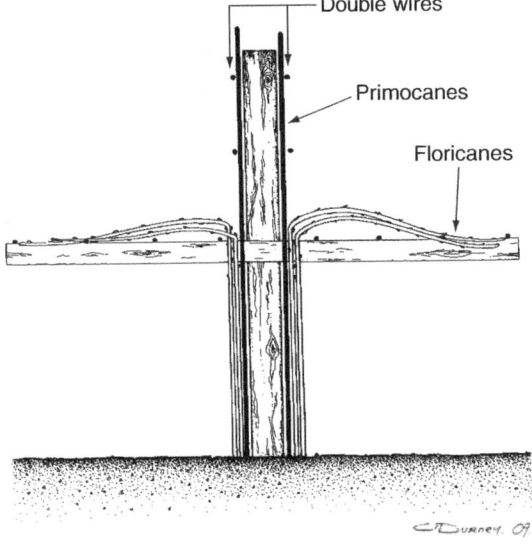

Fig. 6.3. The Lincoln canopy for raspberries.

Sometimes raspberries are grown on a biennial cropping system. In this, canes are mown off in autumn or winter in every other row and these are alternated each year. Thus, one row crops while the next is producing primocanes for the following season's crop.

Autumn-cropping raspberries are simple to manage. Since they crop in autumn on the primocanes, all canes on all rows can be removed after cropping and the following year's primocanes will produce the crop.

Straddle harvesters, using either a finger-wheel shaking device or a beater arm principle, are now in common use for the hedgerow and similar systems. Machines are locally made or imported and yield a product which is satisfactory for jam and other processing uses but generally not suitable for dessert purposes.

Brambles

Brambles, which include blackberries, boysenberries, loganberries and youngberries, have an extensive prostrate habit and require training on to upright posts and wires. This is labour intensive and new methods are being sought to reduce costs. The traditional methods used are illustrated in Fig. 6.4.

The rope method, whereby a handful of canes are tied together on the wires, has gone out of favour. Although the labour cost of training brambles on this rope system is lower, tightly held branches in the rope inhibit flower formation and ultimately the yield is affected. The espalier system is still effective but labour intensive. The 'winding' system is now in general usage. It is proven to provide the best development of flowers and the yield of quality fruit is enhanced.

Primocane growth can be restricted initially by the use of a chemical such as 'Hammer' (carfentrazon ethyl). Chemical application is normally discontinued about the first week in December (southern hemisphere) or the first week in June (northern hemisphere) to allow the primocanes to develop and mature before the onset of winter.

Some thinning of primocanes is often required for raspberries and brambles after removal of the fruited canes, to ensure adequate spray penetration and reduce crowding. The more vigorous canes are selected, especially those which appear to have been exposed to optimum light during their development. Diseased and weak canes are removed. Raspberries on the hedgerow system need about 20 canes per metre of row. Normally about 20–25 canes per bramble plant are trained, and they can be shortened to 2.5 m without significantly reducing yield.

Grapevines

Under natural conditions, vines, such as grapes, kiwifruit or passionfruit, must compete with trees, shrubs and other plants for light, water and nutrients. Instead of a thick trunk to hold them above the ground, vines have evolved other characteristics which enable them to maintain their place in the sun. The first characteristic consists of tendrils, as in grapes, or a twining method of growth, as in kiwifruit. These clasp or entwine neighbouring branches and allow the vine to climb to the top of the canopy of trees and shrubs. The second characteristic is that the growth of the vine is very

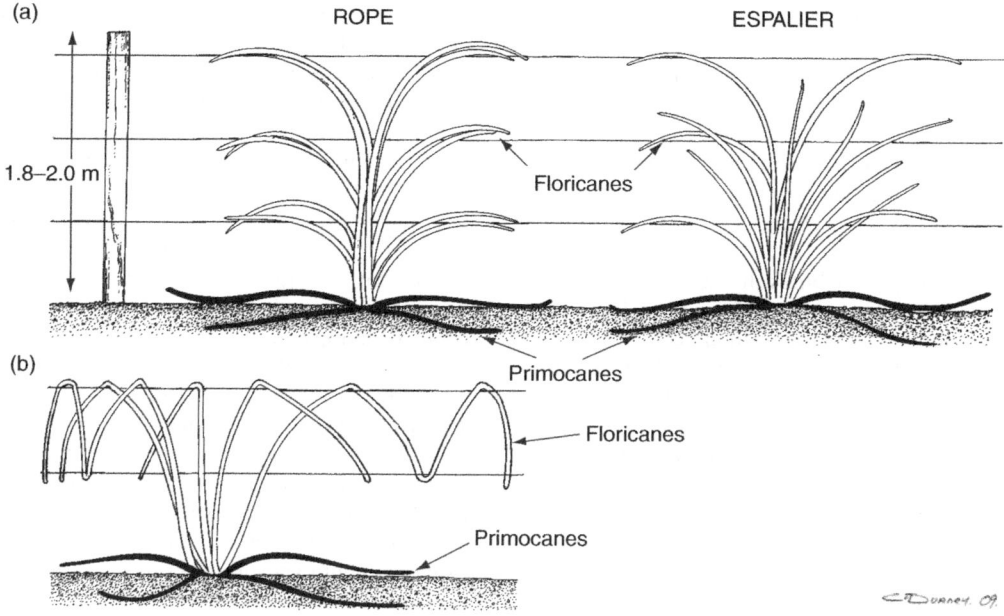

(a)

ROPE   ESPALIER

1.8–2.0 m

Floricanes

Primocanes

(b)

Floricanes

Primocanes

C. Durney. 09.

Fig. 6.4. (a) Two training systems for brambles. (b) Winding floricanes of brambles.

rapid and continues for a long period over spring and summer.

In horticulture, these characteristics have their good and bad points. The lack of a rigid trunk means that some supporting structure must be provided, usually a trellis. This, of course, can be costly. Vigorous growth can likewise be a problem. In a plantation, vines are grown in monoculture with no competition, except from neighbouring vines. Under these conditions, excessive growth has little value and much of it must be removed in the winter and summer. This is a waste of dry matter and may largely explain the lower yields obtained from grapes and kiwifruit compared, for example, with apples.

Pruning and training grapevines

Pruning and training of vines is constrained by the fact that grapes crop on wood that was produced the previous season. The amount of such wood remaining after pruning, which in turn determines the amount of fruitful buds, will have a major effect on the crop next season. Grapevine growers often use node and bud interchangeably when pruning. While high yields will be encouraged by higher bud numbers, too many buds on a vine can be expected to cause an excessive number of shoots to grow, and this can cause congestion and shading of leaves and fruit,

which is a recognized contributor to poor wine quality. Thus pruning can be considered to be a method to obtain the best balance between yield and quality.

Pruning and training aim to:

- Space the shoots so that each will present its leaves to adequate light.
- Space the shoots so that air circulation will be encouraged. This will reduce humidity, which, in turn, lowers disease incidence.
- Space the shoots to allow an adequate penetration of sprays for pest and disease control.
- Provide good replacement shoots for the next winter's pruning.
- Select the length and position of the shoot on which the buds have the best potential for fruiting.
- Achieve an appropriate bud number per plant or per unit length of trellis to give maximum yield of grapes of optimum composition.
- Vary the amount of perennial wood to minimize the hazard of winter freeze injury. Generally in areas with very cold winters a low proportion of perennial wood will increase the likelihood of injury.
- Facilitate movement of people and equipment through the vineyard.

There are many ways to train vines, but most growers use a system which will position last year's canes so that the shoots coming from the buds will intercept maximum light with a minimum of self-shading. The method described here is called the vertical shoot positioned (VSP) canopy system (Fig. 6.5); with modifications it is the major system used for training wine grapes, at least in quality wine-producing areas.

Winter pruning distributes the buds on canes or spurs attached to a base wire, as shown in Fig. 6.5. After buds burst, shoots are trained between parallel wires (Figs 6.5 and 6.6), and when shoots extend so far above the top wire that they begin to fall over, tops are trimmed, usually with a mechanical cutter. In vigorous sites laterals will grow after, and even before, topping. These will need to be trimmed before shading of fruit becomes a problem. Trellises need to be well supported, especially at end posts, two of which are shown in Fig. 6.7.

The geometry of the trellis is also influenced by the distance between rows, which varies from 1 to 3 m. Vines cannot be trained on trellises which are too high, otherwise the base of the vines will be excessively shaded. As a guide, the height of the foliage canopy should be no more than 0.6 of the distance between rows. It is important to emphasize that 'height of foliage canopy' means the height from the lower to upper leaves, not the height above the ground. A general 1:1 rule, however, is sometimes used, where the upper height of trimmed canopy above the ground should be no more than the row spacing.

Narrowly spaced rows generally utilize more sunlight over a given area but may require special tractors and other equipment for effective management. With standard tractors, rows will not normally be less than 2.5 m apart and the height will be in the order of 1.7–1.8 m. North–south rows are generally preferred, since even light exposure on both sides is obtained.

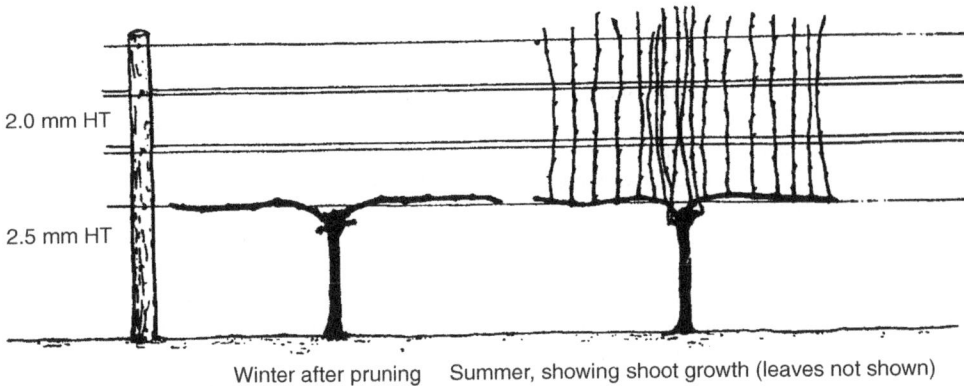

Winter after pruning   Summer, showing shoot growth (leaves not shown)

Fig. 6.5. Vertical shoot positioned vines.

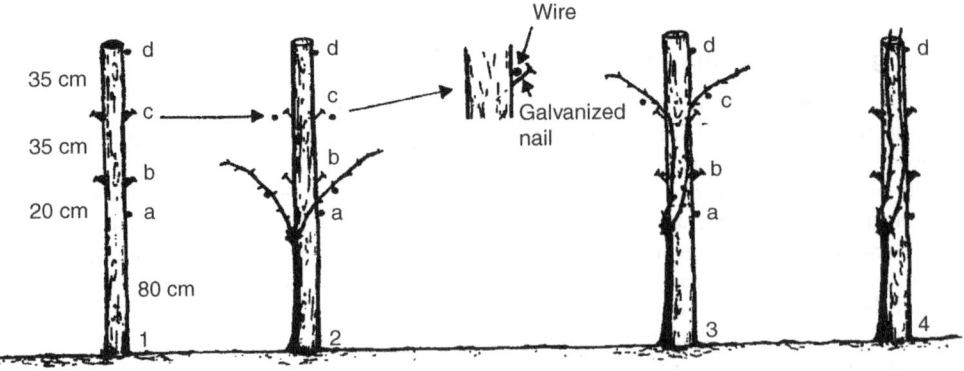

Fig. 6.6. End view along trellis, showing positioning of wires during shoot growth.

G. Thiele *et al.*

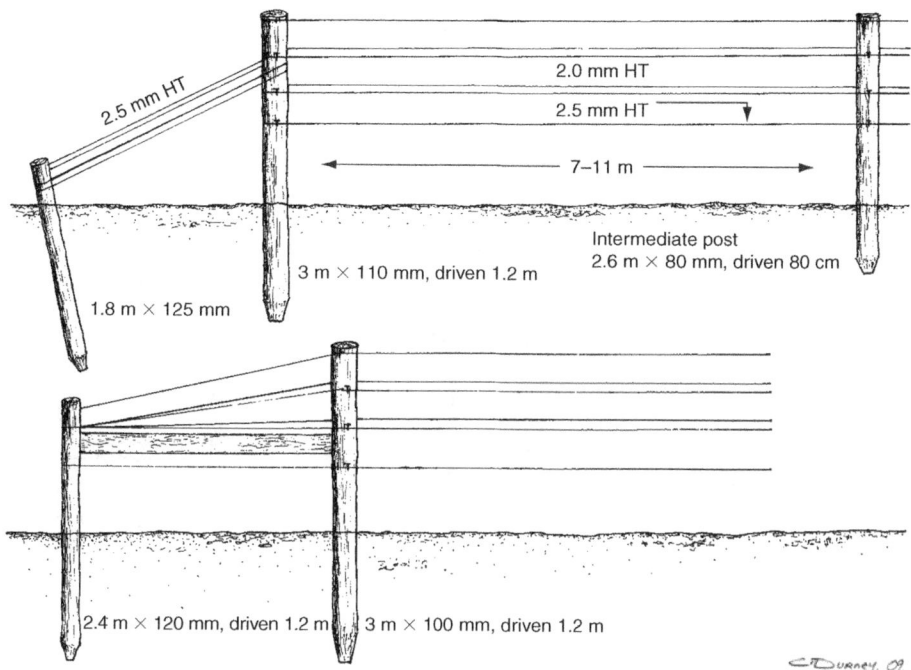

2.0 mm HT

2.5 mm HT

2.5 mm HT

← 7–11 m →

Intermediate post
2.6 m × 80 mm, driven 80 cm

3 m × 110 mm, driven 1.2 m

1.8 m × 125 mm

2.4 m × 120 mm, driven 1.2 m   3 m × 100 mm, driven 1.2 m

C. Durney. 09.

Fig. 6.7. Two end assemblies.

In recent years considerable research on pruning systems has occurred, particularly in Asian areas. The prime motivation for this work has been to cope with excessive vigour or over-wide row spacing. High vigour, found in vineyards with a surfeit of water and/or nutrients, can cause foliage management difficulties and often inferior juice composition. Wide spacing is wasteful of land and may also contribute to vine vigour. Many of the new trellis designs will increase leaf area exposed to sun, reduce foliage congestion and promote juice quality with economic yields. They are not always better than the VSP described earlier but can be superior in the conditions described above. The shapes of some alternative systems are illustrated in Fig. 6.8.

Pruning

Pruning removes 90% of the previous season's growth each winter. It is primarily the 1-year-old shoots, on which the next season's crop will be produced, that are cut off, and it could be suggested, therefore, that 90% of the potential crop is being removed. However, the trellis and the vine could never successfully support this crop level, and removal of shoots is essential.

SPUR PRUNING The spur-pruned vine usually has two permanent arms fixed to the base wire. Every winter, canes are cut back to two to six buds so that the required node numbers are left (Fig. 6.9). The advantage of this system is that it is simple and easy to teach to inexperienced workers. The disadvantage is that, with many cultivars, the lower buds are not particularly fruitful, so the cropping level may be inadequate; low cropping is more of a problem in cool climates, especially if associated with high cloud levels.

Spur pruning is commonly used by home gardeners and is ideally adapted to growing on fences, pergolas, the sides of houses, or in greenhouses, commercial or domestic. It should be noted that these 'spurs' are not true spurs, as found on apples, pears, etc. They are really severely pruned canes, but the term spur is so common it will be retained in our usage here.

CANE PRUNING In a cane-pruned vine there are no permanent arms along a wire or, if there are, they are short, and most fruit is borne on canes which are laid down each winter.

Two or more canes can be laid down per vine and generally each contains six to ten nodes. Spurs are commonly retained in the renewal zone below

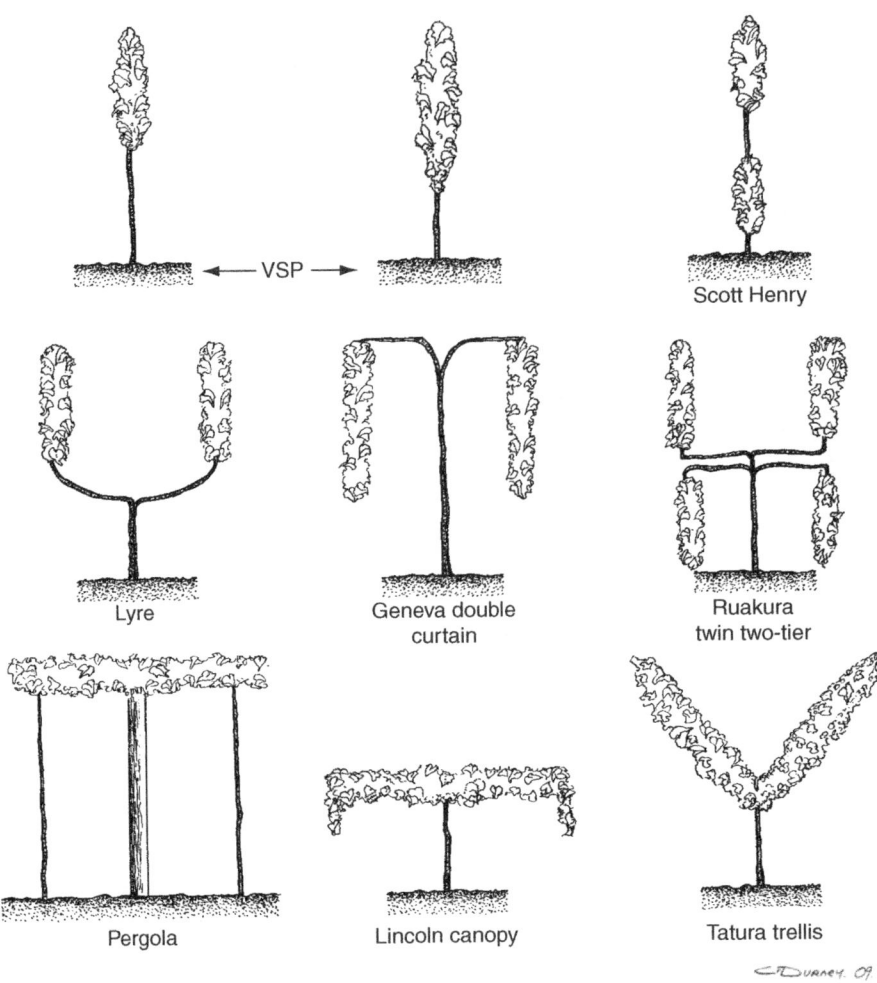

Fig. 6.8. Cross sections of some grape canopy designs. Top row, variations of the VSP system; middle row, double canopies; bottom row, horizontal and angled canopies.

the head of the vine (Fig. 6.10) to ensure there is an adequate choice of replacement canes for the next season. This system of pruning is often called the double-guyot pruning system, while the training system previously described was the vertical shoot positioned (VSP) canopy.

A mature vineyard will normally have 15–25 nodes per metre of row or canopy, whether cane or spur pruned.

Other training systems

The VSP and modifications thereof are common for wine, juice and drying grapes. For table (dessert) grapes, more complex systems are often encountered. There are several reasons for this, including the fact that the value of the crop is often higher than for other grape uses and therefore a higher cost can be justified. Also the cosmetic appeal of the fruit is of considerable importance, and training systems which reduce contact with neighbouring leaves, shoots and fruit will reduce skin blemishes. High-value dessert crops, sometimes grown in protective structures such as greenhouses, may need careful positioning of fruit to enable easy access for intensive labour activities such as bunch thinning or to assist spray penetration to obtain good disease control.

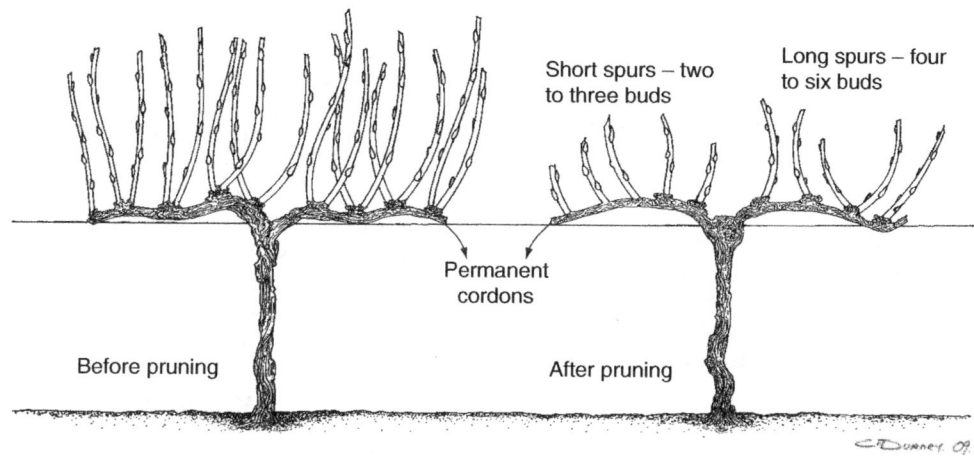

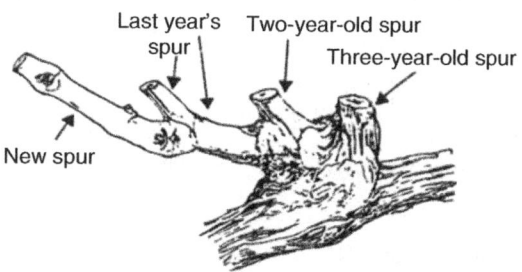

Fig. 6.9. Spur pruning of vines.

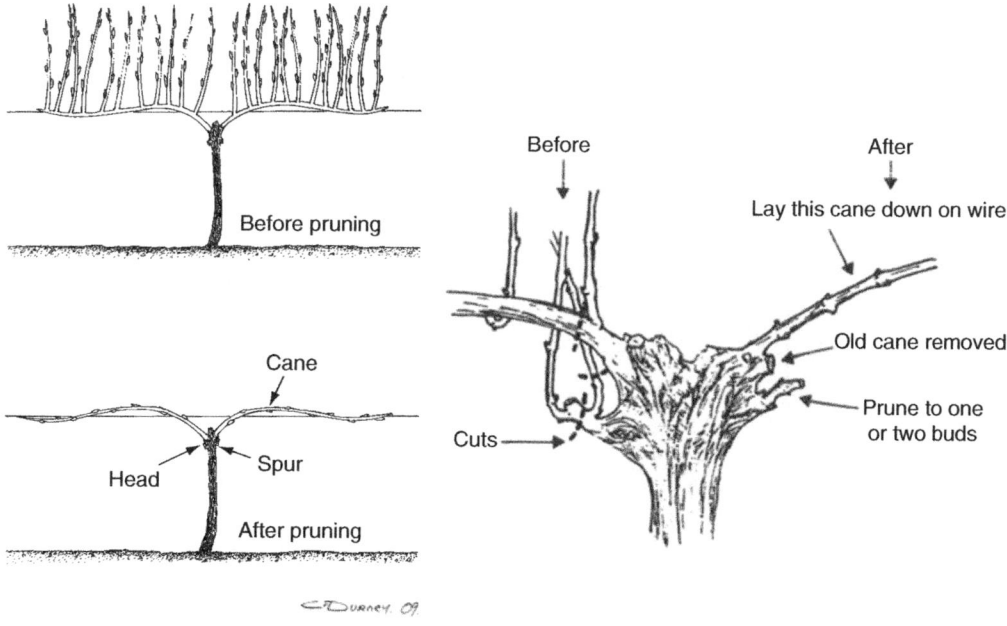

Fig. 6.10. Cane pruning.

Training the young vine

In districts with mild winters, vines can be planted at any time from late autumn to spring, the former often being preferred. Early-spring planting is recommended for regions experiencing cold winters. After growth begins, one shoot only is allowed to grow and this is lightly tied to a stake or grown up a string to the first wire (Fig. 6.11). In districts which are known to induce vigorous growth, two shoots may be left, one being left growing along the ground. The reduced vigour of the trained shoot will result in nodes which are closer together and allow easier choice of shoots at the wire the next season.

The growth of young vines depends on water availability, soil moisture and fertility, plus climate. These must be optimized to encourage early vigour. The alternative methods of training weak, moderate or vigorous vines are shown in Fig. 6.12.

Sometimes, in the first year, the shoot will not reach the wire, in which case it is normally pruned back the next winter to three or four nodes. If growth is very vigorous and reaches the wire early in the season, it may be either trained one way along the wire or cut off below the wire to encourage the shoot to divide. If growth is moderately vigorous and reaches the wire later in the season, it will be cut off in winter and tied to the wire. Subsequent treatments of weak, moderate or vigorous vines are shown in the lower part of Fig. 6.12. It will be seen that vines with moderate to vigorous growth have produced a small crop in the second season after planting. After the second winter (the third growing season), these fruiting vines are nearing full production. Weak vines will need a further season to reach this stage.

Kiwifruit

Readers may like to refer to the earlier sections on grapevine pruning, outlining the aims of pruning and training – they are similar for grapes and kiwifruit.

Support structures

Hitherto, New Zealand growers have normally chosen between the T-bar system (and its modifications)

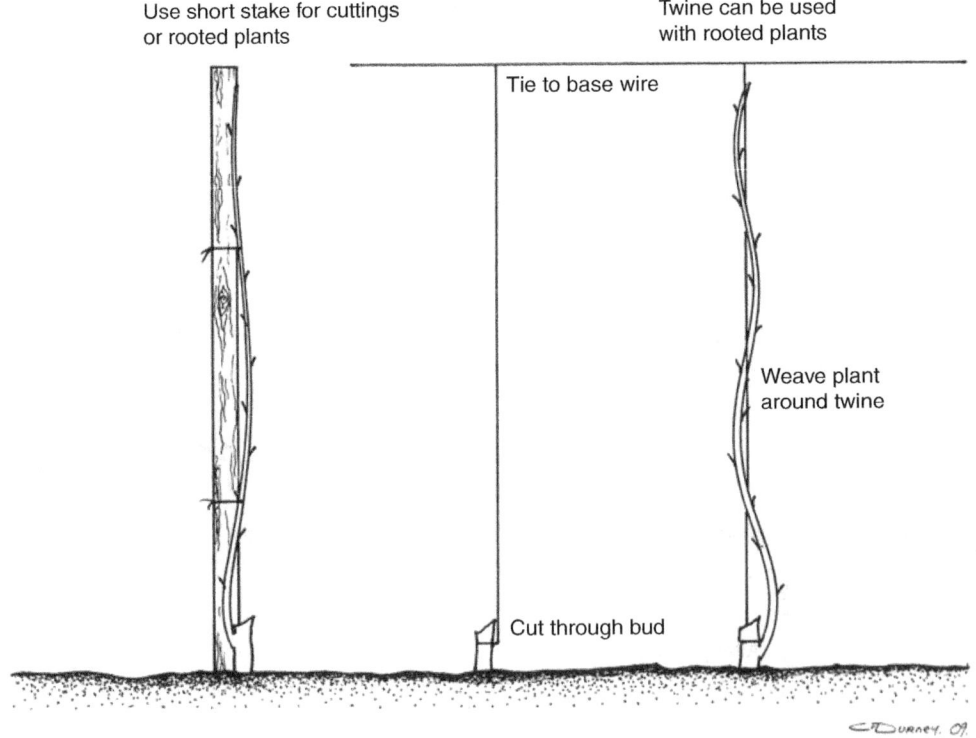

Fig. 6.11. Supporting the shoot.

G. Thiele *et al.*

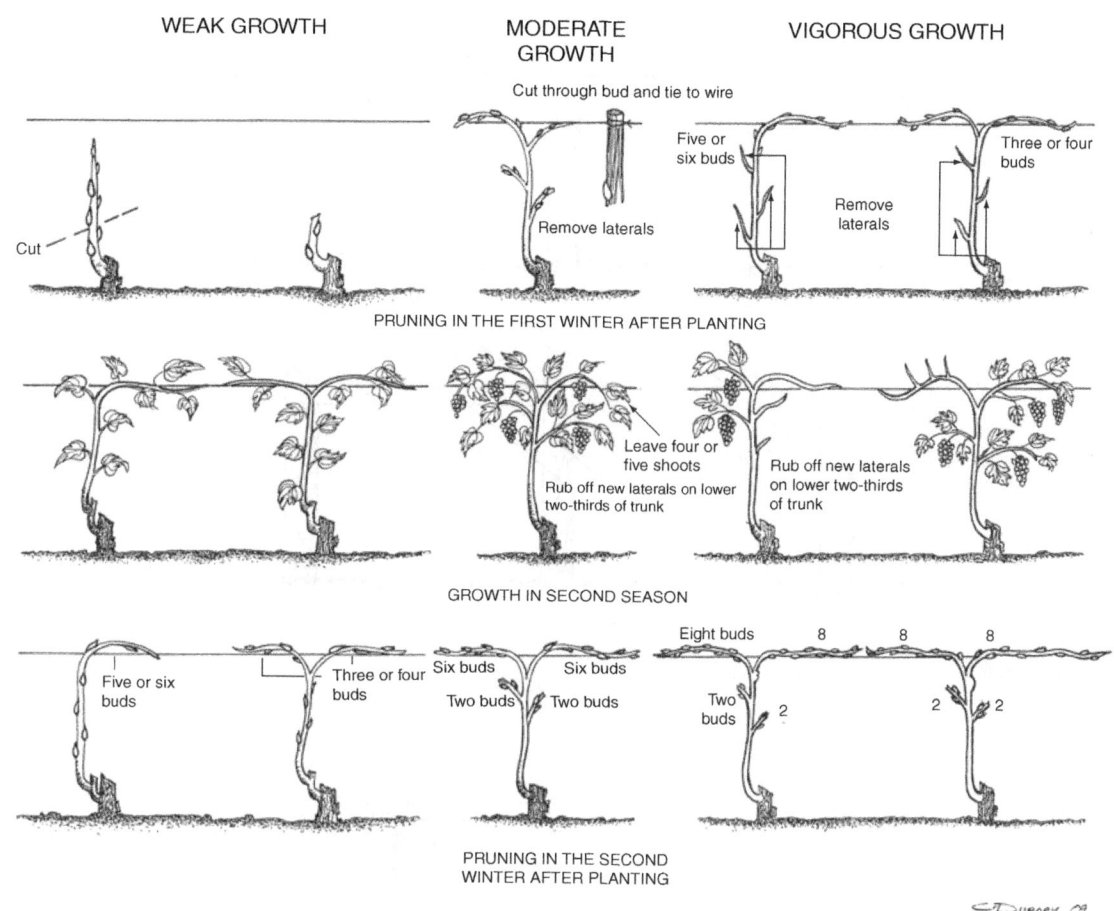

WEAK GROWTH   MODERATE GROWTH   VIGOROUS GROWTH

Cut through bud and tie to wire

Five or six buds

Three or four buds

Remove laterals

Remove laterals

Cut

PRUNING IN THE FIRST WINTER AFTER PLANTING

Leave four or five shoots

Rub off new laterals on lower two-thirds of trunk

Rub off new laterals on lower two-thirds of trunk

GROWTH IN SECOND SEASON

Five or six buds

Three or four buds

Six buds   Six buds

Two buds   Two buds

Eight buds   8   8   8

Two buds   2   2   2

PRUNING IN THE SECOND WINTER AFTER PLANTING

C. Durney. 09.

Fig. 6.12. Early training of vines of different vigour.

and the overhead pergola. These are illustrated in Fig. 6.13. In recent times, more and more growers have adopted the overhead pergola. Other structures have been utilized both in New Zealand and overseas – principally A frames and tall, upright trellises. A recent development is use of the 'bandolier' system with the pergola structure. The bandolier posts and attached strings provide a means for vigorous current-season canes to be separate and above the bearing canopy.

The pergola gives complete overhead cover with leaves and fruit. Relatively little light penetrates a fully established pergola canopy, which results in less need for weed control and mowing. Wind damage of canes and fruit is less of a problem with pergola-grown vines. However, some people have suggested bee visitation of pergola-grown flowers is lower and that botrytis may be more of a problem. The pergola is more expensive to construct compared to T-bar structures. Fruit yields may be lower in pergola plantings compared to T-bar plantings, but the lower fruit number can be offset with lower crop management inputs, improved pack-out percentages, better-quality fruit grades and earlier attainment of full cropping potential.

End posts

Since the weight of the vines and fruit is greater than with grapes, stronger end-structures are needed. Two major strainer assemblies are shown in Figs 6.14 and 6.15. Both are adequate for either the T-bar or pergola.

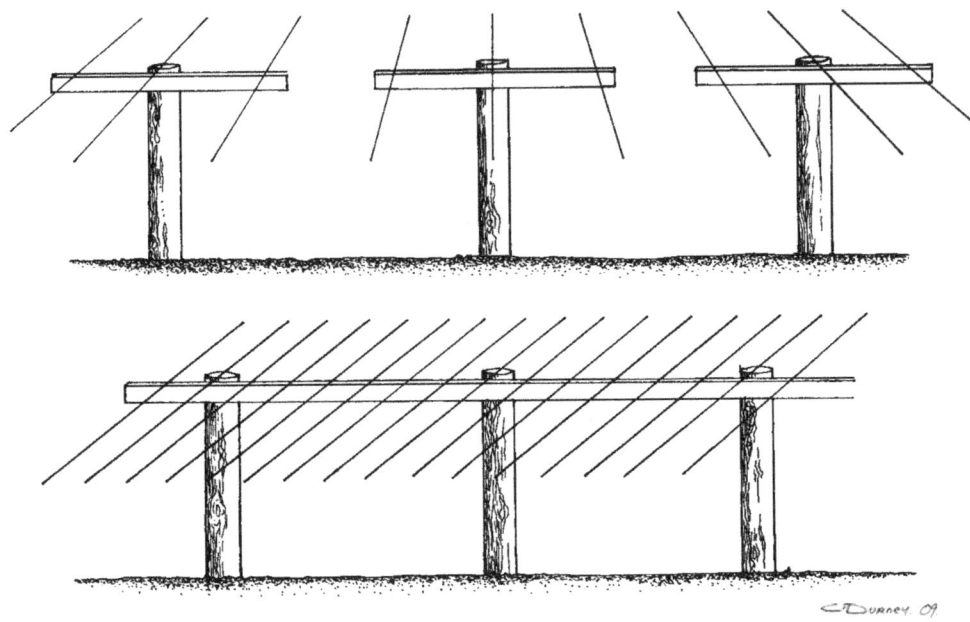

Fig. 6.13. T-bar and overhead pergola.

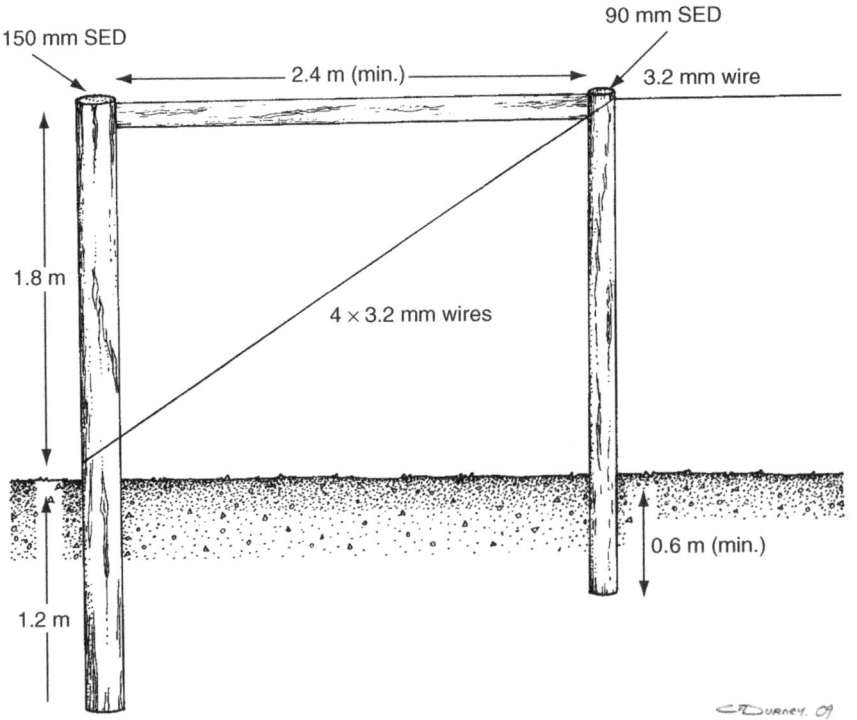

150 mm SED

90 mm SED

2.4 m (min.)

3.2 mm wire

1.8 m

4 × 3.2 mm wires

0.6 m (min.)

1.2 m

Fig. 6.14. Horizontal-stay end assembly. SED, small end diameter of the post – use this figure when ordering posts from the merchant.

G. Thiele *et al.*

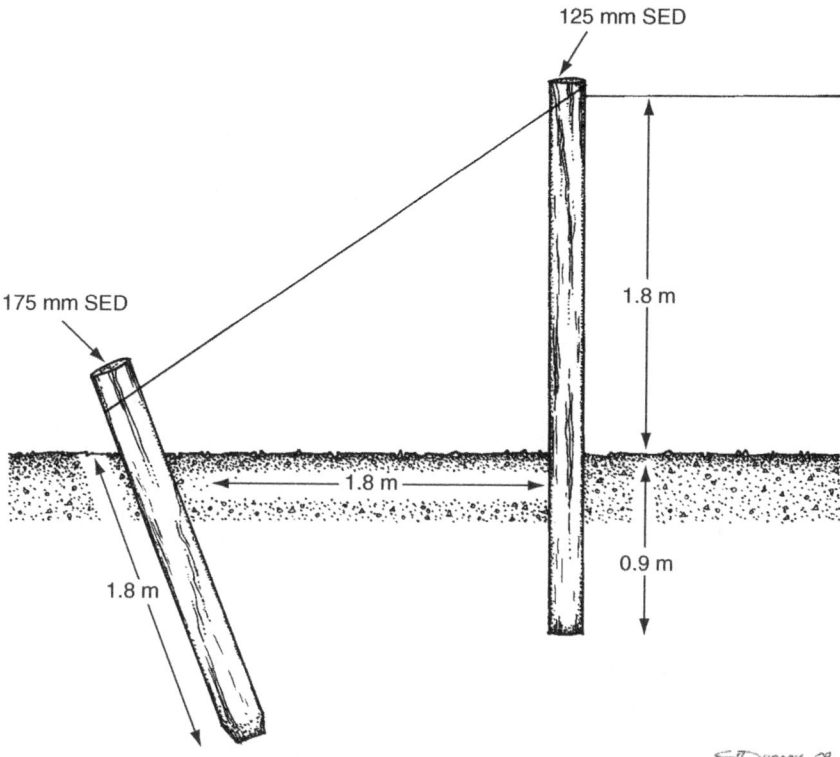

125 mm SED

175 mm SED

1.8 m

1.8 m

1.8 m

0.9 m

C. Durney. 09

Fig. 6.15. Tie-back end assembly.

The T-bar and its modifications

The standard T-bar is 1.8 m tall and 1.5 m wide. This and two common modifications are shown in Fig. 6.16.

The pergola

The construction of a pergola is illustrated in Fig. 6.17.

Pruning and training

Initial establishment

The initial training of the vine in both the T-bar system and the pergola is essentially the same. As shown in Fig. 6.18, vines are planted midway between posts at 5.5–6.0 m apart if traditional spacings are adopted. This figure also shows the arrangement of male and female plants in the two systems. A light stake is placed next to the vine and attached to the central wire on the T-bar or pergola.

The strongest shoot is selected and lightly, but securely, tied to the stake at regular intervals, until it reaches the top. A second shoot can be kept, but on grafted plants, this must not come from below the graft and must not be allowed to compete with the main shoot; it will be removed once the main shoot is established. The main shoot must not be allowed to twist tightly around the stake.

When a grafted plant or rooted cutting reaches the wire, a permanent leader should be allowed to grow in either direction along the wire, as shown in Fig. 6.19.

Sometimes a seedling is planted and will be grafted just below the wire the following winter. In this case, careful training above the wire is not needed before grafting, since this will eventually be removed and the shoot from the bud will take over the top part of the vine. Subsequent training to establish the basic framework is illustrated in Fig. 6.20.

After 3–4 years, there should be two strong leaders running in opposite directions with fruiting arms extending at right angles over the canopy surface, as illustrated in Fig. 6.20. If the main leader(s)

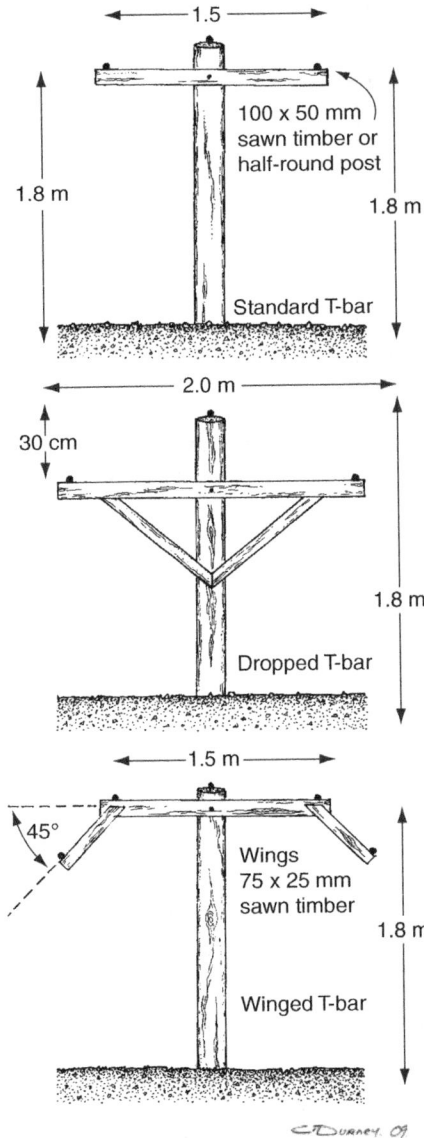

1.5

100 x 50 mm
sawn timber or
half-round post

1.8 m   1.8 m

Standard T-bar

2.0 m

30 cm

1.8 m

Dropped T-bar

1.5 m

45°

Wings
75 x 25 mm
sawn timber

1.8 m

Winged T-bar

C.Durney. 09

Fig. 6.16. Standard T-bar and its modifications.

starts to lose vigour during the establishment, the shoot can be pinched back to a strong bud and a new shoot encouraged to take over the leader role.

Maintenance pruning

WINTER PRUNING (FEMALE VINES) As with grapes, kiwifruit vines fruit on shoots produced on 1-year-old laterals.

- Cane pruning. One method of pruning kiwifruit is similar to cane pruning grapes. The main arms on the central wire are permanent, but the lateral canes, which are taken at right angles across the canopy and allowed to fruit, are replaced each year after fruiting with strong canes produced in the previous season. On the T-bar the new replacement canes are selected at 25–40 cm intervals, hung over the outer wires, secured and cut off at about knee height. On the pergola, the lateral is run out until it reaches its opposite number on the adjacent row. The fruiting canes are removed each winter and replaced by last season's shoots that have arisen in an appropriate position (Fig. 6.21).

- Short lateral pruning. The second method is to retain some of the fruiting lateral after fruiting and encourage secondary lateral growth in the next season. Buds on fruited laterals may produce new fruitful lateral shoots in the following year. The new-season secondary lateral shoots can produce flowers and fruit on the basal nodes (Fig. 6.22). Eventually the density of retained lateral growth is likely to require removal of old laterals and replacement with a new lateral – if possible a 1-year-old lateral from near the base of the originating shoot. Selected laterals should have had good light exposure because floral potential is affected by light. Laterals that have been shaded have low productivity (Fig. 6.23). Laterals that are selected for laying down on the canopy can be shortened to about eight buds for ease of handling. Sometimes short, self-terminating shoots are produced; these are referred to as 'spurs' and do not need to be cut back.

SUMMER PRUNING (FEMALE VINES) Kiwifruit vines may need to be trimmed in summer. Pruning is done with the objective of maintaining a permeable canopy for light penetration and good air movement. The latter should assist with spray penetration and reduce the build-up of disease.

Summer pruning can begin in late spring, when unfruitful shoots, which are not required the following season, are removed. Erect, strong water shoots are cut back to a stub. Shoots starting to curl and tangle (a sign of weakness) can be cut back. During summer, kiwifruit vine growth can be

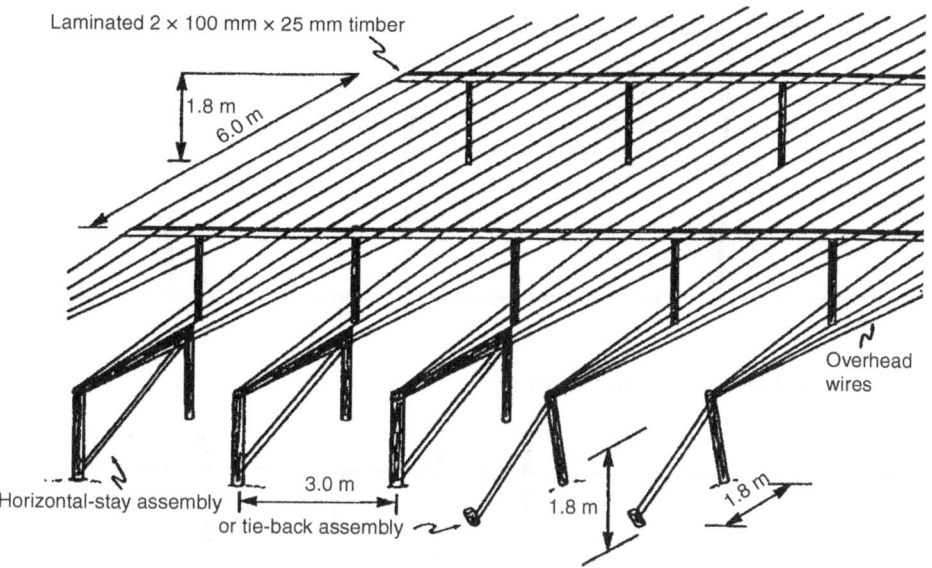

Fig. 6.17. Pergola system.

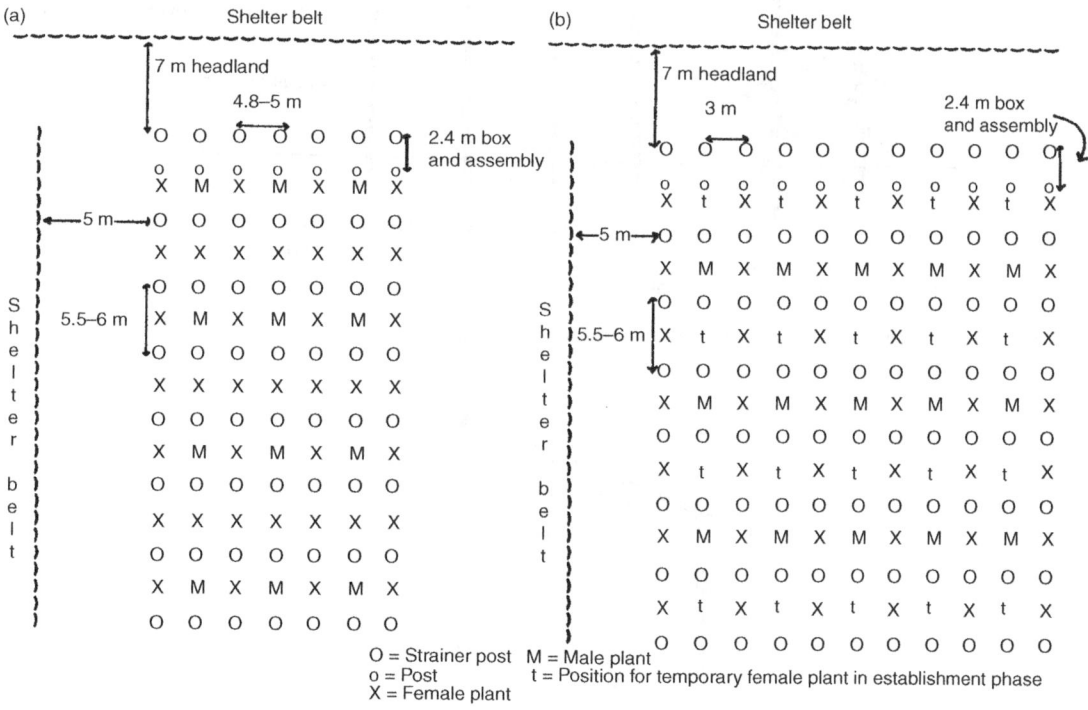

O = Strainer post M = Male plant
o = Post t = Position for temporary female plant in establishment phase
X = Female plant

Fig. 6.18. (a) Diagram of seven-row-wide planting for the T-bar system, with males every second plant in every other row. In this example, males are situated on the ends of the appropriate rows to allow bees access to male pollen as they enter the block from the headland. (b) Diagram of planting system for pergolas. Males are planted between every other post in each alternate row. The aim is to have a narrow band of males between each female row.

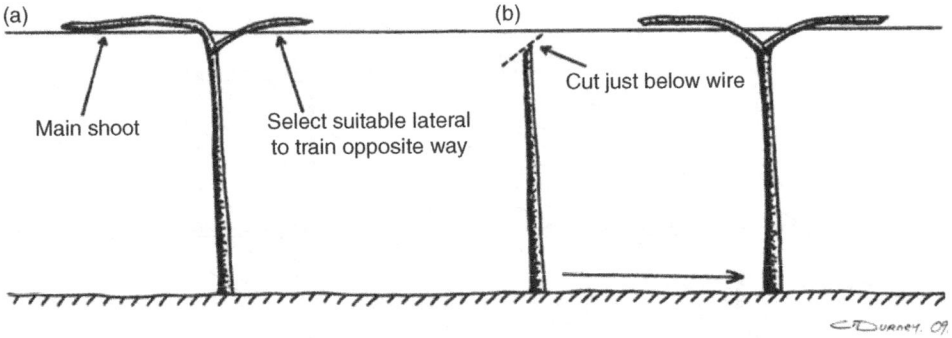

Fig. 6.19. Two methods of selecting the main leader.

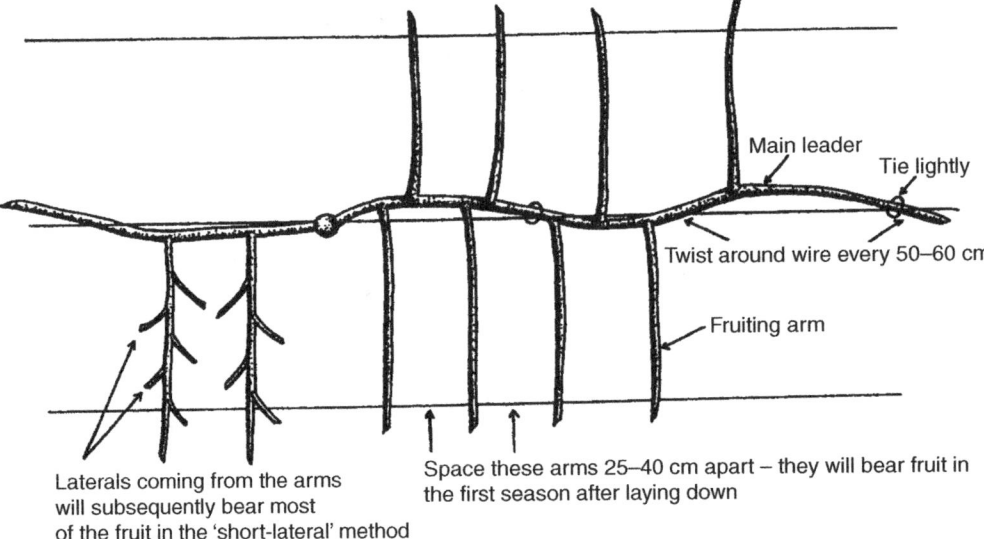

Fig. 6.20. Bird's-eye view of T-bar pergola showing training of leaders and fruiting arms.

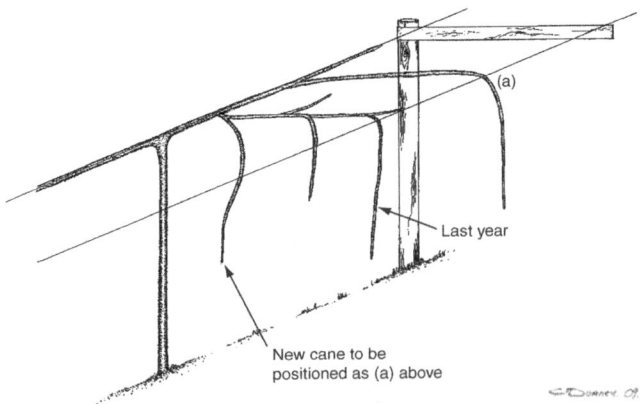

Fig. 6.21. Winter cane pruning.

G. Thiele *et al.*

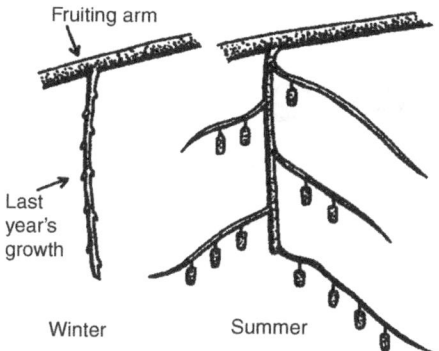

Fig. 6.22. Fruit formation on kiwifruit vine.

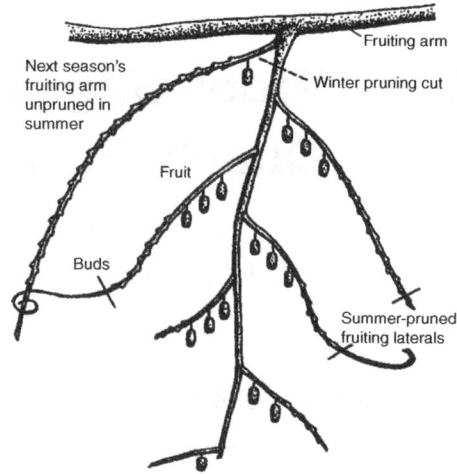

Fig. 6.24. Pruning cuts for summer and winter on a kiwifruit vine.

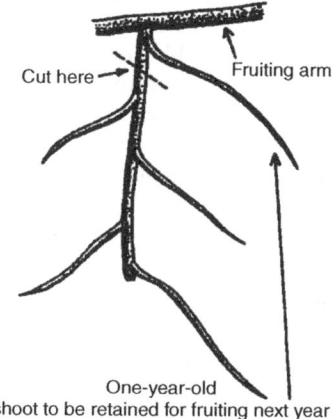

Fig. 6.23. Winter pruning of lateral.

especially vigorous, and several prunings may be needed to achieve the objective of having a permeable canopy. As summer vigour diminishes, growers may cut back the fruiting shoots, with the aim of producing shoots with a length that extends to between three and six leaves above the position of the last fruit, as shown in Fig. 6.24.

Gel pruning – where cut ends of shoots are left with a smear of paste with auxin – has been shown to reduce the number of summer prunings in a season. However, some care must be taken with use of gel pruning with regards to vigour depression and growth deformation. Attention to determining and achieving vine balance, a balance between vegetative vigour, fruit yield and

vine maintenance, will lead to less need for summer pruning.

PRUNING OF MALE VINES Male plants, which do not fruit, need different treatment. While they are grown to a form basically similar to females, their main purpose is to provide pollen – thus the major pruning takes place after flowering has occurred. At this point, the flowering shoots are cut back to a new growth, as close to the base as possible. This new growth is pruned back to 60 cm in January (southern hemisphere), and again in February or March, to 75–80 cm. Any further pruning to maintain a tidy vine is undertaken in winter.

Further Reading

Coombe, B.G. and Dry, R.R. (1992) *Viticulture, Vol. 2, Practices.* Australian Industrial Publishers, Adelaide.

Crandall, R.C. (ed.) (1995). *Bramble Production.* Haworth Press, Binghampton, New York.

Jackson, D. (1997) *Monographs in Cool Climate Viticulture 1. Pruning and Training.* Lincoln University Press, Canterbury, New Zealand.

Warrington, I.J. and Webster, G.C. (eds) (1990) *Kiwifruit Science and Management.* New Zealand Society for Horticultural Science, Ray Richards Publisher, Auckland.

7 Producing and Marketing Quality Fruit

DAVID JACKSON, NORMAN LOONEY,
AND MICHAEL MORLEY-BUNKER

The Process of Maturation

As fruits grow and develop, many important chemical and physical changes occur. These are best understood as an adaptation by the fruit to facilitate reproduction. For example, before the seeds are fully developed, it would be disadvantageous for fruit to be attractive to birds and mammals which consume them and disperse the seeds. Thus, the plant has evolved in such a way that, before ripening, its fruit is hard and lacking in flavour. Immature fruits often have unpleasant characteristics such as excessive acidity or high tannin levels, which make them sour and/or bitter. Once the seeds are mature and ready for dispersal, quite dramatic changes will occur and the fruit assumes a more attractive flavour and texture. The changes in sugar and acids in grape berries illustrate this change (Fig. 7.1). Note how these changes relate to the double-sigmoid growth curve of that fruit. However, many other changes occur that are not shown in this diagram. For example, while sugars are increasing, so also are many other chemical components which give flavour and aroma to the fruit. There may be hundreds of chemical changes during fruit ripening and by no means have all of them been identified by scientists. Changes in fruit colour are nearly universal, usually resulting in the fruit becoming more conspicuous.

All of these changes require chemical energy and, with some fruits, ripening is accompanied by an increased rate of respiration. We refer to such fruits as having a climacteric ripening pattern and these fruits also exhibit an increase in ethylene production during ripening. This phenomenon is discussed below. Figure 7.2 shows some of these changes in developing and ripening apples, a fruit with a single sigmoid growth curve. The following sections detail some of the changes that take place during maturation and ripening of fruits.

Sugars

In apples, starch, which is a more complex carbohydrate than sugars, builds up in the young fruit, but it changes to soluble sugars during ripening. Grapes and stone fruit, on the other hand, do not contain starch and the sugar is transported into the fruit during maturation. Some fruits, such as bananas, contain large quantities of starch at maturity. The following approximate sugar and starch levels have been recorded in different fruits on a fresh weight basis:

- Amount of sugar
Grapes	20%
Pome and stone fruit	10%
Oranges	9%
Kiwifruit	8%
Lemons	2%
- Amount of starch
Bananas	12%
Apples and pears	4%
Grapes and stone fruit	0%

Carbohydrates include starch, cellulose and various sugars. Whereas in animals energy is stored as fat, in plants carbohydrates are the main storage compounds.

The sugar level obviously determines the sweetness of a fruit, but the level of acid also affects our appreciation of sugar. For example, a fruit with 10% sugar and little acid will appear sweeter than one with 20% sugar and very high acid levels.

Acids

There are many different acids in fruit, the three most common being malic, citric and tartaric, the latter most prominent in grapes. At maturity, the level of most acids is below 1%, although lemons

©CAB International 2011. *Temperate and Subtropical Fruit Production*, 3rd Edition (eds D. Jackson *et al.*)

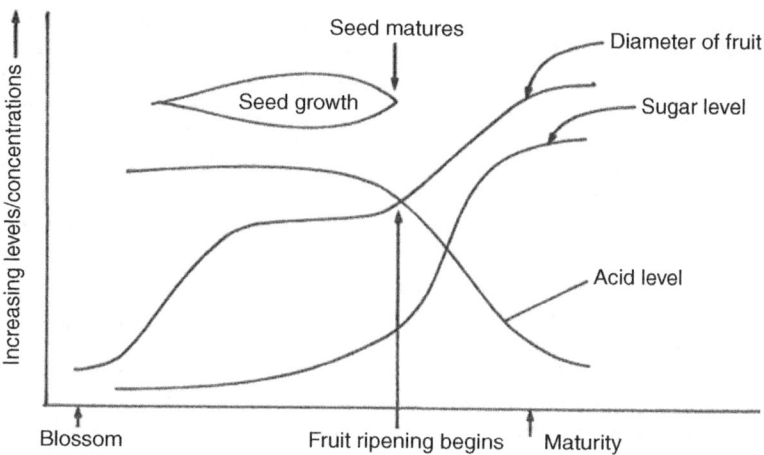

Fig. 7.1. Changes in diameter, sugar and acid levels of grape berries.

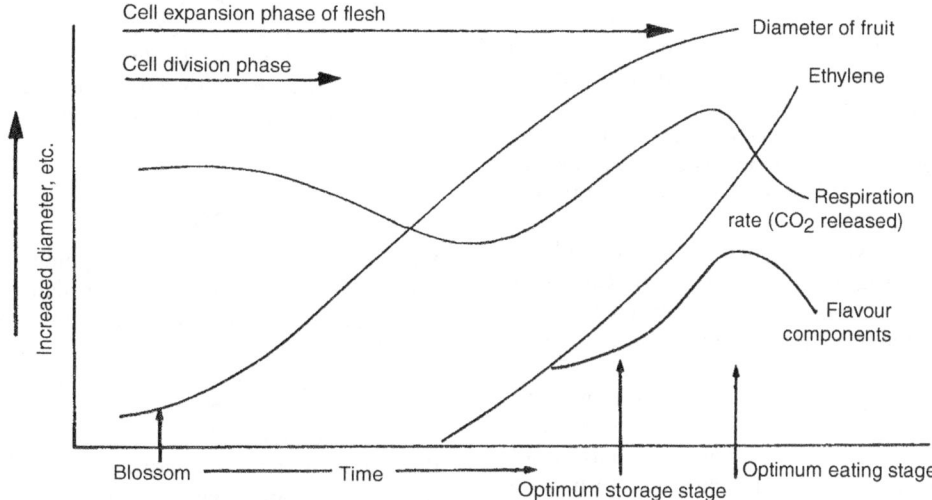

Fig. 7.2. Changes in size, respiration rate and level of flavour components in an apple.

still have 5%. Acids are an important flavour component. A very acid fruit tastes unpleasantly sour, but some residual acid is necessary to give the fruit freshness and interest, otherwise it will taste flat. People vary in their tolerance to acid. Asians seem to prefer fruit with low acid levels, while many Europeans tolerate, even prefer, high acid levels. The Cox's Orange Pippin apple, for instance, is very popular in much of northern Europe, but the Asian market prefers the less acid Delicious apple. Such understanding can obviously have important marketing consequences. Any fruit planted,

selected or bred must be suitable for the market to be served.

Flavour components

Flavour is determined by many compounds, including sugars, starch, acids and specific flavour components. These latter are aromatic compounds, generally esters and alcohols, several dozen of which may give the characteristic flavour to a fruit. As seen in Fig. 7.2, flavour components rise towards maturity. However, they will decline as the fruit becomes over-mature or may be supplemented by other less desirable flavour principles.

Colour

The green colour of fruit is due to chlorophyll contained in special subcellular organelles called chloroplasts. Other colours, such as red, orange and yellow, are due to a whole range of compounds, including anthocyanins, carotenoids and flavones. During maturation and ripening, chlorophyll is degraded and the other compounds are revealed. Often anthocyanins and carotenoids increase in concentration during fruit ripening. A number of environmental factors are known to affect colour development.

- Light. Generally, most fruit develop better colour when exposed to sunlight. The ultraviolet component of light is especially significant in inducing colour formation. Thus, fruit grown at high altitudes or in desert climates are often more colourful. Heavily shaded fruit tend to remain green and have minimum development of other colours. Cherries, strawberries and many grapes will develop colour almost as well when shaded and are therefore an exception to this general rule.
- Nutrition. Excessive applications of fertilizers high in nitrogen will delay the disappearance of chlorophyll from fruit peel, stimulate shoot growth and thus increase fruit shading, and may even directly suppress red colour development.
- Temperature. Colour development in temperate-zone fruits is often better in cooler climates. Apples, for example, will generally have more brilliant colour in cooler climates. With citrus the situation is not so clear. Oranges produced in cool climates may be less coloured than those grown in warmer areas, but in very warm regions colour is often inferior.
- Water. Drier climates are more conducive to colour development than wet climates, which may be explained by temperature and light differences. However, drought conditions can lead to fruit colour or finish that can be described as dull.

Tannins

The term 'tannin' refers to a complex range of phenolic compounds which give the fruit a bitter or astringent taste. This taste is often confused with sourness but will tend to leave a furry sensation in the mouth after swallowing. Tannins are important in the overall taste of the fruit: too much of this flavour principle is unpleasant, but too little can result in the fruit tasting flat. People vary in their ability to tolerate bitterness in fruit. Tannins will often change in form as fruits mature and become less astringent to the taste.

Softening

During the later stages of fruit enlargement there is a considerable increase in the size of cells and a concomitant thinning of cell walls. Such fruit are softer and easier to eat, but if there are large intercellular spaces, or the fruit cells are separating from one another, fruit tend to be mealy and lack crispness. Most tissue softening is due to the breakdown of pectic acid and protopectins, chemicals that bind the walls together. However, cellulose, the structural carbohydrate in all plant cell walls, is also degraded during fruit ripening. This results in thinner cell walls and contributes to our perception of texture.

Respiration

Most stone fruits, berries and pome fruits that ripen early in the summer have a high respiration rate, undergo rapid senescence and cannot be stored for long periods. Others, such as late-maturing apples and pears, kiwifruit and citrus, have a low respiration rate and may be stored for a considerable time. The respiration rate indicates the degree of metabolic activity taking place in the fruit. In general, when this rate is high, ripening and senescence will proceed rapidly.

Respiration rate is reduced during cool and controlled atmosphere storage. This will be discussed more fully later.

Figure 7.2 illustrates the rise in respiration and ethylene production that is characteristic of climacteric fruits. These fruits display a climacteric rise in respiration and this rise generally coincides with the time when the fruit is at optimum condition for eating. However, most fruit physiologists interpret the climacteric rise in respiration and ethylene production as the start of an accelerated senescence process; once the climacteric is under way fruit ripening and senescence are irreversible. Apples, pears, kiwifruit, apricots, peaches, plums, avocados, bananas, passionfruit and mangoes are climacteric fruits. Fruits without a climacteric include cherries, figs, grapes, grapefruit, lemons, oranges, melons, pineapples and strawberries.

Ethylene

As already mentioned, ethylene production increases during ripening of climacteric fruits and it is believed that this gaseous plant hormone is the initiator of the ripening process in these fruits. One unique characteristic of climacteric fruits is that the presence of a small amount of ethylene, either applied or produced by the fruit, stimulates the fruit to produce more ethylene, which further accelerates the ripening process. With non-climacteric fruits, applied ethylene will stimulate some aspects of fruit ripening, such as chlorophyll breakdown in citrus rind, but the fruits do not respond by producing more ethylene. Clearly, the effects of ethylene on fruit ripening profoundly affect our approach to fruit storage.

Factors Affecting the Rate of Maturation

The time of maturation and ripening has considerable significance for fruit growers. It is not uncommon for a popular cultivar to mature and ripen in a time period when markets are oversupplied. In many cases, the highest returns come from an early crop but, in other cases, late crops can obtain premium prices. Another scenario is that a valuable cultivar or species may not ripen early enough to escape the autumn frosts. If the climate or season is marginal for maturing a late-season crop, any factor affecting maturation rate can be important in determining whether a grower succeeds or fails. The most logical way to deal with such problems is to choose early cultivars of the fruits to be grown. However, the following factors, some under the control of the producer, can also affect time of maturity.

Temperature

Heat is the most important environmental factor promoting fruit maturation. As discussed in Chapter 2, temperature in an orchard can be increased by shelter, soil cultivation, choice of correct slope or exposure, or modification of the training system. While the trend has normally been to select warmer sites in temperate countries, there are

situations where the producer may choose a cooler site or manage the planting in such a way as to achieve a late harvest.

Nutrition and water

Frequent watering and high nitrogen tend to promote vegetative growth, which in turn reduces fruit exposure and heating, and therefore delays fruit maturation. In some circumstances this may be desirable, but the price is usually poorly coloured fruit with reduced storage life.

Light

As indicated, shade reduces the temperature of the fruit and delays maturity and colour development. It is modified by the amount of shoot growth and leaf production and by the way trees are pruned and trained. As a general rule, pruning should promote even light penetration within the canopy. The sheer size of the canopy (as influenced by rootstock selection) greatly influences the light climate of fruit trees and vines.

Pests and diseases

Attacks by pests and diseases which cause fruit and leaf damage sometimes advance fruit maturity. Physical damage to an apple or pear caused by codling moth, for example, can cause ethylene to be produced, which initiates ripening. Damage to leaves will sometimes increase light penetration or reduce assimilate flow to the fruit and so affect maturity.

Hormones and plant bioregulators

Hormones within the fruit or plant can affect the rate of ripening. For example, early-maturing apples produce high levels of ethylene and these fruit ripen very quickly. Natural auxins and gibberellins, when in adequate supply in fruit tissues, are believed to 'protect' developing fruit from the effects of ethylene. Avocado fruit seldom ripen while attached to the tree, suggesting that the leaves produce substances that inhibit the ripening process.

Externally applied bioregulators will also modify ripening in a number of cases. For example, very small amounts of a strong synthetic auxin, such as 2,4,5-T, applied to apricots at the beginning of pit-hardening will advance harvest maturity. In western North America, gibberellic acid is used by producers interested in late-harvested sweet cherries to delay colouring and harvest maturity by 2–5 days. This chemical (about 20 mg/l) is applied as a whole-tree spray about 3 weeks before expected harvest.

Gibberellic acid is also used to delay ripening in lemons and can be used to extend the harvest period. Ethephon, a plant bioregulator that breaks down to form ethylene, is used to promote ripening of bananas, apples and other fruit.

All of these plant bioregulator techniques will be discussed in greater detail in the next chapter.

Date of blossoming

If winter conditions are mild in a cool climate, blossoming will often be earlier. However, in a warmer climate, a cool winter may lead to the earlier completion of bud dormancy and bloom may be advanced. While the time of bloom does have an effect on the time of harvest, a much greater influence comes from the air temperatures experienced by the crop during its normal growing season. In other words, it is not unusual for an early bloom to be followed by a normal harvest period.

Determining Maturity

Unfortunately, maturity is an ambiguous term in fruit crop horticulture. It is often used to describe the time when a fruit has optimum flavour and texture for fresh eating, or perhaps for processing, directly after picking. In this manner of usage, mature and ripe are virtually identical terms. This would be the case, for example, for cherries and grapes. However, 'mature' is also used to describe the time when a fruit should be picked so that it will keep for the longest period in cool storage. Thus, 'optimum harvest maturity' for apples and pears intended for long storage does not mean that the fruit is ready to eat at the time of picking. A similar, but not identical, definition of maturity is the stage at which, when harvested, the fruit will develop optimum quality upon ripening. This would apply to banana, avocado, some pears and other fruit that must be ripened after harvest before they can be consumed. Here mature and ripe are distinctly different terms.

To make matters more confusing, we often judge fruit maturity in relation to the market channel to be followed. For example, peaches intended for

export are harvested at a less mature stage than when they will be marketed locally.

Generally, for commercial production, growers and marketers desire a fruit that will store for an appropriate period, yet still ripen to give good flavour during and after storage.

In mild, insular or Mediterranean climates, the maturity date can vary considerably from year to year, due to variability in the weather patterns. This makes it difficult to forecast the optimal date of picking, although such forecasts are obviously valuable for planning the marketing of the crop. Deciduous fruits, which depend on cold weather to overcome dormancy, present an additional problem in milder climates. A characteristic effect of mild winters is a considerable spread of blossoming – 3–4 weeks for apples in New Zealand, South Africa and California, for instance – whereas in the cooler climates of Washington State or northern Europe, most flowers will open within a few days.

The following guides or parameters have been used to determine the maturity of fruit.

Ease of separation of fruit from the tree

This is not very reliable when used alone but can be a useful guide if used with other indicators for judging maturity of apples, pears, stone fruit, raspberries and feijoas.

Fruit colour

In most fruit, colour change is the most conspicuous aspect of ripening and, consciously or unconsciously, this is the aspect most used in selecting fruits to harvest. Colour changes have been described earlier and are valuable for most fruit. Kiwifruit, most nuts and avocados are exceptions to this rule.

Serious attempts have been made to quantify or illustrate colour changes and, in many places in the world, special colour charts are used to make harvesting decisions. Especially good examples exist for pears and apples, where the loss of chlorophyll (i.e. yellowing of the background colour) is the key change. For sweet cherries, colour comparators are used to judge redness.

Seed colour

In apples, pears and strawberries, seeds change from white to brown as the fruit matures. This probably has no advantage over general observations of skin colour changes and is not often considered seriously as a means of assessing maturity.

Sugar or soluble solids

In some fruit, the level of sugar in the expressed juice provides a useful assessment of maturity, although it has been argued that it is a better measure of fruit quality. Normally the fruit is squeezed or pulped and the juice sugar level measured with a refractometer. This is very simple and quick and gives a direct reading in degrees Brix (°Brix), which is more correctly a measure of 'soluble solids'. Since most soluble solids in fruit are sugars, these terms are interchangeable in commercial practice.

Determination of soluble solids is the main method used to assess maturity of kiwifruit, blackcurrants and grapes and has been used for peaches, nectarines, apricots and citrus. There will normally be a distinct recommendation given which states that fruits may not be picked until a certain °Brix reading is achieved. A minimum °Brix for kiwifruit is in the order of 6.2. For grapes it will vary according to cultivar, but will range from 16 to 24.

Starch – the starch–iodine test

As apples and pears mature, starch is converted into sugars, and starch tests may give a useful guide to maturity. Fruit is picked and cut in half on the latitudinal equator. The exposed surface is dipped into a solution of 1 g potassium iodide plus 0.25 g iodine in 100 ml of water for 30 s. The amount and the pattern of black coloration on the exposed surface reveals the extent to which starch has disappeared from the fruit flesh and, when this fruit is compared to a set of standards, a maturity value can be applied.

Acid levels

The decline in juice acidity as the fruit develops also provides a guide to maturity. Acid tests are used for citrus, currants and grapes, usually in conjunction with tests for sugars. The sugar to acid ratio or 'balance' is an important quality measure in some fruits. Since absolute levels of juice acidity can vary greatly from year to year, it is the change in acidity from one measurement to the next that is indicative of fruit maturation.

Flesh characteristics

The softening of flesh as fruit matures and ripens has led to research focusing on flesh firmness as an indicator of maturity or ripeness. There are now many instruments that measure flesh firmness or texture but the most useful is the penetrometer. This device consists of a plunger which is pressed against the flesh (the peel is normally removed) and, when this gives way, a firmness value is recorded on the scale. It has been used to assess the maturity of apples, pears, peaches and apricots.

Fruit size and shape

Fruit size and shape can indicate visually how maturity is progressing. Because of variability between seasons, orchards and cultivars, no categorical size to maturity relationship has been established, but observations of daily or weekly size increase can give an orchardist useful clues to development. A good example of fruit shape changes that signal advancing maturity is found with peaches and nectarines. These fruit exhibit a 'full' or rounder shape, especially along the suture and at the calyx end of the fruit, when they are mature enough to ripen to good quality. Pickers can be trained to recognize those fruits which are ready for harvest.

Date, or time after a reference date

In any district there will be a history of particular fruits maturing around a certain calendar date or within a given number of days after bloom time. With experience, weather conditions influencing this date or this time period are recognized and advisory officers will be able to predict whether, in a specific season, harvest will occur earlier or later.

Theoretically, it is possible to measure heat units (see Chapter 2) after bloom to achieve a more accurate estimate of harvest date. In practice, however, factors other than average temperature are important enough to minimize the use of heat unit accumulation to predict maturity.

Finally, where there is no reliable method, such as sugar, firmness or starch measurements to predict harvest maturity, the following checklist can often be used to make better decisions.

- Observe the blossom date. Is it earlier or later than normal?

- Observe other developmental markers. The date of pit hardening in stone fruits, for example, is a good measure of early-season development.
- Estimate heat unit accumulation. Check the mean monthly temperatures after bloom using local meteorological office figures. Are these higher or lower than the long-term means?
- Observe the maturity of other crops. Are they earlier or later than normal?
- Check for fruit loosening. This will at least indicate a potential problem with pre-harvest fruit drop.
- Observe colour and size development. This will at least predict marketplace acceptability.

Recording these observations over a few years will reveal patterns and help the grower gain confidence in making important decisions about harvest maturity.

Strip verses selective picking

It is much easier for an orchardist to harvest all of the fruit in a single operation, i.e. stripping the trees or vines. However, if the indicators discussed in this section are used by the grower, and the pickers are trained to recognize a fully mature fruit, it is possible to pick selectively to achieve a more uniform and probably more attractive and valuable product.

Selective harvesting generally results in higher profits, despite the extra labour input. The fruit will have a more predictable storage and shelf life and the producer will gain a reputation for high quality. Furthermore, total yield may increase because the late-maturing fruits improve in size and quality by the removal of competing fruits and because they remain on the tree for a longer period.

Harvesting Fruit

The removal of fruit from the tree, bush or vine is a very critical process and incorrect handling can jeopardize an otherwise excellent crop. For convenience, fruits will be grouped into four categories and only representative fruits will be mentioned.

Fruit used primarily for processing and needing least care

- Wine or juice grapes
- Blackcurrants
- Currant and raisin grapes
- Raspberries and brambles (not dessert)
- Apples and pears for juice/cider
- Sour cherries for processing.

These fruits are normally harvested mechanically and the fruit is quickly processed or frozen before chemical degradation occurs.

Nuts

The grower will either allow the nuts to fall to the ground or have the trees mechanically shaken. Nuts are protected by the husk and the hard shell and are seldom damaged. They are picked or swept up off the ground, dried and processed. They should not be allowed to stay for any length of time on the ground, since the shells may stain and they may be eaten by rodents.

Fruit with moderately hard skin or rind

- Gooseberries
- Citrus fruit
- Feijoas
- Passionfruit.

This group is moderately easy to handle and will tolerate more rough handling than others. These fruits could probably survive a 30 cm fall on to a hard surface without too much damage.

Tender fruit

- Strawberries
- Raspberries and brambles – dessert
- Blueberries
- Grapes – dessert
- Apples
- Pears (Asian and European)
- Peaches and nectarines
- Sweet cherries
- Apricots
- Avocados
- Kiwifruit
- Persimmons
- Tamarillos.

These crops need careful attention to harvesting and must not be subjected to dropping or bruising. A 30 cm fall would seriously bruise or cut the fruit. If these fruits are to be bulked up into bins, they must be carefully lowered and not dropped.

If grading is required, surfaces should be padded and it must be remembered that one fruit falling from 20–30 cm on to another could cause bruising or damage, even if the lower one lies on a soft surface.

Pickers of such fruit need to be carefully instructed and it is necessary to emphasize that very little pressure applied by fingers and thumbs will cause a bruise. Sharp fingernails are another source of damage, and padded picking gloves are used to advantage. Fruit should be carefully placed into picking bags or in trays, and these should not be roughly handled. With apples, a major source of bruising comes from climbing up and down ladders with over-full picking bags. Similarly, sweet cherries are bruised by pickers on ladders who reach too far to grasp a fruit. When these crops can be harvested without ladders there is a great deal less bruising. This is a strong argument for small trees.

Harvesting and handling procedures

Figure 7.3 illustrates the many steps involved in getting a fresh fruit from the orchard to the consumer. Although some of these stages may be omitted in any picking and marketing operation, there are still many points at which fruit is vulnerable to bruising or deterioration. Indeed, some of these stages may involve several handling operations; for example, each transfer may involve loading and unloading a truck, moving to a ship or plane and loading, then unloading and reloading on another truck. Cooling at the stages shown in Fig. 7.3 can extend storage life. Fruits that begin this journey in

poor condition are unlikely to survive the trip as marketable produce.

Picking containers

After the fruit has been detached from the plant it will normally be placed in a container carried by the picker. There are two basic types of picking containers.

Picking bags or buckets

Apples, pears, peaches, nectarines, passionfruit and citrus are examples of fruit picked into some form of picking bag or bucket which is carried by straps around the neck and shoulders. This leaves the hands free to pick and place fruit into the bag. The bag has hooks which, when removed, allow the fruit to be lowered from the bag into another container.

The size of the bag will depend on the weight and number of fruit that can be gathered without damage to the fruit or the picker. Smaller bags are used for very sensitive crops like peaches, and some growers of export stone fruit have special straps which hold a wooden or cardboard half-case. These cases are two-thirds filled and stacked on a pallet ready for transport to the packing shed. This cuts down the number of times that the fruit must be transferred from one container to another. In most cases, growers will place large wooden or plastic bins at strategic places throughout the orchard, into which pickers empty their bags. These are carried by fork lifts, sometimes on both

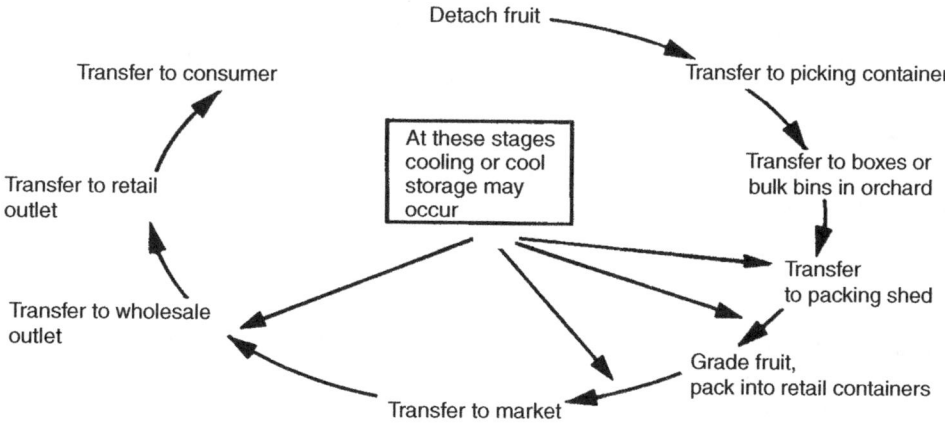

Fig. 7.3. Stages of harvesting and handling fruits.

the front and back of the tractor. Fruit contained in bins may leave the orchard in that form and be transported to a cooperative packing house or processing unit. Alternatively, they may go to the orchardist's packing shed and grader.

Trays or punnets

Strawberries, raspberries, blueberries, gooseberries, brambles, dessert grapes and sweet cherries are fruits that need special handling in shallow containers. For berry fruit, a plastic tray about 200 × 150 × 100 mm is usually used, and a larger tray containing about four of these can be constructed. This will normally have a handle or may even be held in a harness around the neck. Alternatively, pickers may place fruit directly into punnets, also contained on a larger tray. The fruit may be marketed without re-grading if the grower has full confidence in the pickers. Normally, fruit is taken into the packing shed, emptied on to a bench or belt, and re-graded into punnets. The fruit is covered with cellophane, held down by a rubber band.

Table grapes are normally picked into a larger, shallow tray. To avoid loss of bloom, fingers should not touch the berries. Cherries should also be picked into small pails or shallow trays.

Field cooling

With some very perishable fruits intended for distant markets it is highly advisable to cool the fruit before it is transported to the packinghouse. This is accomplished with cold water or with cool air chillers brought to the field. At the very least the grower should harvest during the cool time of the day and provide shade for the fruit waiting to be transported to the packing shed.

Grading and packing

Apart from the fruit destined for processing or packed during picking, most fruit will go to a grading and packing facility for sorting and repackaging for a specific market. This is often done at a large cooperative facility, although many producers sort and pack their own fruit. Specific details about packing will be provided in later chapters but some general points are appropriately made here.

Larger fruits are often machine graded to achieve uniform sizing in the various packs. Apples, pears, stone fruit (with the possible exception of sweet cherries), citrus, kiwifruit and some other subtropical fruits fall into this category. Smaller fruits will usually be graded by eye if required, although this may simply mean discarding those which are excessively large or small.

Various grading machines are available that discriminate on the basis of fruit size or weight. Fruit from the orchard is carefully fed into the machine by gently emptying the boxes or bins. To reduce bruising, some packing houses use water to empty the bins and transfer the fruit on to the grading belts. As it progresses through the grader the fruit is inspected and any rejects are removed. Once the fruit has been sized, it is packed into containers. Where fruit is to be packed for export, government officials or the marketing organization will specify grading standards.

The following descriptions of the packs used to market various fruit crops apply to the New Zealand industry but are representative of those used elsewhere.

- Apples and pears. For local or farm-gate sale, loose wooden or plastic bushel packs or polythene bags are used. For export and 'high-end' domestic markets, 18 kg (1 bushel) cardboard cartons with pressed cardboard trays between layers are used.
- Peaches, nectarines, apricots and plums. For local and farm-gate sale, wooden or cardboard half-bushel packs are often used. If they are to be transported, the fruits will be carefully placed in these packs and a lid applied. For export, stone fruits are packed in single-layer trays. These trays are commonly 90 × 355 × 440 mm for peaches and nectarines and 75 × 355 × 440 mm for apricots and plums.
- Cherries. For high-quality produce they are often placed in special cardboard containers measuring 68 × 220 × 360 mm. A six-pack of these is normally prepared. A single, larger pack is used for the local market. Fruits are not graded for size, but diseased or damaged fruits are excluded, and the top layer may be hand placed to enhance appearance.
- Kiwifruit. Like cherries, special cartons are produced for kiwifruit. These measure 65 × 340 × 440 mm and fruits are carefully placed in special 'Plix' trays (moulded plastic trays).
- Other subtropicals. Similar cartons to those for kiwifruit are used.

D. Jackson *et al.*

- Citrus. A wide variety of containers are used, including bushel and half-bushel boxes (18 kg and 9 kg packs).
- Berry fruit. Most berry fruits are packed into punnets as described above. If these are to be exported or transported for long distances, they are placed in special cardboard boxes. For strawberries, 12-punnet cartons measuring 95 × 355 × 440 mm are common. For raspberries and boysenberries, nine-punnet packs measuring 70 × 355 × 440 mm are used.

Storage of Fruit

Many fruits continue to live and respire once detached from the tree. They will often ripen after harvest, and senescence occurs over many weeks or even months. Death is usually as a result of consumption or disease and decay. As has already been noted, ripening involves the loss of acidity and starch, an increase and then a decrease in sugar, and changes in pectic substances. It involves the production of volatile substances such as esters, acetaldehyde, ethyl alcohol and ethylene. During respiration, oxygen is absorbed, carbon dioxide and water are evolved, and energy is released to drive vital processes. Some of that energy is lost as heat.

When considering the cooling and storage of fruit it is important to consider this heat of respiration and understand that cooling the fruit slows the respiration rate progressively and reduces the heat of respiration. The term Q 10 is used to indicate the relationship between temperature and respiration. The reverse relationship is referred to as the Q 10 of respiration; a doubling of the rate occurs between 0 and 10°C, again between 10 and 20°C, and almost double between 20 and 30°C. As an example of the effect this has on storage life, apples and pears will ripen as much in a day at 20°C as in a week at 0°C.

A number of factors and conditions influence the storage and shelf life of fruit. These factors are discussed in three time-windows: before, during and after storage.

Pre-storage factors

The following genetic and environmental factors have been shown to affect storage and shelf life.

Cultivars and rootstocks

Differences in the ability to store are common between different cultivars. This is especially noticeable in apples, where some, like Scarlet Pimpernel, will store for only a few weeks, while Granny Smith, Fuji or Delicious are stored in controlled atmospheres for up to a year. In apples, late-maturing cultivars generally store better than early ones.

Rootstocks have been reported to affect storage, but this is probably related to tree vigour, crop load and fruit size. Rootstocks have also been reported to influence fruit maturation and, thus, optimal harvest date.

Soils

There is no doubt that the storage ability of fruit can vary from orchard to orchard and a number of factors could contribute to this. It is generally found that orchards on light soils are better balanced with respect to growth and cropping, and the fruit will store better. Some orchards may have specific problems which may be related to nutrition or soil type. For example, some orchards may have consistent problems with the development of bitter pit in storage, whereas others are remarkably free of it. This is probably due to the availability of calcium and other minerals, and, perhaps, since bitter pit is a symptom of moisture stress in the fruit, water-holding capacity may be an important factor.

Fertilizers

As already mentioned, orchard nutrition can profoundly affect storage quality. The amount of fertilizer which produces the best tree growth and highest yields may not give the best-storing fruit. If fruits consistently store poorly, it is generally advisable to obtain a leaf analysis of the trees in the block to see if any element is lacking or over-abundant.

As a general rule, high nitrogen applications depress fruit quality. This effect comes from unbalancing vegetative and reproductive growth, where flowering is reduced and the fewer fruits produced develop in a shaded environment. They may be excessively large and soft. While it has been suggested that adding additional phosphorus and potassium (a 'balanced' fertilizer) will improve fruit quality under high nitrogen conditions, this hypothesis is largely unproven. High potassium and high magnesium are known to compete with calcium uptake and transport, thus reducing fruit calcium levels. Applying phosphorus to many soils does not dramatically improve phosphorus availability, since this nutrient is quickly absorbed on to soil particles.

The amount of nitrogen and most other nutrients available to the fruit crop (magnesium appears to be an exception) can be reduced by planting a ground cover. Conversely, these nutrients are released by tillage and the application of herbicides. Thus, any change in cultivation should be considered in relation to the effect it can have on nutrient availability and fruit quality.

Water

Under non-irrigated conditions when water is in short supply, trees often produce small fruit of excellent keeping quality. However, under these conditions a year with exceptional amounts of rainfall may result in large fruit with poor colour and poor storage potential. These fluctuations may result from the release of reserve nutrients in these non-irrigated soils. In areas which normally have moderate drought stress, the grower must balance irrigation with natural rainfall. Where natural rain is very low and irrigation is mandatory, soil structure and mineral nutrition are more important factors influencing fruit quality.

Still, some specific fruit-quality problems are known to be related to the water relations within the tree or vine. Bitter pit, often seen in large fruit on light crop trees, is induced when the fruit is unable to compete with the foliage for water and calcium. Drought spot is a special disorder of apple relating to boron deficiency. It is most likely to appear when irrigation has been less than optimal.

Fruit size

Large apples seldom store as well as small fruit and, for this reason, any measures which promote excessive size, such as pruning or thinning, should not be overdone. Light crops and fruit from young trees should never be stored for long periods and these fruit should always be separated from the rest.

Diseases and pests

A good spray programme will reduce the carryover of diseases into storage. During harvesting, damage by bruising and puncturing should be kept as low as possible, since it is through cracks and skin abrasions that many disease agents enter the fruit.

Delays between harvesting and storage

These should always be as short as possible, since any delay can reduce storage life. It must be remembered that the weather is warm during the harvest period for most fruits. Two days at ambient temperatures following harvest can be much more damaging than 2 days later in the year when the fruit comes out of cool storage to be marketed, often in winter. This aspect is especially important for berries and stone fruit, which have a limited storage life under the best of circumstances. Two hours in the hot sun will dramatically reduce the storage life of peaches, nectarines, cherries, plums, apricots, strawberries, raspberries, etc. Producers are often advised to find a way to remove the field heat from such fruit crops before they are transported to the packinghouse, cold storage or market.

Having mentioned field cooling, it is important to understand that coolers differ widely in capacity and effectiveness. The most appropriate coolers are those capable of rapidly cooling a large quantity of warm produce. Some of these coolers rely on the rapid movement of very cold air, but this air must not be too dry or fruit dehydration will occur. If it is too cold, freeze damage will occur. The technique of hydro-cooling, where fruits are immersed in water between 2 and 4°C for a limited period, is highly desirable for some crops but not recommended for others.

Maturity

Picking at the correct stage of maturity is one of the best ways to ensure optimum storage life. This aspect has been discussed earlier but the difficulties it poses need to be re-emphasized. Growers must liaise with local experts to ensure that their practices accord with those demanded by the people marketing their fruit.

As a general rule, climacteric fruits will be picked between the time the first colour changes begin – yellowing of the green background colour, with increases in other pigments – and the time when the

D. Jackson *et al.*

fruit is at optimal eating quality. Fruit picked before such colour changes begin may fail to ripen to acceptable quality. Fruits harvested early in this period will be hard and are excellent for handling and storage. However, the flavour will suffer to a certain extent and seldom reaches the level of tree-ripened fruit. Late-harvested apples or peaches may be excellent for immediate consumption but cannot be expected to store for more than a few weeks.

Keep in mind, however, that many fruits ripen little or not at all after picking. These non-climacteric fruits include grapes, raspberries, currants, gooseberries, blueberries, cherries and citrus. They are picked at optimum maturity for fresh eating when the period from field to consumer is short. They are usually picked somewhat earlier than this when they must withstand a long trip to market.

Storage conditions

There are three common types of storage used for fruit: non-refrigerated storage, cold storage and cold storage with a controlled atmosphere (CA storage). Hypobaric storage, another way of controlling the atmosphere, may prove valuable in some situations.

Non-refrigerated storage

In some parts of the world the term 'common storage' is used to describe this practice. It is only appropriate where fruits are stored for a limited period of time or if there is no other choice. Growers who sell from the farm will often keep their fruit in this way, and anyone, including retailers, holding fruit prior to sale without the availability of cool storage is, of course, operating a similar system.

The aim is to provide an environment that is as cool as possible. In many parts of the developing world, caves or caverns are used because they maintain a constant cool temperature and protect the fruit from freezing. If a structure is built for this purpose it should be located to avoid the winter sun. It should be well ventilated, but not excessively draughty, and it should be insulated to prevent freezing if the temperatures drop that low.

Cold storage

Cold stores aim to reduce the temperature of fruit to as low a level as is possible in order to reduce metabolism and respiration. Most fruits have lower limits to which they can be cooled before damage of some sort is sustained. Freezing injury occurs when the temperature of the peel or flesh drops below a critical point. A physiological disorder known as chilling injury is observed with a number of fruit species and cultivars. It occurs when the fruit is held for a few days or weeks at temperatures below a critical threshold. Depending on the crop, this injury can be induced by temperatures between -1 and $+4°C$.

There is much to know about cold-storage construction and management. The following subsections note a few salient points.

AIR MOVEMENT AND HUMIDITY CONTROL Cold stores operate by the circulation of cool air, which removes the heat from the fruit and then maintains a steady temperature. Rapid air movement is needed to remove field heat quickly, but once the temperature is at the correct level, too much air flow can be a problem. It can cause excessive moisture loss from the fruit, which becomes wilted and less marketable.

Thus, maintaining a high relative humidity is very important for ensuring high fruit quality. This can be achieved by applying water to the floor of the storage room and by ensuring that the cooling coils do not remove moisture by freezing. The secret for good humidity control is to have adequate, even excessive, cooling capacity, so that air movement does not have to be high and the surface of the cooling coils can be at a temperature not greatly different from that desired in the room.

There should also be adequate provision for the exchange of air within the storage room. The build-up of ethylene gas can accelerate certain senescence processes, even at the low temperature of most storage environments. This can be a special problem for those fruit held at slightly higher temperatures to avoid low temperature injury problems.

All of these considerations require a well-designed store which can be appropriately manipulated and careful stacking of the fruit so that adequate air circulation is achieved. It is wise to become familiar with refrigeration equipment, storage design and storage management before deciding to take on this challenge. It is often better to engage the services of a commercial or cooperative fruit-shipping company with professionally managed storage facilities.

TEMPERATURE Later chapters will discuss the recommended temperatures for storing various fruit crops. There are normally firm guidelines available, based on research and experience. However, with some of the newer fruits, or new cultivars of well-known fruits, the recommendations are still being developed. It is wise to seek out the most recent information. Keep in mind that if a temperature range of, say, 1–2°C is given, it is especially important not to go below this range, since injury may occur. Slightly higher temperatures than those recommended will simply shorten storage life. There can be no doubt that good temperature management is the most important ingredient of successful fruit storage.

MONITORING, CLEANLINESS AND OTHER ASPECTS Fruit in cold storage must be regularly removed and inspected to determine shelf life. When shelf life is no more than 10–14 days it is time to empty the store. The store should be kept clean so that fungal spores do not build up. This is achieved by not overestimating storage life and by thoroughly cleaning the walls and floors of the store when the room is empty. Most commonly, different fruits are stored in different rooms. This is primarily because the optimum storage temperature is likely to be different, but there are other good reasons. For example, the ethylene produced by apples dramatically promotes ripening of kiwifruit. It may also be advantageous for different cultivars to be stored separately, since they may require different conditions. This is especially important for apples and pears.

Controlled atmosphere (CA) storage

With this technique, not only is the fruit cooled to reduce respiration, but oxygen, carbon dioxide and nitrogen gases in the storage atmosphere are carefully regulated. When the oxygen content is lowered and the carbon dioxide level increased, a considerable reduction in respiration will occur, and long-term storage becomes quite feasible. For apples, the oxygen is maintained at between 1.5 and 3% and carbon dioxide in the range of 2–5%. Each cultivar has specific limits, or tolerances, and if these are not adhered to very carefully serious physiological disorders may occur. Thus, while the benefits of CA storage are substantial, there are also important risks that must be understood.

In the early days of CA storage, growers attempting to use this technology experienced difficulty in achieving a gas-tight store. Now, many operators accomplish CA conditions within an existing cold store by erecting an inside tent of polythene. Fruits are stacked in the room over a film of polythene. When cool, a wrapper is placed around the top and sides and sealed to the base film.

In a 'passive' CA system the respiring fruits reduce the oxygen level (initially at about 20%) over a period of a few weeks. When it reaches the desired level the tent is ventilated to maintain that level. Similarly, carbon dioxide builds up from the initial level of 0.35% to the required 2–5% and is maintained at that level by passing the air over lime (calcium hydroxide), which 'scrubs' carbon dioxide from the air. Equipment for monitoring gas levels and adding or removing oxygen and carbon dioxide can be obtained commercially in various degrees of sophistication.

An advancement in CA technology is called 'rapid CA'. Using this technique, the operator reduces the oxygen level in a large CA room from 20% to 1–2% in a matter of a few hours. This is often preceded by a very rapid pull-down in fruit temperature. The result of using this technology is very good fruit quality and even longer storage life, but the equipment is expensive and highly sophisticated. It is more likely to be used in stores operated by large packing and shipping firms.

Hypobaric storage

This technology involves the storage of fruits at atmospheric pressures well below normal. It has been shown to be successful for apricots, peaches, cherries, pears, apples, strawberries and limes. For most fruit, a pressure of 0.1–0.2 atmospheres is used, together with normal cold storage temperatures. While the technology appears to be sound and the concept is similar to that of CA storage, the equipment is very expensive. Hypobaric storage technology could be used in shipping very valuable commodities to distant markets.

Post-storage factors

Once fruits have been removed from storage, deterioration can be quite rapid. As previously discussed, it is the storage manager who must judge for how long the fruit will retain its quality once removed and ensure that it has the shelf life required for the marketing channel selected. The grower's or commercial storage manager's responsibility will normally cease when the fruits leave the store. It then becomes the responsibility of the wholesaler or retailer to ensure that fruits are kept cool properly.

Further Reading

Bartsch, J.A. and Blanpied, G.D. (1990) *Refrigeration and Controlled Atmosphere Storage for Horticultural Crops*. Publication NRAES-22, Cornell University, Ithaca, New York.

Ben-Arie, R. and Philosoph-Hadas, S. (eds) (2001) IV International Conference on Postharvest Science. *Acta Horticulturae* 553.

Beaudry, R.M. (ed.) (2010) IX International Controlled Atmosphere Research Conference. *Acta Horticulturae* 857.

Florkowski, W.J., Shewfelt, R.L., Brueckner, B. and Stanley, E. (2009) *Postharvest Handling, 2nd edition, a Systems Approach*. Academic Press (Elsevier), San Diego, California.

Hellickson, M. L. (2003) Fruit storage systems. In: Heldman, D.R. (ed.) *Encyclopedia of Agricultural, Food, and Biological Engineering*. CRC Press Inc., Boca Raton, Florida, pp. 422–424.

Herppich W.B. (ed.) (2010) III International Conference Postharvest Unlimited 2008. *Acta Horticulturae* 858.

Kader, A.A. (ed.) (2002) *Postharvest Technology of Horticultural Crops*, 3rd edn. ANR Publications, California.

8 Use of Bioregulators in Fruit Production

NORMAN LOONEY AND DAVID JACKSON

In this chapter a range of orchard practices will be considered where plant growth-regulating chemicals (i.e. plant bioregulators) are used to regulate cropping, harvesting and fruit quality. With the exception of fruit thinning by hand, these practices would not be possible without an appropriate chemical. Some plant bioregulator-based practices that influence tree growth and development in the nursery and in the orchard will also be introduced.

Fruit and Flower Thinning

Many fruits do not require thinning to achieve good quality and annual cropping. For those that do, however, this is one of the most important of all orchard practices. If done badly the producer will pay the price in low fruit quality (small, poorly coloured fruit), limb breakage and the risk of biennial flowering. The amount of thinning required varies with species and cultivar, between fruit-growing districts and between seasons, depending mainly on the level of set in a particular year. It may also be conditioned by climate and market requirements. Growers may thin less in areas subject to late frosts, heavy winds or hail, although it could be argued that such an area should not be used for commercial fruit production unless effective control measures can be devised. Market requirements are more often an important determinant of thinning intensity. Consumer preferences do change, and at one stage large fruit may offer the greatest returns whereas smaller fruit may be preferred at other times or in other places.

Thinning always reduces tonnage in the season that it is applied but it is necessary to produce the fruit size and quality demanded by the marketplace. More importantly, successful thinning will ensure a return crop. When viewed with all of these benefits in mind, the crop loss due to thinning is an absolute necessity for profitable fruit growing.

The higher the value of the crop, the more time can usefully be spent reducing crop load. At a size which gives 23–25 fruit per tray of nectarines, for example, this fruit is twice as valuable as when there are 36–40 fruit per tray. Thinning to achieve this size is well worthwhile. Similarly, Fuji apple production in Japan aims for very large, well-coloured fruit for the gift market. Thinning (at least twice), fruit positioning and, finally, selective leaf removal may take place sequentially during the growing season to achieve this high-value product.

Timing of the thinning operation

The earlier that crop reduction can be achieved the better the result, from the point of view of both optimizing fruit size and overcoming biennial bearing in those crops where this is a problem. However, a number of factors may mitigate this practice. Natural drop (poor fruit set) or frost, wind or hail damage may occur after thinning and cause over-thinning. A better idea of the amount of thinning required can be obtained if it is delayed for a few weeks. Certain crops, such as apples and some stone fruit, can be thinned with chemicals applied as late as 4 weeks from full bloom.

A key to developing successful fruit thinning practices for a given orchard is to ensure that pollination is not a limiting factor. This is achieved by planting enough pollinizer trees, providing the bees needed during pollination and establishing the microclimate that encourages bee activity (e.g. shelter where needed). It is only in those orchards where reliable fruit set can be expected that early blossom and fruitlet thinning can be used with confidence.

For those crops where hand thinning is the only option available, thinning is usually delayed until the wave of natural fruit drop is complete or at least well under way. This 'June/December drop' normally occurs within about 6–8 weeks from bloom. This avoids over-thinning and removing fruit that would drop naturally. However, with some stone fruit like apricot and peach, hand thinning of blossoms is a common practice. There is a stage just before bloom when flowers can be rubbed off quite easily. Furthermore, foliage development is such that the flowers are easily visible. Note, however, that flower thinning of peaches, nectarines and apricot is not done as rigorously as fruitlet thinning, because one cannot judge the quality of the fruit that will arise from each flower retained. Thus, flower thinning is usually followed up with some touch-up hand thinning about 6 weeks later.

Hand thinning

Hand thinning is a very labour-intensive practice and, when applied with the degree of detail and concentration required to do a good job, can account for as much as 20% of the total cost of production. For apricots and peaches it is usually the largest single cost of production.

Depending on the species and the market requirement for fruit size, the general practice is to space fruit along the branch, leaving one fruit every 10–12 cm and choosing the largest and most well-formed fruit available. Note that it is better to compromise the spacing objective than to select small fruit that will not develop good size and quality (i.e. (a) rather than (b) in Fig. 8.1).

The orchardist is always faced with the problem of gauging the most economical number of fruit per tree or vine. One method is to relate fruit number to the size of the tree or vine. With some experience with a given cultivar in a given location, it can be determined how many fruit can be carried per centimetre of trunk or vine circumference, taken at a point 10–20 cm above the soil surface. The producer would then thin a few trees or vines to demonstrate to the thinning crew the fruit spacing required to achieve this final crop load.

Chemical thinning

Chemical thinning of apples is a well-established and highly successful practice. Some stone fruit

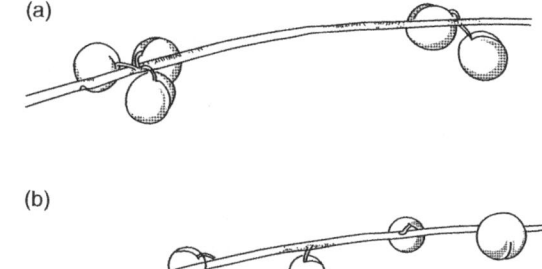
(a)

(b)

Fig. 8.1. Two thinning methods. (a) Largest fruit. (b) Undesirable smaller fruit.

can be thinned successfully with chemicals, and with pears and grapes chemical thinning is possible, although problematic. Few other fruit crops are commercially thinned with chemicals. The following materials are available in various countries and regions. Details about how to make best use of them in a particular region can be obtained from local extension publications.

Desiccants like DNOC, ammonium thiosulphate and endothall

For apples and stone fruits, the only blossom thinners available for many years were dinitro-O-cresylate (DNOC) and dinitrobutylphenol (DNBP or dinoseb). However, these chemicals have been replaced with less toxic materials. An aquatic herbicide, endothall, applied at low rates to apples has resulted in promising flower-thinning activity. It might also prove to be a good blossom thinner for some stone fruit. The nitrogen- and sulphur-containing fertilizer, ammonium thiosulphate (ATS), has proven to be a reasonably good replacement for DNOC in apple orchards in western North America. Lime sulphur applied during flowering, sometimes combined with a plant-based oil, is also known to reduce the number of apple flowers that set fruit. This approach has found favour with organic growers.

All of these chemicals cause thinning by desiccating the stigmatic surface and the pollen that may be on that surface before pollen germination and fertilization occurs. Thus timing is the key to success. Since most apples are cross-pollinated, the idea is to judge bee movement and daily mean temperatures, and when one is quite sure that the king flower has been pollinated and fertilized, the spray is applied to disable the remaining flowers on that cluster. This strategy is successful because the side flowers in an apple blossom cluster open later than the king flower.

Note that all of these desiccants will cause some leaf burning, but this side effect has not proven to be a problem worthy of serious concern.

Ammonium thiosulphate is effective when applied at about 60% bloom to apricots and about 80% bloom to apples. It has not yet been shown to be effective for other stone fruits or for pears. A common rate is 1% ATS applied with 1000–2000 l/ha of spray solution. Note that fertilizer-grade ATS comes as a saturated solution containing about 60% ATS in water.

Naphthalene acetic acid (abbreviated NAA or ANA)

NAA is a synthetic auxin long used for apple thinning. It is applied at petal fall to 18–21 days later. When applied at the later time it is often combined with carbaryl (see below). NAA stimulates ethylene production by fruit tissues, which in turn slows the development of the youngest and weakest fruits more than the older fruits in a cluster. The result is that the weaker fruits cannot compete for resources and are quickly shed.

The best time to use NAA depends quite a lot on cultivar. Early applications are recommended for Gala and Delicious, cultivars that will often retain 'pygmy' fruits if treated later in the fruit development period. With all cultivars there is a tendency for NAA-treated fruit to grow more slowly for some days or weeks after treatment. However, the end result is often satisfactory thinning and good return flowering. A common recommendation for NAA usage is to apply a 5 ppm solution of NAA at about 2000 l/ha within about a week of full bloom. However, it is sometimes used as a blossom-time spray or combined with carbaryl and applied as late as 3–4 weeks after bloom.

Naphthalene acetamide (NAAm or NAD) is the amide salt of NAA and is considered a safer chemical to use in some situations. It has less immediate hormonal action but remains active over a longer period. It has been used successfully to thin some cultivars of European pear. Effective rates are usually in the 20 ppm range.

Ethephon (ethrel ®, CEPA, 2-chloroethylphosphonic acid)

This ethylene generator has been used successfully to thin apples in a number of regions. It can be used alone or combined with NAA and carbaryl. The timing is also quite flexible, with good results obtained from petal fall to several weeks after bloom. The principal concern about using ethephon is that it is highly dependent on air temperature for its effectiveness. Thus, if hot weather occurs following treatment one may observe excessive thinning activity, the opposite being the case if the weather is unusually cool. None the less, ethephon has been used successfully with Golden Delicious and other cultivars that set excessive crops and are prone to biennial flowering.

6-Benzyladenine (6-BA)

This chemical represents another family of natural plant growth substances, the cytokinins. 6-BA will selectively reduce fruit set of apple when applied during the post-bloom period, again eliminating the weaker fruit. Available commercially as Maxcel® (Valent BioSciences Corporation), this plant bioregulator also contains a small amount of gibberellins. Experience with MaxCel® in some regions, especially in eastern North America, indicates that it will also increase the size of the remaining fruits to an extent greater than would be expected from the reduced competition for resources. This is explained by the effect that cytokinins have on extending the cell division period in apples and other fruits. Apples with more cells are likely to be larger at maturity. However, MaxCel® alone is not considered a strong chemical thinner in most regions and is usually combined with carbaryl to achieve the desired amount of fruit set reduction. A rate of 50 g of MaxCel® active ingredients per hectare is a common recommendation but may vary from region to region. It is usually applied with 4.5 l of 50% a.i. carbaryl when the king flower is about 10–12 mm in diameter. Again, however, the optimum rates and timing will vary from location to location. It is important not to combine Maxcel® and NAA or use both chemicals in the same season. The retention of 'pygmy' fruit is increased by this combination.

Carbaryl (1-naphthyl-N-methyl carbamate, Sevin)

This interesting chemical is primarily known as an insecticide but was shown to have plant bioregulation properties in the 1950s, particularly as an apple fruitlet thinner. It appears to favour carbohydrate retention by young leaves, resulting in a temporary stress that sorts out the weak from the stronger fruit. Carbaryl has the great advantage of reducing set on individual clusters to a single fruit. Furthermore, where initial set in a cluster is already a single fruit, carbaryl is unlikely to remove that fruit. Thus, from the standpoint of thinning efficacy, carbaryl is the safest and most effective thinner presently available.

However, being an insecticide, it must be used with caution and with great attention to the effects that it can have on the integrated pest management (IPM) strategy in place in an orchard. Obviously, there will be times and places where carbaryl cannot be considered. For example, while it is effective when applied from full bloom to some 4 weeks later, carbaryl should not be applied when flowers are still on the tree. Even the preferred formulation for thinning (Sevin XLR, an oil-based suspension with very fine particle size) will kill bees if they are in the vicinity. For the same reason, flowering weeds under the trees should be mown prior to spraying. In regions where carbaryl has been in continuous use for chemical thinning for nearly 40 years, it has been demonstrated that the insect ecology of the orchard has adjusted remarkably well. The explanation seems to be that mite predators and other beneficial insects have developed sufficient resistance to carbaryl to maintain their populations. Despite this encouraging development and its obvious value to apple producers everywhere, only time can tell whether carbaryl will continue to have a place in apple production systems that are increasingly dependent on biological insect control strategies.

Gibberellic acid (GA₃)

Gibberellic acid represents a family of natural plant hormones that influence flowering and fruit set. GA_3 is used to reduce the number of berries per cluster in some cultivars of table and wine grapes and, by an entirely different mechanism, to reduce the amount of hand thinning in peaches and other stone fruit. These technologies are often rather complicated and specific for a given region and cultivar. Thus, it is necessary to seek out local advice when considering GA_3 as a tool to regulate crop load.

With one unique table grape cultivar, Delaware, grown widely in Japan, GA_3 is used to turn a normally seeded cultivar into one that is seedless. This is accomplished with treatments applied to individual clusters by dipping twice during the 2–3 weeks surrounding full bloom. The resulting seedless berries are encouraged to grow larger by follow-up GA_3 dips in the post-bloom period.

Cultivars vary widely in their response to GA_3 treatments and problems with poor return flowering prevent GA_3 usage in many situations.

On the other hand, this tendency to reduce flowering can be used to advantage with stone fruit species that flower primarily on l-year-old wood. GA_3 applications made early in the growing season (often at about the time of pit hardening) will reduce the number of nodes producing flowers for the following year (Fig. 4.1). If this treatment is combined with a complementary pruning strategy so that an adequate number of flowering nodes are retained, a full crop can be achieved the following summer and very little hand thinning is required. Furthermore, because fruit to fruit competition is lower than is the case with normal hand thinning, these fruits will be larger and develop better internal quality.

The amount of GA_3 required to achieve this result with peaches, nectarines and some plums is in the range of 20–75 ppm in a spray applied to runoff. The effect on the crop in the season of application can also be beneficial (larger, firmer fruit) if the producer is willing to accept a slight delay in fruit maturation.

Removal of unwanted crops

Sometimes a grower will wish to remove all of the fruit from a tree – for example on a young tree, where tree growth and development is valued more than the crop. A good practical example can be found in sour cherry production. With this mechanically harvested crop, fruiting is considered undesirable until the tree is large enough to withstand mechanical shaking. Gibberellic acid is routinely used to inhibit flowering until the tree reaches a certain size.

With apple and many other fruit crops, however, defruiting is more difficult to achieve. Blossom thinners such as ATS will remove much of the crop but the remaining fruits may not respond to the post-bloom thinners normally available. A relatively high rate of ethephon (250–500 ppm) at petal fall may do the job but more likely it will prove necessary to remove these fruits by hand.

Environmental factors affecting thinning

Since over-thinning is very easy with many chemical thinners, Table 8.1 summarizes the conditions under which sprays may over-thin.

Some practical aspects of chemical thinning

- Directing the spray. Because flowers are weak and natural set is lower on the inside and lower parts of the tree canopy, direct the bulk of the spray to the top and exterior sections.
- Spray volume. Most chemicals used for thinning are best applied with a high volume spray that achieves thorough coverage of the fruit and leaves.
- Spray timing. In desert climates it is best to apply chemical thinning sprays at dusk or in the early morning. The spray dries more slowly and more chemical is absorbed.
- Leave some check trees and keep records. The experience gained after several seasons will greatly improve the success and reliability of chemical thinning in any orchard block.

Stop-drop sprays

Some cultivars of apple, particularly the early-season dessert varieties like McIntosh, Jonathan and Delicious, are susceptible to crop loss from pre-harvest fruit drop. The problem may not occur every year and seems to vary widely from region to region. Some early-season pear cultivars are also prone to pre-harvest drop.

With susceptible cultivars, an early symptom of ripening is the initiation, by ethylene produced by the fruit, of cell wall breakdown at the abscission zone. This zone is a specialized layer of cells between the pedicel and the spur. If the fruit contains enough natural auxin it can override this ethylene effect but, as we have already learned, the natural tendency during fruit maturation is for auxin to disappear.

Thus, there are two approaches to controlling pre-harvest drop with plant bioregulators. The first is to apply a synthetic auxin, such as NAA, shortly before normal harvest. In some countries a stronger auxin, 2,4-DP (dichlorprop), is the preferred chemical. The second approach is to inhibit ethylene production within the fruit. This was the basis for the outstanding effectiveness of daminozide (Alar®), a plant growth regulator no longer available to fruit growers in most countries. The modern equivalent is aminoethoxyvinylglycine hydrochloride (AVG; ReTain®).

NAA and dichlorprop

NAA is applied 7–10 days prior to the anticipated harvest date at 10–20 ppm active ingredient, depending on cultivar and region. It will control fruit drop for a period of about 10 days and should not be used twice. It is important to apply NAA before pre-harvest drop becomes a problem. It is applied as a full-volume spray, thoroughly wetting fruit and leaves. Dichlorprop is used at a lower rate, about 5–10 ppm, and the effect persists for at least 2 weeks.

Both of these auxins have the important disadvantage of stimulating ethylene production in the flesh of apples and pears. This means that, even though abscission is being prevented, the fruits are in fact ripening at an accelerated pace. This is why

Table 8.1. Factors contributing to over-thinning by chemicals.

Tree factors	Environmental factors
Weak trees with weak spurs and thin wood	High humidity for several days preceding application
Trees badly pruned	High humidity on day of spraying, leading to slow rate of drying
Trees too closely spaced or shaded by shelter belts	High maximum temperatures
Inadequate pollination	Moderate rain within several days following application
Young trees	Frost-affected foliage

N. Looney and D. Jackson

a second spray is seldom recommended and why treated fruits are not, in some cases, suitable for long-term storage. This is also why NAA usage on Winter Cole or Winter Nelis pears may cause severe calyx-end cracking on a small percentage of the crop.

Aminoethoxyvinylglycine hydrochloride

AVG, the active ingredient of ReTain® (Valent BioSciences Corporation), is a natural inhibitor of ethylene production obtained from microbial fermentation. To effectively control pre-harvest drop of apples, AVG is applied as a full-volume spray 4 weeks before expected commercial harvest. The recommended application rate is 125 g/ha active ingredient, and adding a wetting agent is considered important. Unlike the auxinic drop-control chemicals, AVG does not promote fruit ripening. In fact, it suppresses fruit softening before harvest and during fruit storage. In apple cultivars susceptible to developing 'water core' and storage 'scald', AVG is known to reduce the incidence of these disorders. In some seasons and on some cultivars, this AVG treatment can slow the natural development of red colour. This affect can be countered by delaying harvest a few days.

Miscellaneous Bioregulators Used in Fruit Production

NAA and other auxins

It has been said that NAA, the oldest of all widely used plant bioregulators, is one of the most versatile agricultural chemicals. We have seen how it acts as a chemical thinner, yet, later in the season, it will prevent fruit abscission. It also can be used to prevent sucker growth from around large branch-removal cuts or, applied as a basal trunk spray, to prevent the development of rootstock suckers. The amount of active NAA needed to prevent bud break and shoot growth, however, is very high compared to these other uses. The material is painted on to the cuts at 1% concentration (10,000 ppm) in water or is mixed with bituminous or latex paint. A commercial NAA product designed for sucker control on apple, pear, avocado, citrus and olives is Tre-Hold®, a product of Amvac Chemical Corporation, Los Angeles. Although time consuming initially, this treatment can avoid considerable later effort devoted to pruning off the

shoots which arise after such cuts or which arise from the rootstock.

The phenoxy auxins, such as 2,4,5-T, 2,4,5-TP and 2,4-DP, have also played an important role in fruit production and continue to have a place in some countries. Even though they have been used in agriculture primarily as herbicides, they can have very beneficial effects in horticulture when used at low concentrations. For example, 2,4,5-T has been used to advance harvest maturity by 3–6 days and increase fruit size of apricots when applied just before pit hardening. The correct timing is just when resistance is being felt by a knife cutting through the apex of the developing stone. The appropriate concentration of phenoxy auxins depends on cultivar and location, but care must be taken to ensure that this practice is fully registered and available for commercial use.

Gibberellins

The gibberellins have a number of uses in fruit growing. Gibberellic acid (GA_3) has been used to increase fruit set in pears under conditions where the flowers were damaged by frost. It will increase fruit size in seedless grapes, such as Sultana and Black Corinth. As we have already learned, for Delaware table grapes it can be applied as cluster dips before full bloom to induce seedlessness and after full bloom to increase size. It is used to delay rind senescence of oranges and grapefruit and prevent yellowing of limes.

GA_3 can be used on lemons to delay fruit maturity by 4–6 weeks, increase the percentage of fruit with long storage life and decrease small tree-ripe fruit. It is applied at 5–15 ppm just as the colour is changing from green to silver. Where permitted, it may be more effective if applied with 2,4-D at 8–12 ppm.

Gibberellic acid is used in sweet cherry production to delay fruit colouring, increase fruit size and improve postharvest quality.

A mixture of gibberellins A4 and A7 is used to reduce russet development on apples (ProVide®, Valent BioSciences). The procedure is to apply four sprays containing 10 ppm of gibberellins at weekly intervals, starting at petal fall. When combined with 6-BA (Promalin®; Valent BioSciences), these same gibberellins are used to enhance elongation of Delicious apples in climates where warm temperatures (especially night-time temperatures) following bloom result in apples that lack the typical shape of this cultivar. Here the useful concentration is about 20 ppm of both active ingredients, applied during the bloom period.

Ethephon

We have already discussed the effects of this 'ethylene generator' on thinning and mentioned its important effects on fruit ripening. However, this useful plant bioregulator also influences shoot elongation and flower initiation if applied at the correct time during the growing season. These two effects are probably related.

Fruit ripening

One of the most important uses for ethephon (Ethrel®; Bayer CropScience) in apple production is to stimulate fruit colouring and ripening, with the aim of advancing harvest by as much as 3 weeks. This practice is well developed for McIntosh apples in North America, where a portion of the crop is treated in this manner every year. Whole trees are sprayed with about 500 ppm ethephon and 20 ppm NAA (the latter to prevent excessive fruit drop) about 2 weeks before the anticipated picking date. The result is well-coloured ripe apples suitable for immediate sale and consumption. These fruits have a short shelf life and are certainly not suitable for long storage. Average fruit size will be smaller than if the fruits had remained on the tree until normal harvest maturity. It is also important to understand that not all apple cultivars respond to ethephon with improved red coloration.

Blueberry harvest can also be advanced and facilitated (see below) by the use of ethephon.

Fruit loosening

Because ethylene causes fruit loosening in many fruit crop species, the introduction of ethephon in the late 1960s provided many opportunities to consider mechanical harvesting in fruit crops horticulture. A good example of success is found with sour cherries. When the fruits approach full size and maturity, the trees are sprayed with 300–500 ppm ethephon to induce loosening, and mechanical harvesting starts about a week later. Harvest involves trunk or limb shakers and catching frames. Less energy is required to remove the fruit (less potential for tree damage) and a greater proportion of the crop is removed by shaking

(improved efficiency). Furthermore, because the fruits separate from the stem cleanly, there is less loss of juice during transport and less postharvest decay. The only disadvantage is some phytotoxicity to the tree, which is revealed by the appearance of 'gum' exuding from pruning cuts throughout the canopy. While this detracts from the appearance of the tree it has not been shown to have economic importance.

Growth control and improved flowering

It has long been known that ethylene induces flowering in a wide range of crops. With the advent of ethephon this effect could begin to be used in production horticulture. Today we find that ethephon is used in pineapple production to synchronize flowering of whole plantations so that once-over harvesting is possible. It is also used in mango production with much the same aim. With these crops ethylene is the hormonal signal for flower initiation.

With apple, the effect of ethylene on flowering is less direct. Ethephon sprays early in the growing season temporarily suppress shoot elongation, resulting in more of the products of photosynthesis being available for bud development. A technique suggested to Washington State apple growers in recent decades was to apply a whole-tree spray to vigorous apple trees about 5 weeks after full bloom. When used at this time it has little effect on fruit retention and does not greatly advance fruit maturity. The principal effect is to check shoot elongation during the period when flower initiation and development is most critically influenced by competition for carbohydrates from growing shoot tips. The concentration of ethephon often suggested for this purpose is 500 ppm in a full-volume spray.

An alternative approach that has shown promise in eastern North America is to apply a series of sprays using very low concentrations of ethephon (100 ppm or less) once a week for 8–12 weeks. Suppressed shoot elongation and enhanced return flowering have been observed, but the effects on fruit quality are such that this treatment is mainly used on young non-bearing trees.

Ethephon also has possibilities for the grape grower. It is an effective inhibitor of shoot growth and can thus reduce the amount of summer pruning required. For this purpose, it is applied at 300 ppm 3 weeks after fruit set to the foliage above the flower cluster. A repeat spray 4–6 weeks later may be required. It must not be allowed to touch the flowers or the adjacent leaves, since this can seriously reduce fruit set. Applied closer to harvest to the whole grape canopy, ethephon will hasten ripening and increase the colour of some red cultivars, such as Pinot Noir. It is used at 250 ppm at a stage when berries are about 50% coloured.

Despite these promising results, ethephon is not widely used on grapes and growers should experiment on a small part of the vineyard before fully adopting these technologies.

Anti-gibberellin growth retardants (CCC, daminozide, paclobutrazol, prohexadione-Ca and others)

Growth-suppressing and flower-enhancing chemicals in this category have long been important tools for floriculture and the ornamental horticulture industry around the world. They are classed as anti-gibberellin compounds because they interfere with natural gibberellin biochemistry and physiology, and many of the effects of these chemicals can be reversed by applying gibberellic acid.

Success with woody ornamentals led to research with fruit trees and vines, and several of these chemicals are presently, or have been in the recent past, important in fruit production. Daminozide, known in the ornamental horticulture trade as B-Nine® and in tree fruit horticulture as Alar®, is one such chemical. Alar® was widely used in apple production to control shoot elongation, enhance flowering and improve fruit quality. This formulation of daminozide was withdrawn by the original manufacturer in about 1990 but remains available in some countries.

Cycocel (CCC) is another growth retardant with wide usage in other branches of horticulture and agriculture. It is very important to control growth of many ornamental plants and to prevent lodging of grain crops. Its use in fruit crops horticulture, however, is now limited to improving berry set of grapes for wine and raisins in a few countries. It had been used very successfully in northern Europe to control vegetative growth and promote fruit set of pears, but that practice is no longer supported by the product label.

Paclobutrazol (Cultar®, Syngenta) is a very potent anti-gibberellin. When applied as a soil drench, as little as 1 g of paclobutrazol can suppress shoot elongation for several years without adverse effects on flowering or fruit quality. It is also applied as a foliar spray with less dramatic growth-suppressing effects.

Cultar® is registered in South Africa for use on a range of tropical and subtropical fruit species (litchi, avocado, mango) and also on macadamia, pecan, peaches and plums. Research in Europe has shown that cropping and fruit quality of apple can be enhanced and the need for pruning is virtually eliminated. With sweet cherry, the use of Cultar® permits high-density plantations of compact trees that can be economically protected from rain and birds by covering. However, an important feature of paclobutrazol is its very slow rate of breakdown in the soil, which has blocked approval by registration authorities in many countries.

A related chemical registered widely for use on apple is prohexadione-Ca (Apogee®; BASF Corporation). This anti-gibberellin is applied by foliar sprays and its effect is relatively short-lived. It may find wide application in apple production systems involving larger trees, excessive shoot growth and problems with shading and fruit quality. Agogee® is known to reduce the incidence of fireblight, the bacterial disease caused by *Erwinia amylovora*. Thus, it has proven to be a dual-purpose production tool, where reduced shoot growth and fireblight protection are important objectives. Reduced shoot growth improves light and spray penetration into large canopies, reduces pruning costs and often improves fruit coloration. Commonly recommended rates range from 75 to 125 mg/l of active ingredient, with treatments starting shortly after bloom and continuing at bi-weekly intervals during the first 2 months of the growing season.

Branching agents

In the fruit tree nursery industry there is strong interest in being able to provide customers with a well-branched or 'feathered' tree. Such trees are believed to be more productive in their early years in the orchard.

One traditional approach to producing such trees has involved hand pinching of the central shoot tip to induce lateral branching, but this practice is laborious and not always successful. Another is to cut the tree back to the ground after the first full year in the nursery and then grow the tree for an extra year to produce a well-branched tree. This European system, called 'knipboom', works very well but adds great cost to the nursery tree. Thus, there has long been interest in finding a plant bioregulator that would stimulate branching in a 1-year-old nursery tree.

A number of chemical 'pinching agents' have been tested with varying degrees of success, but the approach presently favoured is to apply a hormone that forces axillary buds to develop into shoots. The hormone commercially available for this purpose is 6-benzyladenine (6-BA) and it can be purchased as Promalin® and Accel® (Valent Biosciences) or in Europe as Paturyl 10 WCS. Promalin contains GA_4 and GA_7 in addition to 6-BA. The technique is to apply the 6-BA-containing spray (about 500 ppm) to the top and upper portion of the nursery tree where branching is desired. Tree height in the nursery row should average about 60 cm and the trees must be growing very strongly if the desired result is to be achieved. Success is measured by the number of feathers greater than about 10 cm long that develop at a height that will make good fruiting branches once the tree is planted in the orchard.

Another plant bioregulator shown to induce branching of apple and cherry trees is cyclanilide. This chemical is registered for use, together with ethephon, as a harvest aide for cotton but is not yet registered for use on any other crop. None the less, recent research shows promising results for improving the branching of young apple and cherry trees and registration is being sought in North America.

Further Reading

Davies, P.J. (ed.) (2004) *Plant Hormones: Biosythesis, Signal Transduction, Action!*, 3rd edn. Kluwer Academic, New York.

Janoudi, A. (ed.) (2000) International Symposium on Growth and Development of Fruit Crops. *Acta Horticulturae* 527.

Kang, S.M., Bangerth, F. and Kim, S.K. (eds) (2004) IX International Symposium on Plant Bioregulators in Fruit Production. *Acta Horticulturae* 653.

Webster, A.D. and Lee J.M. (eds) (2008) Proceedings of the International Symposium on Endogenous and Exogenous Plant Bioregulators. *Acta Horticulturae* 774.

Webster, A.D. and Ramirez, H. (eds) (2006) X International Symposium on Plant Bioregulators in Fruit Production. *Acta Horticulturae* 727.

N. Looney and D. Jackson

9 Soils, Nutrients and Water

David Jackson

Selection of soils

A wide range of soils is used for fruit production. Few are entirely unsuitable, but others need so much effort and money spent on them that they are best avoided. The broad functions of soils have already been considered in the section on roots (in Chapter 3). They supply the following general plant needs.

- Providing a suitable environment for root growth and development.
- Storing and supplying plants with water.
- Storing and supplying plants with essential nutrients.

The first two relate to the broad chemical and physical properties of the soil, while the last is a reflection of its fertility. For most fruit-growing soils, the chemical fertility, if below optimum, can be remedied by the application of essential nutrients as fertilizers, either to the soil or perhaps by foliar sprays.

However, undesirable physical features in the soil are often impossible, difficult or expensive to change. For this reason, particular attention should be given to a thorough, in-the-field examination of the soil's physical properties. It is worth remembering that if the soil has failings that cannot be remedied, yields will always be limited, but the costs of production will not be similarly reduced – indeed, they may be increased. Thus the profitability of the enterprise will always be less than that of a similar enterprise on a non-limiting soil.

Soil: physical characteristics

The nature and depth of the subsoil, soil drainage and the depth of the water table are three important physical characteristics of soil. The growth of many fruit plants and their productivity are determined more by the subsoil than by the kind and quality of the surface soil. An ideal subsoil should be well drained and loose enough to permit root development. A sandy loam subsoil is preferred to heavy clay or compacted soil.

Depth of the soil is the depth of soil to the weathered rocks, stone lines, pans or other physical impediments to root penetration. The restriction of roots to a shallow depth or a small volume of soil will result in poor tree or vine anchorage and poor growth. Supporting such plants with a post or trellis system will improve performance but growth will still be restricted by the inability of the roots to fully explore the subsoil for nutrients and water.

Thus, the depth of a root system has an important influence on the productivity and longevity of trees and vines. A deep soil can hold a much greater reserve of nutrients and moisture than a shallow soil. The soil depth for an orchard should be at least 1.2–1.8 m for optimum root and top growth; bush fruits require less anchorage and can adjust to a shallower soil of 0.7–1.2 m. Many plantations are established on soils with less than optimum depth, but they place greater demands on the grower to supply water, nutrients and drainage. Occasionally, however, shallower soils can be advantageous when they control excess vigour. Grapevines, for example, will often be easier to control and wine quality may improve under soil-imposed de-vigoration.

Water table and drainage

If a hole is dug into the ground and water flows into the hole from the surrounding soil, the level at which the water remains is termed the 'water table'. Fruit trees and vines may not survive if their roots are submerged in water for an extended period. Water replaces the air in the soil environment and the roots die from lack of oxygen. 'Wet feet' can be tolerated for a longer time in the winter because the tree is virtually dormant, soil temperature is low and root respiration is at a minimum rate. It is very important, however, that the water drains away before spring growth commences. Other undesirable consequences of wet feet are root disease and decay, and elements such as manganese may increase in the soil solution to toxic levels.

Methods of assessing the water table and drainage requirements are discussed below. If the problem is serious, an underground drainage system may be required. If it is minor, it may be overcome by less costly ridging (Fig. 9.1).

To form ridges or hills, topsoil is moved from alleyways. This can be done by ploughing towards the eventual tree row, rotary hoeing (rototilling) to loosen the soil and finishing up with a blade on a tractor. This effectively doubles the depth of topsoil immediately available to the tree or vine and helps to overcome waterlogging of the roots. The trees or vines are planted in the centre of the ridge. Drip irrigation or mini-sprinklers are a very important part of the system, since, after heavy rain or irrigation, water will flow from the ridge to the alleyway, leaving those roots near the tree dry. One potential disadvantage of this system is difficulty in mowing the alleyway.

Growers should think twice before buying land that needs extensive tile drainage. However, there are certainly situations where a drainage system may be considered both desirable and economic.

The methods of drainage and installation details will not be discussed here. In most regions where this practice is deemed necessary there will be consultants and engineers with specific knowledge, skills and services.

Specific soil types

Some types of soil and their suitability are shown in Fig. 9.2.

Texture and structure

The term 'soil texture' refers to the particle size distribution of the mineral constituents of the soil. Laboratory methods can be used to accurately determine the various soil textural classes, which show the distribution of different proportions of sand, silt and clay. The classes commonly recognized are detailed in Table 9.1. It should be noted that pure sands, silts or clays are seldom found. People referring to a 'clay' will normally mean a clay loam.

For the practical grower, field determination of soil texture is often adequate. This is carried out by putting some soil in the palm of the hand and moistening and working it into a smooth paste before recording the 'feel' of the soil by rubbing some of the paste between the thumb and the forefingers. A sandy loam produces a faint rasping sound and a slightly gritty feel. It can be moulded into a cohesive ball, which fissures when pressed flat. A clay loam

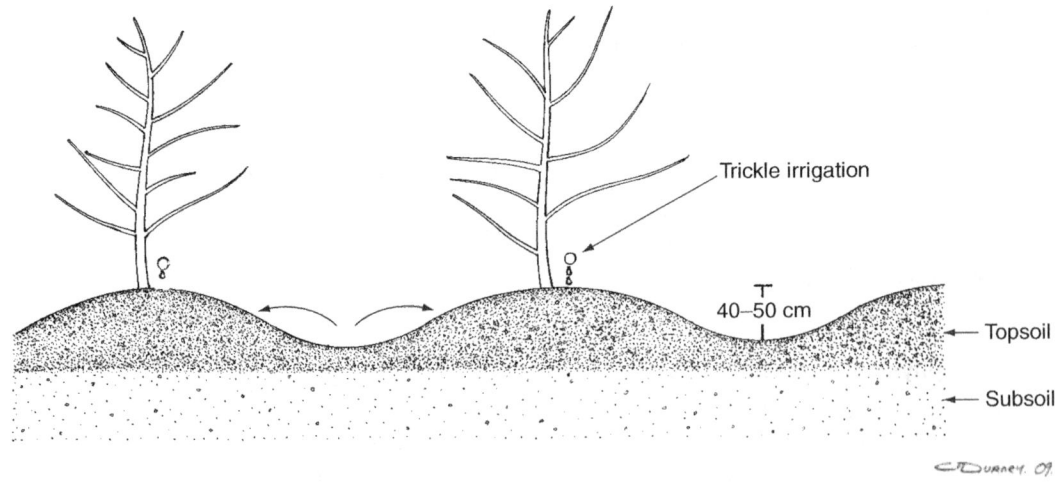

Fig. 9.1. Ridging used to increase soil depth and reduce water-logging.

D. Jackson

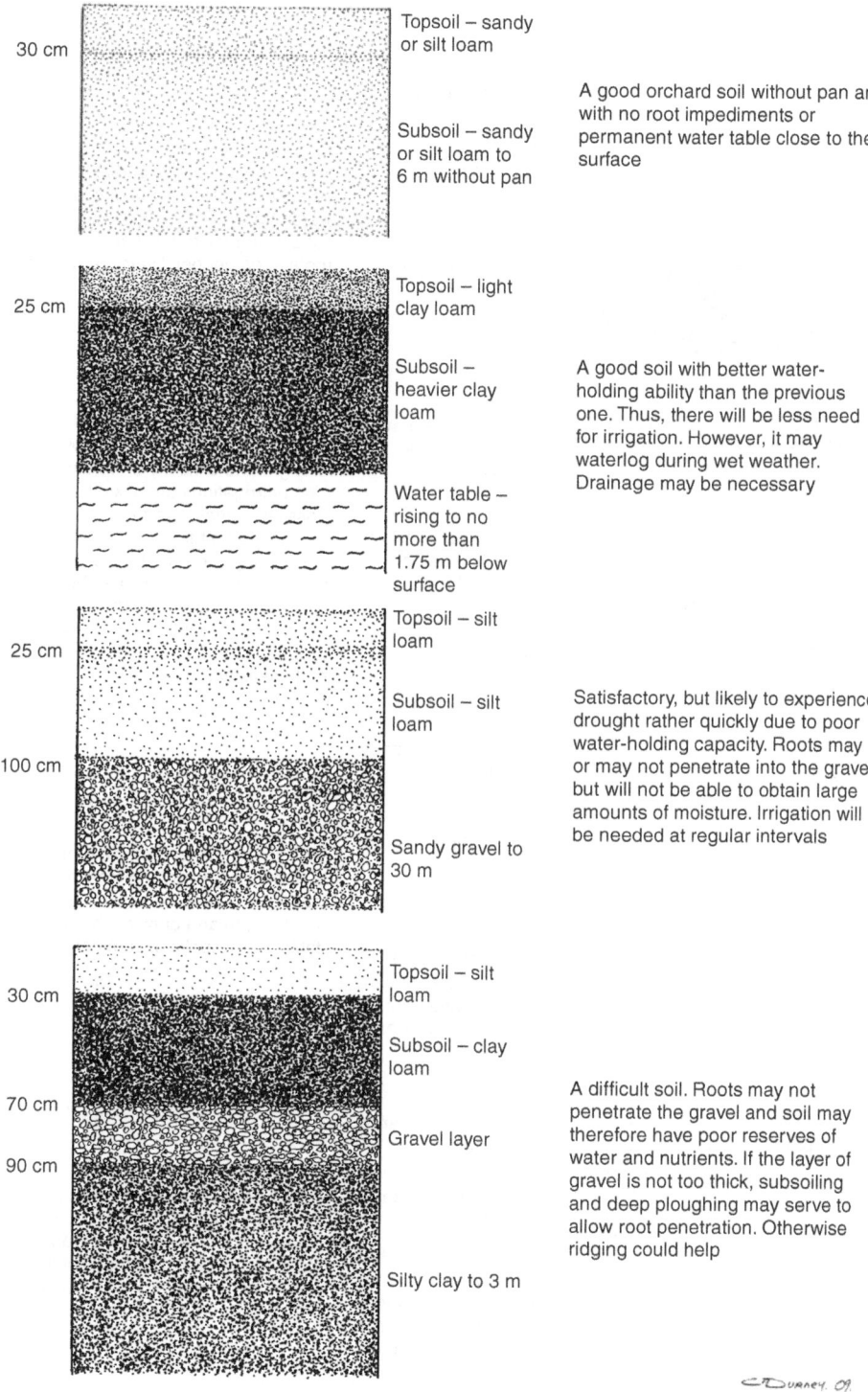

30 cm — Topsoil – sandy or silt loam

Subsoil – sandy or silt loam to 6 m without pan

A good orchard soil without pan and with no root impediments or permanent water table close to the surface

25 cm — Topsoil – light clay loam

Subsoil – heavier clay loam

Water table – rising to no more than 1.75 m below surface

A good soil with better water-holding ability than the previous one. Thus, there will be less need for irrigation. However, it may waterlog during wet weather. Drainage may be necessary

25 cm — Topsoil – silt loam

Subsoil – silt loam

100 cm

Sandy gravel to 30 m

Satisfactory, but likely to experience drought rather quickly due to poor water-holding capacity. Roots may or may not penetrate into the gravel but will not be able to obtain large amounts of moisture. Irrigation will be needed at regular intervals

30 cm — Topsoil – silt loam

Subsoil – clay loam

70 cm — Gravel layer

90 cm

Silty clay to 3 m

A difficult soil. Roots may not penetrate the gravel and soil may therefore have poor reserves of water and nutrients. If the layer of gravel is not too thick, subsoiling and deep ploughing may serve to allow root penetration. Otherwise ridging could help

Fig. 9.2. Some examples of good and bad soil profiles.

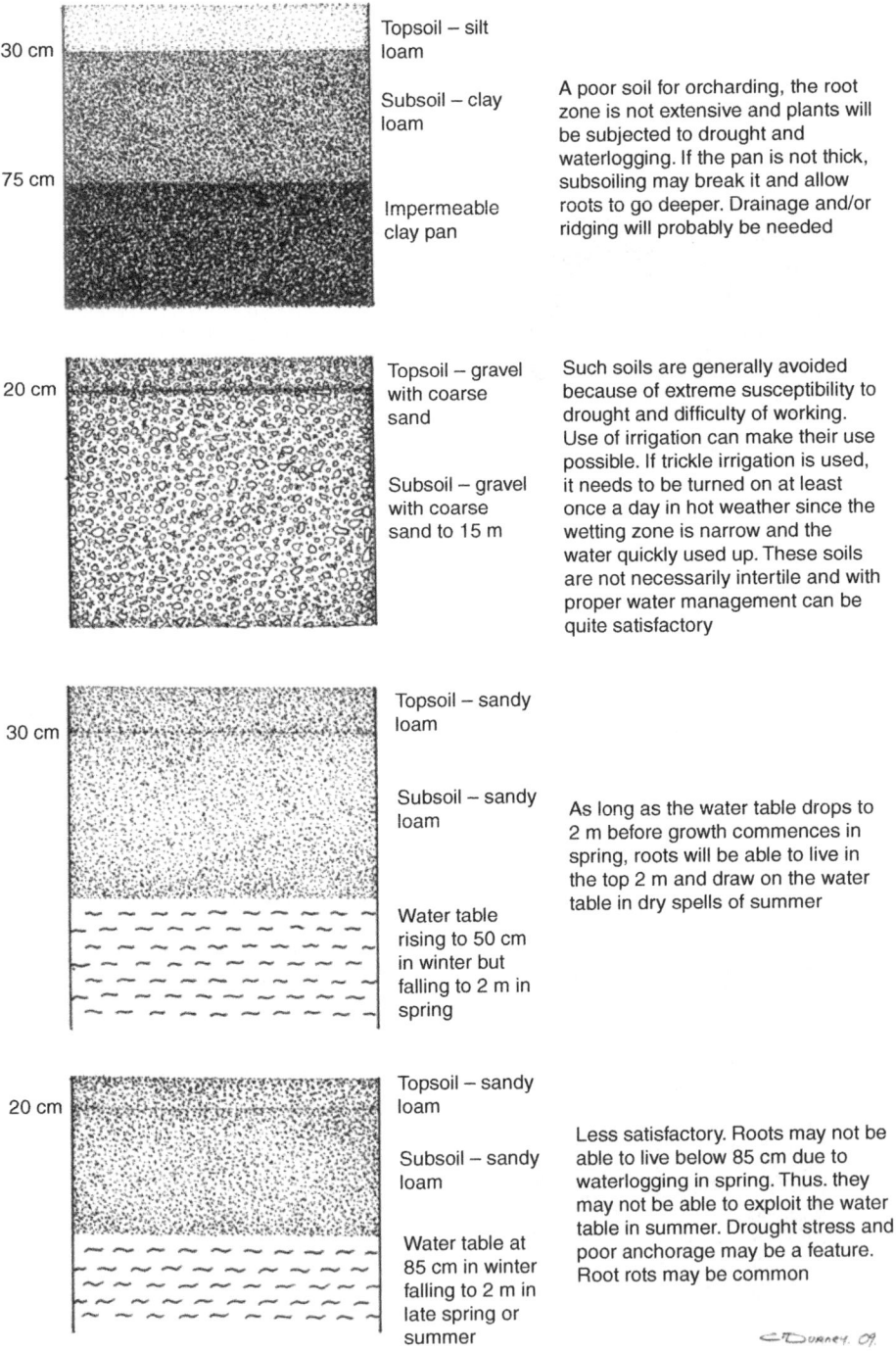

30 cm

Topsoil – silt loam

Subsoil – clay loam

75 cm

Impermeable clay pan

A poor soil for orcharding, the root zone is not extensive and plants will be subjected to drought and waterlogging. If the pan is not thick, subsoiling may break it and allow roots to go deeper. Drainage and/or ridging will probably be needed

20 cm

Topsoil – gravel with coarse sand

Subsoil – gravel with coarse sand to 15 m

Such soils are generally avoided because of extreme susceptibility to drought and difficulty of working. Use of irrigation can make their use possible. If trickle irrigation is used, it needs to be turned on at least once a day in hot weather since the wetting zone is narrow and the water quickly used up. These soils are not necessarily intertile and with proper water management can be quite satisfactory

30 cm

Topsoil – sandy loam

Subsoil – sandy loam

Water table rising to 50 cm in winter but falling to 2 m in spring

As long as the water table drops to 2 m before growth commences in spring, roots will be able to live in the top 2 m and draw on the water table in dry spells of summer

20 cm

Topsoil – sandy loam

Subsoil – sandy loam

Water table at 85 cm in winter falling to 2 m in late spring or summer

Less satisfactory. Roots may not be able to live below 85 cm due to waterlogging in spring. Thus. they may not be able to exploit the water table in summer. Drought stress and poor anchorage may be a feature. Root rots may be common

C Durney. 09.

Fig. 9.2. Continued.

D. Jackson

Table 9.1. Soil classes.

Soil texture	Distribution (mean %)		
	Sand	Silt	Clay
Sand	90	5	5
Sandy loam	70	15	15
Silt loam	40	40	20
Clay loam	35	30	35

Size of sand particles, 0.02–2.0 mm diameter; size of silt particles, 0.002–0.02 mm diameter; size of clay particles, <0.002 mm diameter.

has a very smooth, slightly sticky or sticky feel; it is plastic in nature and, when it is moulded into a cohesive ball, it can be flattened without fissuring.

A sandy loam is generally cited as the ideal soil texture for perennial fruit trees and vines since it contains the most appropriate proportion of sand, silt and clay. The sand 'lightens' the soil and provides good drainage, but too much sand leads to excessive drainage and droughty conditions. Too much clay in a soil makes it very 'heavy' and difficult to work, especially under wet conditions.

In the soil, particles of sand, silt and clay are usually clumped together to form larger aggregates or 'peds'. These form the structure. Types of soil structure can vary from plate-like, prism-like or block-like to smaller crumbs. Crumb-structured soils are friable, with high porosity and water-holding capacity, while blocky-structured soils are firm with low porosity. Naturally the former are preferred.

The aggregation of soil particles is promoted by humus, derived from organic matter in the soil. Hence, improvement in soil structure can be brought about by soil management practices that increase organic matter content.

Assessment of soil suitability in the field

A quick assessment of the soil can often be made by observing the plants growing in the area. Good, healthy growth of trees or bushes suggests a good, deep soil. In late spring or summer, grass in some areas will be observed to lose its greenness sooner than in others. This indicates either light or shallower soils. The parts of the property that remain green over most of the summer may have a good water-holding capacity but they may also be areas that are waterlogged in spring and therefore unsuitable for permanent crops like trees. These spots should be observed after heavy rain to determine the water table.

Exposed soil profiles, as in road cuts or ditches, will help a prospective grower to assess the type of soil on a property. Soil maps may be available, which show the dominant soils in an area. The key to their value is the detail supplied, whether it is, for example, adequate to provide a good idea of the soil on the property under consideration or is only appropriate for getting a general picture of the district as a whole.

The colour of successive layers throughout the soil profile provides clues to the organic matter content, drainage and moisture conditions in the soil. Black is usually indicative of the presence of humus and is generally restricted to the topsoil. A uniform brown soil profile is an indication of well-drained land, while soil profiles with grey, especially bluish-grey colours, or a pronounced mottling pattern containing specklings of orange and rust shades are imperfectly drained. Permanently wet soil horizons are uniformly bluish-grey. If mottles occur close to the soil surface, the soil is generally too poorly drained to be a good fruit production site.

If a further assessment of the water table is needed, this can be achieved by digging six to eight evenly spaced holes per hectare to a depth of 1–1.5 m using a tractor-driven or hand auger. A drain pipe should then be inserted into each hole and the water table observed throughout one season, and especially after each rainfall. In a well-drained orchard soil, the water table may be visible within a few centimetres of the soil surface for an hour or so after a heavy rain, but it should fall rapidly and, within a day or so, drop to a metre below the surface. In an imperfectly drained soil, the water table remains at a depth of about 30–60 cm below the surface for a week or more after such a rain. This soil is unsuitable for fruit production unless a drainage system is installed to lower the water table.

If tile drains are already installed in the land, it is valuable to inspect the outruns after rainy periods to check that they are running freely, as they may clog in silty soils.

The depth of rooting of other plant species can be very revealing and, especially if deep-rooted plants like lucerne (alfalfa) have been grown, it will indicate the likely volume of soil available to plant roots. Deep root penetration to 1–2 m indicates that there is no impermeable layer or persistent water table.

Chemical analysis of soils

After the field examination, it may be useful to collect samples of topsoil and subsoil from the field to

Table 9.2. Availability of nutrients at different pH levels. (After Nelson, 1968).

Nutrient	pH values at which nutrient becomes less available
A. Those minerals more likely to become deficient at low pH	
Phosphorus	Below 6.0; in addition there is some reduction between 8.0 and 8.5
Potassium	Below 5.5
Sulphur	Below 5.5
Calcium	Below 6.0
Magnesium	Below 6.0
Molybdenum	Below 6.0
Boron	Below 4.7, some reduction between 7.7 and 8.5
B. Those minerals more likely to become deficient at low and at high pHs	
Nitrogen	Below 5.0–5.5 and above 8.5
Copper	Below 4.7 and above 7.5–8.0
Zinc	Below 4.7 and above 7.5–8.0
Manganese	Below 4.7 and above 7.0
C. Those minerals more likely to become deficient at high pH	
Iron	Above 7.5

be sent to a soil-testing laboratory to assess chemical fertility. These laboratories provide the soil test results and usually give recommendations for improving the soil nutrient status. However, in order for the soil test results to reflect the fertility of the entire field, it is imperative that the samples are representative. An analysis of soil collected from between 10 and 30 cm of the soil surface may provide a sufficient estimate of fertility, especially if the production site is to be irrigated. However, some experts consider that the soil depths sampled should be continuous from the soil surface right down and through the subsoil to a depth where it is considered the roots will be able to penetrate. A soil auger is the best way to obtain such a sample.

For a fairly flat piece of land showing no obvious evidence of fertility patterns, samples should be taken at regular intervals while walking in a W-pattern across the area. These samples are then thoroughly mixed and a subsample sent off for analysis. For a field with obvious fertility patterns, it is advisable to subdivide the land into these apparent sections, obtaining an analysis for each area. Do not underestimate the number of samples needed to gain an accurate estimate of the fertility of each section. Samples should be sent for analysis as soon as possible after collection.

Major deficiencies identified by the soil analysis should be corrected before planting. Note that nitrogen cannot be accurately assessed by soil test, thus no nitrogen analysis will be provided. Indications of N status can be obtained by leaf analysis (see later).

The soil analysis will also indicate the pH of the soil. While apple and other fruit trees have been grown on soils as acid as pH 4.0 or as alkaline as pH 8.5, for many fruit crops the best level is considered to be about 6.0–6.5. This reflects the effect of pH on the availability of different nutrients (Table 9.2).

For acid soils, correcting pH involves the addition and incorporation of lime. This is best done before planting. Alkaline soils are more difficult to adjust to a lower pH, but the addition of organic fertilizers, irrigation and the use of acidifying fertilizers such as ammonium sulphate will lower pH over time.

Correcting Nutrient Deficiencies

To deal appropriately with nutrient deficiencies it is necessary to understand the plant's need for specific minerals.

Minerals needs and sources

The minerals required by plants can be divided into two groups: major nutrients and minor (or trace) nutrients. The major nutrients are nitrogen (N), phosphorus (P), potassium (K), sulphur (S), calcium (Ca) and magnesium (Mg). N, P and K are the ones most commonly deficient and these are included in many fertilizer mixes.

D. Jackson

Minor nutrients include manganese (Mn), boron (B), iron (Fe), zinc (Zn), copper (Cu), molybdenum (Mo) and cobalt (Co). Most soils are adequately supplied with these minerals, but deficiencies do occur, sometimes as a result of an excessively acidic or alkaline soil. These deficiencies are usually recognized by inadequate growth and typical foliar symptoms. However, the symptoms are not always easy to distinguish and expert advice may be required to obtain or confirm a diagnosis. The information in Table 9.3 can serve as a guide, but only the most common deficiencies are listed.

Common fertilizers available and the major elements they supply are shown in Table 9.4.

Table 9.3. Guide to nutrient deficiencies.

Mineral	Symptoms	Treatment	Comments
Nitrogen (N)	Plant appears light green-yellow with short shoots and small leaves	Nitrogen fertilizer may be broadcast, added to the irrigation line ('fertigation'), or urea may be sprayed over the foliage for a quick response (use 6 kg/1000 l)	Easily leached from soil by heavy rain or excess irrigation. Probably the most common deficiency in neglected orchards
Phosphorus (P)	Small leaves which may have a reddish-purple colour and show autumn colour earlier than normal	Usually applied as fertilizer to soil, but may use ammonium polyphosphate through trickle system	Is quickly bound in the soil and may become unavailable to the plant. Any deficiency in the soil best treated prior to planting
Potassium (K)	Grey margins on the older leaves, which may dry out	Use fertilizers applied to soil. May use potassium nitrate in trickle or sprays of potassium sulphate (10 kg/1000 l) to leaves	Excessive potassium applications may lead to magnesium or calcium deficiency
Magnesium (Mg)	Older leaves have yellow margins and yellow tips. Sometimes red-brown coloration occurs in centre of leaf	Apply magnesium sulphate to soil or use serpentine superphosphate. For quick response, spray with magnesium sulphate (20 kg/1000 l) in spring	Three to four spray applications at 2-weekly intervals may be made if deficiency is severe. Most common in citrus
Calcium (C)	This seldom occurs as soil-induced deficiency but, due to slow movement in the plant, may sometimes become deficient in the fruits (see bitter pit).		
Manganese (Mn)	Yellow leaves, which are sometimes difficult to distinguish from N or Mg deficiency	Spray in spring with manganese sulphate 6 kg, 8 kg hydrated lime, 1000 l water	Mainly seen on peaches and nectarines
Boron (B)	Young leaves die, buds may fall; distorted, pitted fruit	Use 100–200 g borax/tree or spray leaves with 1–2 kg borax/1000 l	Only use if boron deficiency is confirmed; do not over apply – excess is as bad as or worse than a deficiency
Iron (Fe)	Pronounced yellowing in young leaves	Spray iron chelate (1.5–2 kg/ 1000 l) once or twice per season. Alternatively, water around tree with 1–2% solution at 1 l/m^2, followed by irrigation	Most common in alkaline soils
Zinc (Zn)	Young leaves usually bunched like a rosette; they may be thin and pale green or yellow	Use sprays of zinc chelate (1–2 kg/1000 l) soon after leaf emergence	Mainly a problem on peaches, nectarines, grapes and citrus

Table 9.4. Analysis of common fertilizers.

	Percentage composition					
	N	P	K	Mg	S	Ca
Nitrogen						
Di-ammonium phosphate (DAP)	18	20	–	–	–	–
Mono-ammonium phosphate (MAP)	11	20	–	–	–	–
Ammonium nitrate	34	–	–	–	–	–
Ammonium sulphate	21	–	–	–	24	–
Calcium ammonium nitrate	26	–	–	–	–	8
Potassium nitrate	14	–	36	–	–	–
Urea	46	–	–	–	–	–
Phosphorus						
Di-ammonium phosphate (DAP)	18	20	–	–	–	–
Mono-ammonium phosphate (MAP)	11	20	–	–	–	–
Serpentine superphosphate	–	7	0.1	5	8	15
Superphosphate	–	9–10	–	–	11	20
Potassium						
Potassium nitrate	14	–	36	–	–	–
Potassium sulphate	–	–	39	–	18	–
Potassium chloride	–	–	48	–	–	–

Growers often prefer to use commercially mixed fertilizers, which are often in an easily managed, granular form. However, they can often be paying an excessive amount relative to what is required. For example, applying an NPK fertilizer to a soil adequately supplied with K means paying as much as 30% more than necessary.

It is important to remember that nutrients are also available from organic sources such as animal manures and meat industry by-products. Table 9.5 shows the composition of some of these materials. While they are not rich sources of N, P and K, they do have important advantages. Bulky organic manures, such as deep-litter poultry manure, supply a significant amount of organic matter to the soil. Organic manures contain trace elements as well as major elements. It is not uncommon to use organic manures as mulching material. Finally, they can often be obtained cheaply as an unwanted by-product of animal production systems.

Organic manures tend to reduce pH, a feature that can be advantageous in some situations.

However, a major disadvantage is the cost of transport to the site and distribution to the fruit crops. Some organic material, such as fresh poultry manure, can be toxic if applied at too high a concentration, and some manures will introduce weed seeds to the orchard.

Animal by-products may be processed, for example blood and bone, and are sold as organic fertilizers. Their organic content is low but they are moderately high in minerals.

Manures originating from plant sources, such as straw or sawdust, will have less readily available minerals. Furthermore, they must rot before the minerals are released. Incorporating these manures into the soil will usually create transient mineral deficiencies. This occurs because microorganisms need minerals, especially nitrogen, to help them break down the organic matter. Initially, these elements must come from the soil, creating the deficiency.

Thus, if vegetable materials are applied alone, it is advisable to add an additional dressing of about 15 kg N/ha to prevent any denitrifying effect on the soil. If these materials have been used as litter for

D. Jackson

Table 9.5. Average composition (%) of some organic manures and fertilizers. (After Goh, 1985).

	Nitrogen	Phosphorus	Potassium
Organic manure			
Cow manure	0.6	0.3	0.5
Pig manure	0.5	0.3	0.5
Sheep manure	0.9	0.3	0.9
Horse manure	0.7	0.2	0.6
Poultry manure (deep litter)	1.0–4.0	0.8–1.6	0.5–1.5
Seaweed (kelp)	0.6	–	1.0
Sawdust or crushed bark	0.1	–	0.6
Straw – cereal	0.6	0.1	0.6
Straw – leguminous	1.6	0.1	1.0
Organic fertilizers			
Dried blood	12–14	–	–
Blood and bone	6.1	6.9	–
Bone dust	3–4	7–8	–

animals, the presence of the animal manure promotes rotting and helps reduce these deficiencies.

Amounts of nutrients to apply

The report of a soil analysis will often come with specific recommendations about how to correct the identified deficiencies. Advisers have reference sheets which give suggested soil levels of minerals for specific crops. While these vary slightly with crop, soil type and even analytical techniques, they normally fall into the following range:

- Phosphorus (P) 25–35 ppm
- Potassium (K) 12–15 ppm
- Magnesium (Mg) 15–20 ppm
- Calcium (Ca) 10–12 ppm
- pH (acidity or alkalinity) 6.0–6.5.

To raise phosphorus levels by 10 ppm, an orchardist needs to apply superphosphate at 250 kg/ha on a sand or sandy loam, 500–700 kg/ha on silt loam and 750–1250 kg/ha on clay or peat soil.

To raise the potassium level by 5 ppm, application of potassium sulphate needs to be 60–125 kg/ha on sand, 250–350 kg/ha on a sandy or sandy-silt loam, 500 kg/ha on a silt loam, 750 kg/ha on a heavy silt loam and 1000–1250 kg/ha for clay or peat.

Twice these amounts of magnesium sulphate are required for raising magnesium levels by 5 ppm.

The addition of lime (ground limestone, i.e. calcium carbonate) is used to raise pH and calcium levels. Normally, if the pH is in the correct range, the calcium level will be adequate. To raise pH by half a unit, lime should be added at 1.25 t/ha on a sandy loam, 2.5 t/ha on a silt loam and 5.0 t/ha on a clay or peat loam.

The Use of Fertilizers in Established Fruit Crops

Deciding what to use

If fruit trees or vines are planted in a soil with good depth and structure, fertilization has corrected nutrient deficiencies, and soil pH is in the satisfactory range, these plants will be well set to make good growth. However, over time, the soil will gradually lose fertility as minerals are lost by leaching, are incorporated into the tree and are removed with the crop and prunings. Thus, except in the most fertile soils, nutrient replacement is normally an annual event.

Replacement of nutrients is still a difficult process to quantify. Various estimates of nutrient removal have been made for different crops, and fertilizer application rates have been devised to replenish these losses. Although this is a useful guide, it is subject to considerable error, since not all soils retain minerals equally, i.e. the loss of nutrients by leaching can mean that greater than estimated levels are needed. Furthermore, crop yields affect nutrient removal. A high-yielding cultivar or a good crop could remove significantly larger amounts of nutrients than a poor-yielding one on the same soil. Some research work suggests

Table 9.6. NPK removal (kg) per hectare, per year by mature trees.

Crop	Nitrogen (N)	Phosphorus (P)	Potassium (K)
Apples[a]	39	10	71
Kiwifruit[b]	24	4	48
Peaches[a]	76	11	96

[a]Yield of trees not specified. From Westwood, M.N. (1978) *Temperate Zone Pomology*. W.H. Freeman & Co, San Francisco; [b]Yield 16.5 t/ha. (From Ferguson and Eiseman, 1983).

that for apples and peaches, the amounts of NPK listed in Table 9.6 are removed per hectare per year by mature trees, and orchardists would normally attempt to supply replacement N, P and K each year.

The following recommendations of a mixture for deciduous trees can be used as a guide for annual fertilizer applications.

- 150 kg urea, which supplies 72 kg N.
- 300 kg superphosphate, which supplies 24 kg P.
- 150 kg potassium chloride, which supplies 72 kg K.

The total is 600 kg, which is applied to 1 ha of orchard.

It will be noticed that the level of phosphorus is twice that of the amount estimated to be removed. Excess fertilizer is required because phosphorus is quickly 'fixed' in the soil, i.e. is converted into insoluble complexes with the soil.

When deciduous trees or vines are young, the same fertilizer mix is recommended, but (assuming 600 trees/ha) 200 g of the mixture is applied per tree for each year of its growth up to 5 years, i.e. 200 g in the first year, 400 g in the second and up to 1 kg in the fifth. After this stage, the fertilizer is broadcast generally under the tree row. Other crops would be treated differently, the difference basically related to the speed with which the crop covers the allocated area. For example, strawberries in a 1-year rotation will fill the area in 1 year and there would be no build-up time.

Soil analysis

Such general recipes for fertilizer application are very rough and ready. Other estimations are possible, one of which is to obtain soil analyses at regular intervals, say every second or third year. Soil applications then attempt to overcome any deficiencies that are appearing. An annual rate is determined and maintained until the next analysis, after which it may be modified to accommodate changes that have been found. Unfortunately, soil analyses do not provide an estimate of nitrogen levels. Furthermore, sample collection can be tricky. For example, sampling too soon after fertilizer application can give misleading results.

Leaf analysis

Leaf analyses are more reliable and are increasingly used. Leaves can be analysed for major and minor elements, and this gives an accurate measurement of the status of the tree. This is then checked against standard figures for the crop concerned and fertilizer rates are recommended. Because the levels of nutrients within leaves vary throughout the growing season, it is important that leaf samples are taken at a specific period so that results can be compared, both with standards and between successive analyses. Generally, the best time is in mid- to late summer. Consultants may be approached for further information on leaf analyses.

Application rates and times

Winter is not a good time to apply fertilizers since leaching may remove a substantial proportion of the minerals before plant roots are active; this is especially the case with nitrogen and potassium. The best application time is just before growth starts in the spring, although organic manures could be spread a little earlier (the use of organic mulches in winter delays soil warm-up and may increase frost risk.) Sometimes two-thirds is applied in early spring and one-third in late spring or even late summer (which may help the build-up of over-wintering reserves). Fertilizers are distributed evenly within the zone covered by the branches of the tree. This usually coincides with the area kept weed-free by herbicides, and it is in this area that most roots are found.

While the majority of nutrients are spread in solid form to the soil, liquid fertilizers can also be used.

They are applied by two methods. Foliar application is an excellent way to supply trace elements to the tree and zinc, iron, boron and magnesium will quickly be absorbed in this way. The appropriate chemical is simply added to the normal spray, so long as the ingredients are compatible. However, advice as to the timing, the chemical and its formulation, and the concentration is best sought from advisers. Nitrogen, as urea 6 kg/1000 l, is sometimes used in spring, after leafing, to provide a quick response. A number of commercially prepared foliar mixtures are available which contain major and/or minor elements. Foliar-applied fertilizers can be useful if a deficiency is suspected to be causing poor growth or yield – a response to such an application can provide a clue as to the cause of the problem.

Fertigation

The second way of applying liquid fertilizers is via the trickle irrigation system ('fertigation'). Theoretically, this is an ideal method, since the nutrients are placed in a dilute form directly in the root zone of the tree and there is little waste. Nitrogen and potassium salts are commonly used in fertigation since they are easily dissolved. Many phosphate-supplying fertilizers are relatively insoluble but ammonium polyphosphate is a soluble source.

One practical approach is to apply superphosphate to the soil in spring and apply nitrogen and potassium throughout the season by fertigation. Simply divide the total amount needed by the number of irrigations to determine the amount to apply with each irrigation. Urea and/or potassium salts are dissolved in water, which is injected into the system over a specified period of time. Commercially produced injectors are available, but many fruit growers have designed their own injection system, based on the Venturi principle. Note that most water districts require a back-flow preventer as part of this system.

Managing the Soil and Orchard Floor Vegetation

The proper care of soils and effective management of the orchard floor vegetation ensures rapid establishment of a young plantation, early fruit bearing and a long period of fruit production. The system chosen should maintain or increase the level of humus in the soil, maintain soil structure, protect earthworms and other beneficial soil fauna and flora, prevent compaction, conserve soil moisture, prevent erosion, minimize leaching and maintain an appropriate level of soil fertility. It should also contribute to safe and comfortable conditions for carrying out the various orchard operations.

Grassing down and herbicide strips

Several different methods of soil and orchard floor vegetation management have been used. One common practice is to use a winter cover crop followed by clean cultivation in the first few years (3–5), followed by grassing down permanently for the remainder of the life of the plantation. Herbicides are used to eliminate weeds and other vegetation along the tree rows, usually in a strip about 2 m wide (Fig. 9.3).

This management system has a number of advantages.

- Easy working conditions for people and machinery (once the grassed alley is established).
- Reduction of soil erosion on sloping land.
- Improved soil structure and increased earthworm and other biological activities, leading to better water penetration and percolation.

It is claimed that fruit quality and storage may also be improved. If clovers are used, they may also fix nitrogen, but evidence suggests this is mostly used by the grass species. In order to avoid any detrimental effects, the grassed area should be mown frequently; this will reduce competition with the fruit trees for moisture and nutrients. Regular mowing and the return of grass clippings increase the cycling of nutrients, especially phosphorus, potassium and magnesium. Furthermore, frequently mown short grass is less likely to harbour orchard pests and diseases, which seek refuge in tall grass. Fertilizers should be applied to the herbicide strips and not to the grassed portion.

Total herbicide use (zero-tillage)

This system involves spraying the entire orchard floor with herbicides and completely removes the competition between fruit trees and other vegetation. This system produces more vigorous apple tree growth and higher yield than under a grassed-down system, although the fruit quality may be marginally inferior.

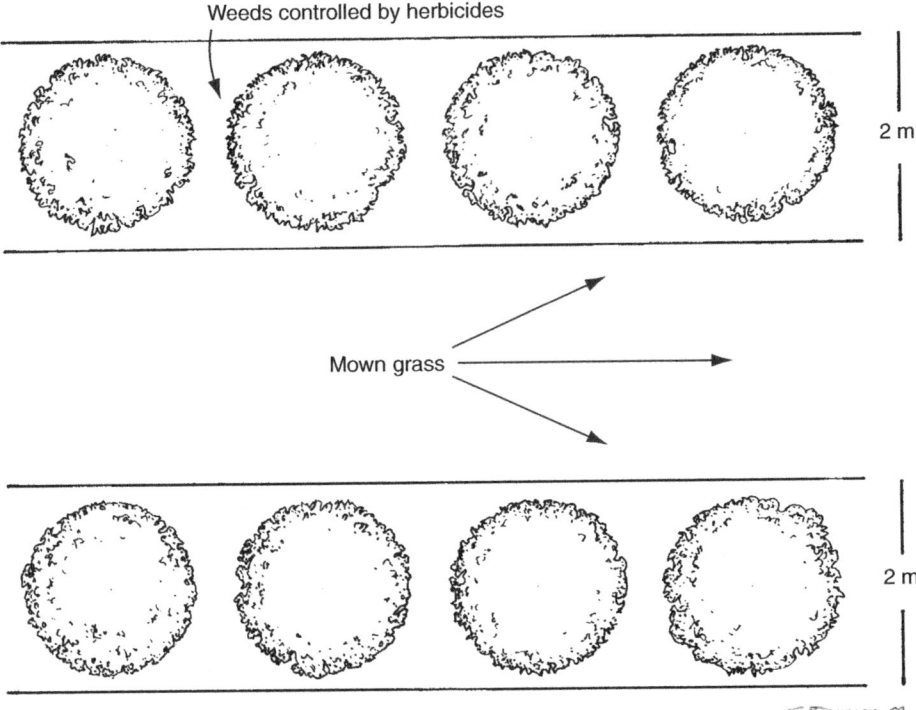

Weeds controlled by herbicides

Mown grass

2 m

2 m

Fig. 9.3. Soil management by herbicides and mown grass.

However, the increased yield and higher financial returns are normal for overall herbicide, relative to the strip system. A moss vegetation cover usually results from the zero-tillage system, and this has been found to be satisfactory for machinery passage. Surface compaction of the soil often occurs, but it is only on heavy soils that it leads to reduced water penetration into the soil profile. Excessive runoff and erosion, however, are likely in sloping orchards. Herbicide mixtures normally contain a chemical such as glyphosate, paraquat or amitrole, which will kill existing weeds, together with a persistent material, such as simazine, which will kill germinating seeds (Chapter 10). In deciduous orchards, herbicides should be sprayed just before buds burst in spring.

Mulching

Instead of using herbicides, a mulching system uses a mulch of straw, hay, sawdust or other organic material over the entire orchard floor or beneath the trees. It is seldom used because of the costs involved in obtaining and distributing the material.

A depth of at least 8 cm when settled is required. It may improve soil structure, but it can also be a fire hazard when dry and can harbour rodents. Orchards with mulches are usually cooler than weed-free, soil-exposed establishments; frost may therefore be more severe.

Irrigation

Irrigation is an important practice in fruit production. It is even being used in regions with 900–1400 mm of rain; a rate of precipitation long considered adequate for commercial production. The justification is that even a 1- or 2-week period of water stress can seriously affect crop yields, fruit quality or plant growth. Still, the decision to install irrigation is not always easy. The grower must consider the total amount of rainfall expected, its distribution and reliability during the season, the estimated evapotranspiration from the soil and plant, the capacity of the soil to retain and supply water, and the cost of the system in relation to the expected returns. This decision is dealt with in more detail below.

D. Jackson

Estimation of available soil moisture

Different soils have different capacities to absorb and retain water, and an understanding of the basic principles involved can be very helpful. First of all, it is important to be familiar with a number of terms.

Saturation

After rainfall or irrigation, water fills the spaces between the particles of soil. When all the air is replaced by water, the soil is said to be saturated.

Field capacity

After a period of 24 h of free drainage following saturation, the soil will retain a certain amount of water – a volume which can be held against gravity. This is called the field capacity.

Evapotranspiration (ET)

While more water may be lost by slow drainage from the field capacity, most of the further loss is due to evaporation from the surface of the soil and transpiration from the leaves of plants. This is called evapotranspiration.

Permanent wilting point

After a period of evapotranspiration, plants may begin to wilt and, if they are unable to recover fully, the soil water content is said to have reached the permanent wilting point.

Available water

This is defined as the difference between the water content at field capacity and that at the permanent wilting point. The amount of available water in soil varies according to its texture and structure. This is clearly indicated in Table 9.7.

A soil with good structure will hold more water than a soil with poor structure. Thus, a newly planted site previously in grass or clover is likely to have better structure than one which grew annual crops for a number of years with no addition of organic manures. The available moisture in the soil of the previously grassed site will be at the high end of the normal range.

Estimation of water loss

Most meteorological stations have a 'Pan Evaporimeter – Type A', which measures daily evaporation. In summer, these values will vary from about 3 to 18 mm per day. This value can indicate the amount of irrigation required to replace that lost to evapotranspiration.

In a fully established orchard with grass between the rows, evapotranspiration can often be at parity with the figure from the evaporimeter, and this is the figure that the orchardist will aim to replace. If the ground between the trees or vines is free of grass and weeds, losses due to evapotranspiration are reduced substantially and should be considered when making irrigation decisions. For example, if 36 mm of evapotranspiration occurs in 7 days and only half of the ground is covered by vegetation (i.e. the tree or vine canopy), then $36 \times 0.5 = 18$ mm of irrigation will be needed. This same calculation can be used for trickle irrigation, even if grass exists between the trees, since the trickle does not water this grass. In this case the water-needed value is converted to volume of water per tree or vine. This is given by the formula:

Water needs per tree in litres = ET in mm × proportion of leaf cover × 10,000/trees per hectare

Thus, if evapotranspiration was 5 mm in 1 day, half the ground was covered by leaf area and there

Table 9.7. Available moisture in soils. (From Goh, 1985).

| Soil texture | [Millimetres of available water per metre depth of soil] | |
	Range	Most common
Sand	0–66	42
Sandy loam	93–126	110
Silt loam	177–186	182
Clay loam	160–177	169

were 650 trees/ha. The daily water needed would be:

$$5 \times 0.5 \times 10,000/650 = 38.5\,l \text{ per tree}$$

Another formula which is used is:

Water needs per tree in litres = 2 × diameter of tree canopy in metres × evapotranspiration in mm

Thus, if the evapotranspiration was 5 mm and trees have an average canopy diameter of 3.5 m, water use would be 2 × 3.5 × 5 = 35 l per tree per day.

Deciding on irrigation

Information about available soil water, seasonal rainfall and evapotranspiration can be used to decide when to commence irrigating or whether an irrigation system is even required. If, for example, the soil is a clay loam with poor structure over a solid clay pan at 1 m, such a soil would hold about 165 mm of available moisture (Table 9.7). In a fully established orchard, the roots would effectively fill the root volume to a depth of 1 m.

If daily evapotranspiration in midsummer averaged 8 mm, 21 days without rain would reduce soil moisture to the permanent wilting point. Even 1–2 weeks would begin to reduce growth and yield, especially if the soil moisture was not at field capacity to begin with. Such 'droughts' are not uncommon, even in the wetter horticultural districts, and irrigation would be very desirable.

However, under the same evaporation conditions, a silt loam of good structure with 2 m of penetrable soil would hold 185 × 2 = 370 mm and the plants would survive 370/8 = 46 days before permanent wilting occurred. Six to seven weeks without adequate rain might be a rare occurrence and irrigation may not be required.

In many areas, water deficit is a normal feature of most summers. In these conditions irrigating will ensure good fruit yields and is essential for successful fruit production. Using data for soil capacity, evaporation and rainfall over the summer, a water balance sheet can indicate when to start irrigating and how much water to apply. The individual orchardist could do this by obtaining an evaporimeter and rain gauge or arranging with the local meteorological station to provide daily figures. In many areas these data are published by the local newspaper.

As an example, take the water budget for a fully established orchard of the silt loam mentioned above. The water capacity is 370 mm at the beginning of the season and changes as follows:

- April (October, southern hemisphere) – total evapotranspiration = 60 mm, rainfall 50 mm; available water = 360 mm.
- May (November) – total evapotranspiration = 180 mm, rainfall 20 mm; available water = 200 mm.
- June (December) – total evapotranspiration = 240 mm, rainfall 40 mm; available water = 0 mm.

Clearly, a serious water deficit will begin to occur in June/December. If the orchardist has an irrigation policy designed to prevent the lost moisture causing substantive soil moisture deficits, then the orchardist would need to commence irrigation during this month. If sprinklers or flood irrigation are used, the first irrigation should attempt to return the soil to field capacity, and subsequent applications should bring it back to this level from no less than half the figure. With trickle irrigation, daily replacement of water loss would need to begin in early June (December).

It may seem complicated to begin water budgeting, but, in fact, once the routine of daily assessments is established, the process is quite simple. Unneeded irrigation wastes water and electricity and washes away nutrients from the soil. By budgeting water consumption, one saves money while achieving optimum yields. Overall, the effort is well worth while.

The effect of soil moisture deficit on fruit tree growth and yield varies with the particular phase of growth and development. Research has shown that shoot growth can be restricted in both its duration and extent by water stress. However, fruit growth may be largely unaffected if the water stress is alleviated during the period of fruit growth and maturation. Thus the strategy termed RDI – regulated deficit irrigation – has been adopted by some growers. Amongst the advantages proposed for RDI is the restriction of shoot growth, which impacts on pruning and other tree management tasks. Restricted shoot growth can lead to more fruit exposure to light, which may in turn have positive effects on fruit colour and maturation.

Allowing some water stress carries some risk for tree and fruit development. The demonstration that a stress response can be caused if part of the root system is exposed to drying soil conditions while the rest of the root system is adequately supplied with moisture has led to a strategy termed PRD – partial root (zone) deficit. This approach requires the tree to have a dual irrigation supply – typically one drip emitter will supply water to one side of the tree and a second emitter the other side. The response to a water stress on the 'dry side' is not sustained indefinitely while the other side of the tree is kept well supplied with moisture. As a consequence the grower will alternate which drippers will be in use and which side of the tree is moist or dry.

Types of irrigation

Government agencies and private-sector advisers are often available to supply information on irrigation systems and designs appropriate for fruit crops. Only the main factors influencing the choice of each type of irrigation will be covered, not the technical aspects.

Flood irrigation

Flood irrigation involves the diversion of water from irrigation channels to the orchard. It is useful on flat or carefully terraced fruit sites. This method wastes water in that it cannot be adjusted to meet individual tree needs, but, once established, it is effective and inexpensive to use.

Sprinkler irrigation

As discussed in Chapter 2, sprinklers will serve both to irrigate and to protect trees from heat or frost – although the rate of water delivery is quite different, depending on the purpose. They are of two main types (Fig. 9.4) – those applying water overhead (both irrigation and temperature control) and those sprinkling underneath the trees (mainly irrigation).

Over-tree sprinklers increase humidity and wash off sprays, with both effects contributing to pest management problems. For over-tree irrigation it is normal to use larger sprinkler heads, each wetting a large area of the orchard. This saves on the number of sprinklers but, under windy conditions, water loss from evaporation can be very high. Worse still, the windward side of the plantation may not be adequately irrigated.

Sprinklers which direct the spray below the canopy overcome these various problems to a

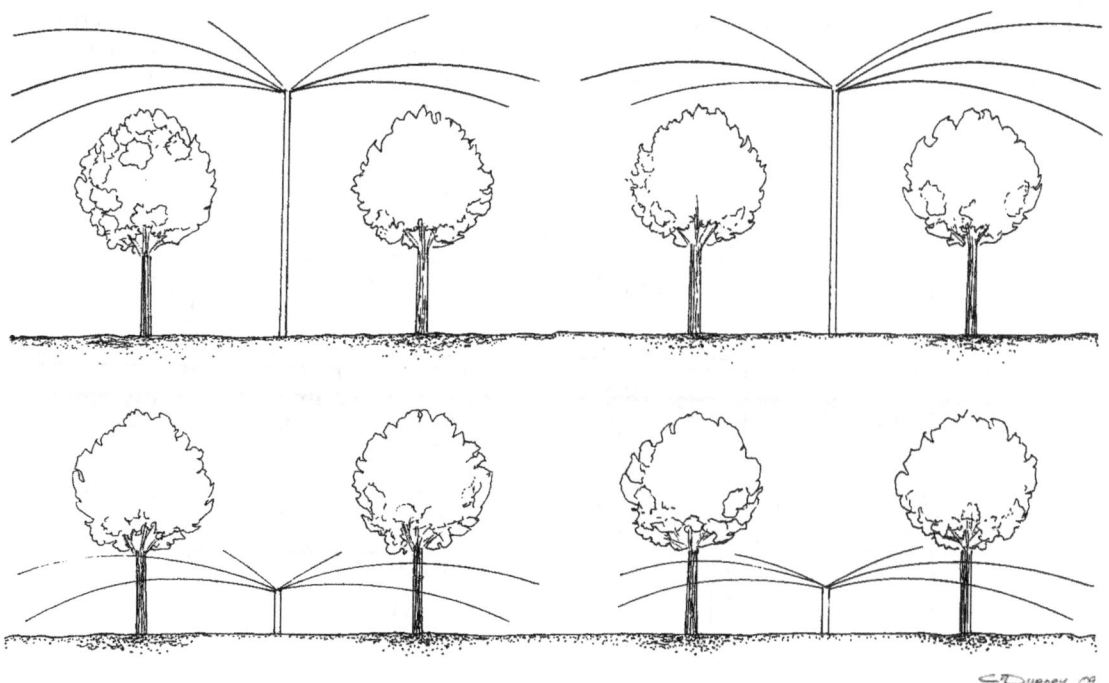

Fig. 9.4. Overhead and beneath-tree sprinklers.

considerable extent, but more sprinklers are needed and they are not intended for frost control. However, applying water in this manner during a frost event will result in a small temperature lift when the water freezes. Evaporative cooling by under-tree sprinklers on a hot day may also slightly reduce heat damage. Sprinklers are expensive to install and need a high pump capacity to operate. An additional problem is that they interfere with other orchard activities during the hours of operation. Because they can bring the whole soil area to field capacity, the interval between irrigations will be longer than for trickle irrigation.

Trickle (drip) irrigation

The use of low-pressure drip or trickle irrigation to supply water to trees has reduced the costs of installing irrigation systems and this has been partly responsible for the increased use of irrigation in fruit crops. Lines of polythene tubing carrying the water along the row dispense water in drops to individual plants through specially designed proprietary emitters. The key feature of trickle is that only the soil in the tree or vine row is wetted. This reduces the amount of water required and helps to reduce weed and grass growth in areas where water is not applied. However, the margin for error is greatly decreased; a blocked emitter can result in serious water stress conditions in a very short time.

Figure 9.5 shows how the water is distributed on different soil types. On light, easy-draining soils, lateral spread is reduced, and water available to a plant after each irrigation is limited. Water must be applied once, or even twice a day in hot weather. On the heavier silt, wider distribution occurs and the interval between irrigations can be increased. Distribution is even greater when an impervious pan occurs below the surface. In fact, trickle irrigation can make such soils much more useful in horticulture, provided that drainage problems can be overcome at other times of the year.

The rule for trickle irrigation is to apply water little and often. Daily irrigation is often recommended, even on heavy soil, with the amount being determined by the previous day's evapotranspiration.

Other emitters

A number of developments have occurred with water emitters, which offer alternatives to the traditional drippers. Some are in-built in the lateral while others soak rather than drip, thus ensuring a wider water distribution.

The problem of limited water spread on light soils with trickle irrigation can be reduced by the use of microsprinklers (microjets). These are placed into the laterals instead of emitters and act like small sprinklers covering an area of 1–3 m diameter.

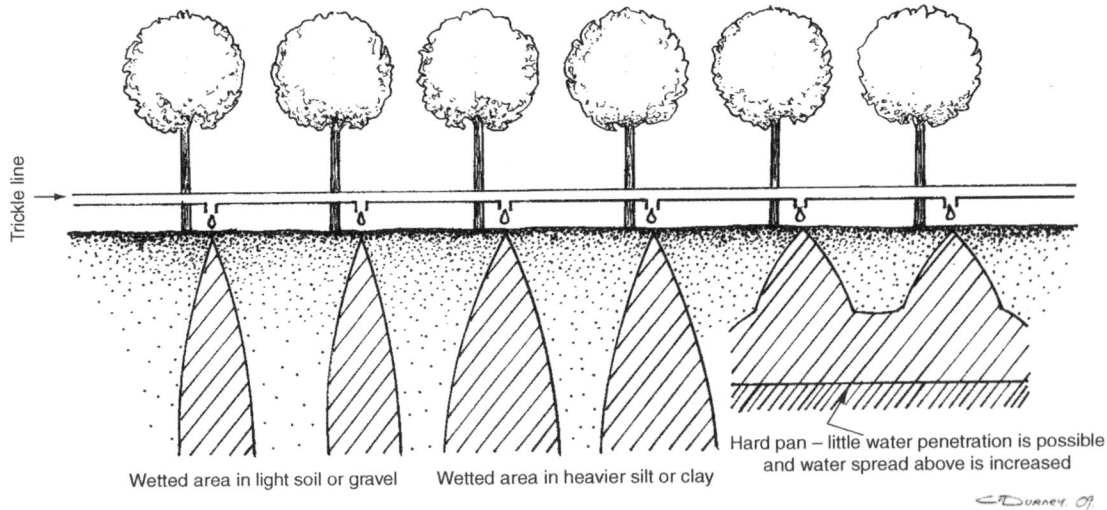

Trickle line

Wetted area in light soil or gravel   Wetted area in heavier silt or clay

Hard pan – little water penetration is possible and water spread above is increased

Fig. 9.5. Water distribution with trickle on different soils.

D. Jackson

Mini-sprinklers wet a significantly greater volume of soil and reduce the need for such frequent irrigation.

Instrumentation

There are a number of instruments which can help assess the water status of the soil or the plant. Growers are following the lead of research scientists and are beginning to choose various methods and forms of equipment that indicate moisture content in plants and soils. Tensiometers (Fig. 9.6) can be inserted into the soil profile and provide readings of soil moisture tension, although the volume of soil that is monitored by tensiometers is small relative to the volume occupied by most fruit tree roots. Neutron probes can provide readings of soil moisture presence for larger soil volumes. Skilled operators are needed to work with neutron probes. Following soil moisture readings over a period of time will help in equating the values obtained with the actual conditions. This will assist in predicting when stress conditions start to develop. Other soil moisture probes may use the

principles behind what is know as time domain reflectometry (TDR) and time domain transmission (TDT).

Changes in plant tissue water potential can be related to the development of water stress. One method for following changing plant water potential is to regularly remove (sample) leaves for testing in a 'pressure bomb'. The values for plant water potential (normally measured in kilopascals (kPa)) are related to water stress in the plant, and the grower will determine the acceptable level of stress at any time during the growing season. Other means of determining leaf water status include observing leaf temperature and observing leaf water conductance. New opportunities are developing with new instrumentation technology. Monitoring the condition of a plant and the surrounding environment on a continuing basis is something that is becoming more practicable. Consultants can assist growers in this technology and especially the management of water resources, plant water stress and irrigation technology in general.

Further Reading

Dris, R., Niskanen, R. and Jain, S.M. (eds) (2003) *Crop Management and Postharvest Handling of Horticultural Crops. Volume III: Crop Fertilization, Nutrition and Growth.* Science Publishers, Inc., Enfield, New Hampshire.

Ferguson, A.R. and Eiseman, J.A. (1983) Estimated annual removal of macronutrients in fruit and prunings from a kiwifruit orchard. *New Zealand Journal of Agricultural Research* 26, 115–117.

Ferreira, M.I. and Jones, H.G. (eds) (2000) III International Symposium on Irrigation of Horticultural Crops. *Acta Horticulturae* 537.

Goh, K.H. (1985) *An Introduction to Garden Soils, Fertilizers and Water.* Bascands Ltd, Christchurch, New Zealand.

Goodwin, I. and O'Connell, M.G. (eds) (2008) V International Symposium on Irrigation of Horticultural Crops. *Acta Horticulturae* 792.

Jones, J.B. Jr (1998) *Plant Nutrition Manual.* CRC Press, Boca Raton, Florida.

Neilsen, D., Fallahi, B., Neilsen, G. and Peryea, F. (eds) (2001) IV International Symposium on Mineral Nutrition of Deciduous Fruit Crops. *Acta Horticulturae* 564.

Nelson, L.B. (ed.) (1968) *Changing Patterns in Fertilizer Use.* Soil Science Society of America, Madison, Wisconsin.

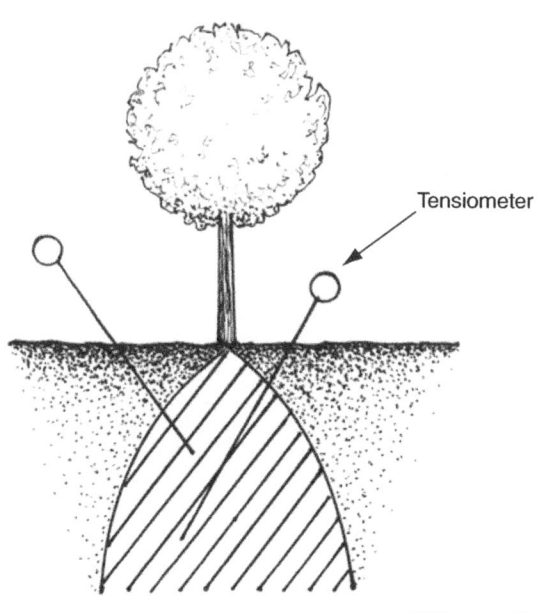

Tensiometer

Fig. 9.6. Use of tensiometers.

Peterson, A.B. and Stevens, R.G. (eds) (1994) *Tree Fruit Nutrition Shortcourse Proceedings*. Good Fruit Grower, Yakima, Washington.

Retamales, J.B. and Lobos, G.A. (2006) V International Symposium on Mineral Nutrition of Fruit Plants. *Acta Horticulturae* 721.

Snyder, R.L. (ed) (2004) IV International Symposium on Irrigation of Horticultural Crops. *Acta Horticulturae* 664.

Tagliavini, M., Toselli, M., Bertschinger, L., Brown, P., Neilsen, D. and Thalheimer, M. (eds) (2002) International Symposium on Foliar Nutrition of Perennial Fruit Plants. *Acta Horticulturae* 594

Vallone, R.C. (ed.) (2004) International Symposium on Irrigation and Water Relations in Grapevine and Fruit Trees. *Acta Horticulturae* 646.

Westwood, M.N. (1978) *Temperate Zone Pomology*. W.H. Freeman & Co, San Francisco, California.

Wild, A. (ed.) (1988) *Russell's Soil Conditions and Plant Growth*, 11th edn. John Wiley & Sons, New York/ Longmans, Harlow, UK.

D. Jackson

10 Crop Protection

DAVID PENMAN AND BRUCE CHAPMAN

Fruit crops are attacked by a complex of organisms, which are, in turn, controlled by a multiplicity of strategies. An understanding of the pest organisms and their control is therefore essential if profitable fruit production is to be maintained. In this chapter 'pest' is used in the general sense to cover those organisms (insects, mites, nematodes, fungi, bacteria, viruses and weeds) which can affect fruit quantity and quality either directly or indirectly. 'Crop protection' is defined as the integration of information on crops, pests and control methods into sensible control or management programmes.

Marketing strategy is often a key factor regulating pest control inputs. The fresh fruit market expects blemish-free fruit of good size and colour. Export grading may be even more stringent, requiring complete freedom from insects, insect damage or blemishes due to disease pathogens. In order to maximize production and meet these market expectations for top-quality fruit, the use of pest control products can seldom be avoided. Of course, pesticide residues in or on harvested fruit must fall below the specified tolerance level.

It is impossible in a book of this nature to provide complete recommendations for pest control in all fruit crops and all regions. Thus, pest control in apple in New Zealand will be used in several sections to demonstrate the principles of management and control. References given in recommended reading will enable readers to widen their understanding of crop protection.

Pest Status

Key pests are those requiring continuous control in order to grow a crop which gives maximum returns and has consumer appeal. The tolerance of damage from these pests is very low, effectively zero, for export-quality fruit. Control of these pests will often regulate the design of control programmes for other pests.

Occasional pests are those requiring only periodic control to prevent economic damage. Interestingly, some of these pests may become more of a problem when a chemical is no longer used to control a key pest. In other words, the chemical was controlling the occasional pest without the knowledge of the producer.

A third, and sometimes related, category is the *secondary-induced* pests. These are organisms which, in the absence of control inputs for the key pests, are kept under economic control by natural predators, parasites and pathogens. If the chemicals used to control the key pests kill these natural enemies, the secondary-induced pest may become a problem. Table 10.1 lists the key, occasional and secondary-induced pests of apple in New Zealand.

Damage

While pest monitoring is intended to identify increasing population levels before economic damage occurs, a damage symptom is often the first indicator of an insect or disease problem. Recognition of such symptoms is therefore an important part of effective crop protection. The linking of symptoms to a particular organism allows the selection of a suitable control strategy. There are several ways of categorizing pest damage symptoms but they are usually based on the organ attacked or type of damage. Table 10.1 illustrates the way that the key, occasional and

Table 10.1. Apple pests classified according to plant parts attacked, pest status or damage types. Similar tables can be prepared for other fruit crops.

Plants parts attacked	Pest status	Damage type
Fruit (direct damage)	*Key pests*	*Chewing*
Leafrollers	Leafrollers	Leafroller
Codling moth	Codling moth	Codling moth
San Jose scale	San Jose scale	
Scab (black spot)	Scab (black spot)	*Sucking*
Powdery mildew	Powdery mildew	Woolly apple aphid
Virus blemishes	Storage rots	Spider mites
Storage rots		Scales
Fireblight	*Occasional pests*	Mealybugs
	Woolly apple aphid	Leafhoppers
Foliage (indirect damage)	Mealybugs	Leafcurling midge
Leafrollers	Leafhoppers	
Mites	Leafcurling midge	*Leaf abnormalities*
Scales	Fireblight	Scab (black spot)
Leafcurling midge	Virus diseases	Powdery mildew
Mealybugs	Crown and root rots	Viruses
Scab (black spot)		Leafcurling midge
Powdery mildew	*Secondary-induced pests*	
Fireblight	Spider mites	*Cosmetic*
Virus mottles and distortion	Scales	Scales
		Woolly apple aphid
		Mealybugs
Branch/shoot/root		Fruit blemishes due to viruses
Scales		and fungal diseases
Woolly apple aphid		
Virus distortion		*Storage rots*
Crown and root rots		Due to various pathogens
Fireblight		

secondary-induced pests of apple are described in terms of damage. Each category will be examined in more detail.

Plant parts attacked

The most significant pests in fruit production are often those directly attacking the fruit itself. These are called *direct pests*. Downgrading, and hence price reduction, can occur from surface blemishes or the presence of young insects, even though the fruit may be internally sound. Foliage feeders, called *indirect pests*, do not directly affect fruit quality and quantity but, by reducing photosynthesis, they decrease the subsequent uptake of assimilates by the fruit and thus reduce fruit size. Some indirect pests affect tree vigour and distort shoot growth; for example, scales, woolly apple aphid and viruses. Some disease organisms have a major effect on plant growth by infecting the roots and crown of the plant. Phytophthora crown rot and various basidiomycetes root diseases are common examples.

Damage type

The feeding actions of insect and mite pests will give different symptoms, largely dependent on the mouth parts and feeding habits of the organisms.

Chewing damage is particularly common. Leaf tissue is consumed by those insects with chewing mouth parts. Chewing damage to fruit can be particularly troublesome. Even small blemishes from a brief period of surface feeding may be enough to downgrade fruit. Some insects, such as codling moth, chew into the fruit and complete their larval stages inside the apple or pear.

By contrast, many insects and mites have *piercing* and *sucking* mouthparts. Chloroplasts are damaged and leaves take on a mottled appearance. Some insects also infest branches and twigs, sucking nutrients from the phloem. During the feeding process, enzymes may be injected which induce a reaction in the plant. This is manifested by galls on branches, such as those caused by the feeding of woolly apple aphid.

D. Penman and B. Chapman

Diseases, particularly viruses, may be transmitted by sucking insects, and some, such as aphids and scales, may ingest comparatively large quantities of liquid, the surplus of which is excreted as 'honeydew'. This sugary substance is an excellent medium for the growth of sooty moulds, ruining the appearance of the fruit. The downgrading caused by pear psylla is largely explained in this way.

Biology of Pests

In order to protect fruit crops against pests, it is important to have some information about the biology and life cycle of the organisms concerned. It may then be possible to identify key points in their life cycle when they are especially vulnerable to simple but effective control methods. To devise suitable control strategies, the following aspects of the pest's life cycle need to be known.

- *Survival from one season to the next.* In general, this means the pest's survival from autumn to spring. Many fruit crops are deciduous perennials, have an annual period of growth and are more subject to damage during the spring–summer–autumn period. The growth cycle for apple is shown in (Fig. 10.1). Similar growth diagrams can be prepared for the other fruit crops and the biology of pests related to such cycles.
- *Spread after overwintering.* Some fungal diseases spread by means of airborne spores coming from the overwintering stages on dead leaves or on the tree itself. Insects and mites may crawl from ground cover into trees, while some may fly into trees from sites either within the plantation or from neighbouring areas.
- *Time of initial spread.* This is essential information in order to be able to time the use of appropriate control measures.
- *Spread during the growing season.* If the pest spreads during the growing season it is important to know how and when this spread occurs.

All of the above information will help the producer formulate a plant protection programme. Some understanding of the usefulness of these data can be obtained from studying Figs 10.2 and 10.3 and Table 10.2. The figures illustrate the life cycle of two major diseases of apple. With apple scab (black spot), primary spread is in the spring, by means of airborne spores released from overwintering perithecia found in dead leaves on the ground. Within-crop spread, called 'secondary spread', happens once primary infection becomes established. It can occur at any time suitable weather conditions arise, from early summer to harvest, by means of splash-dispersed conidia.

On the other hand, powdery mildew survives as mycelium within buds, and in spring the fungus grows from that source to infect emerging leaves. Subsequent spread to other leaves and to the fruit

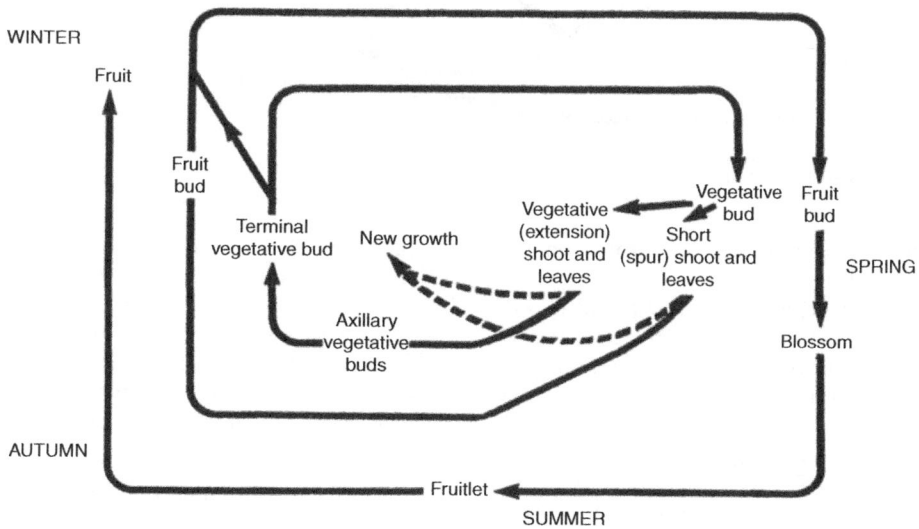

Fig. 10.1. The apple growth cycle.

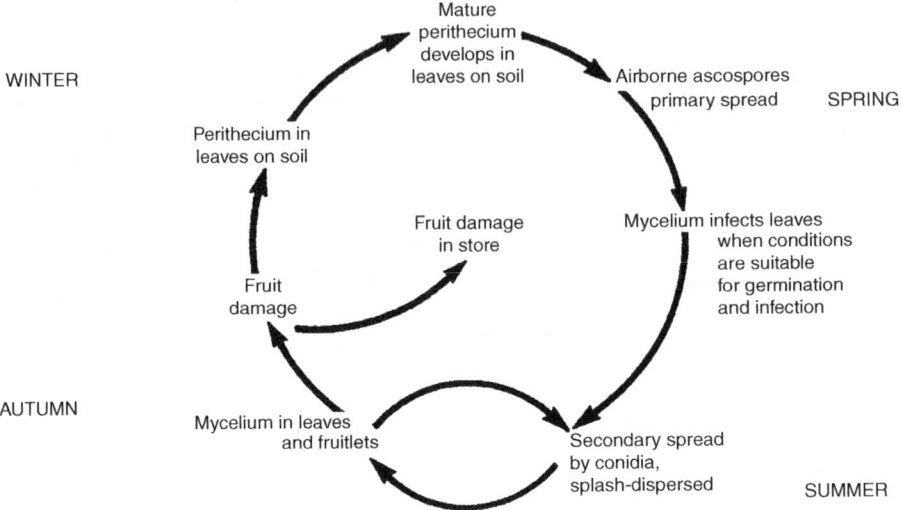

Fig. 10.2. Scab (black spot) of apples.

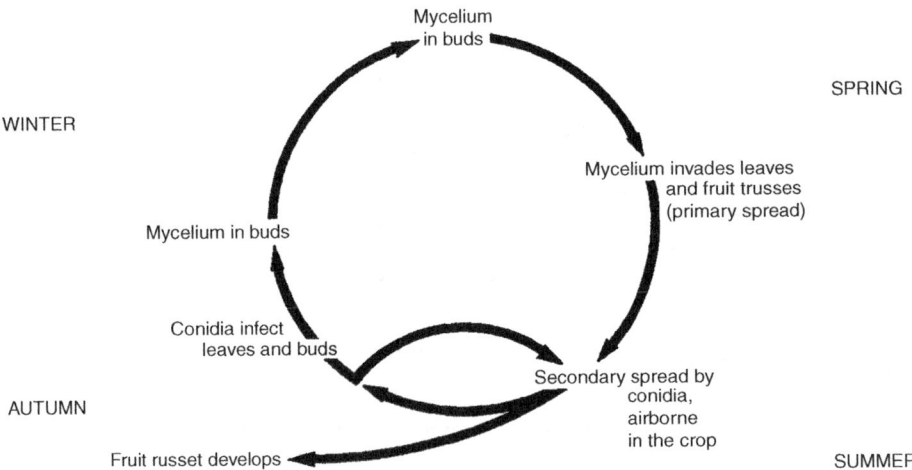

Fig. 10.3. Powdery mildew of apples.

is by means of airborne conidia formed on the initially infected leaves.

Table 10.2 also depicts these events but relates them to stages of development of the tree. It also provides similar information on the major insect pests and European red mite.

Control

With the continuing high value placed on blemish-free fruit for export and even for domestic markets, most producers feel they have little choice but to try to eliminate damage-causing pests from the plantation. In many cases this can only be achieved with pesticides or a combination of biological and chemical control agents. This is despite the fact that there is very concerted research under way in most countries aimed at devising alternative control strategies that minimize the use of chemical pesticides.

Chemical control

Chemicals are often very effective agents for pest control. The choice of chemical is linked to the target

Table 10.2. A time profile for apple tree growth and the occurrence of selected insects, mites and diseases.

System component	Time of year					
	Winter	Spring	Summer			Autumn
Tree	Wood/buds	Wood/leaf/ buds/ blossom	Wood/leaf/ buds/ fruitlets		Wood/leaf/ buds/fruit	Wood/buds/fruit
Codling moth	Larva on bark	Pupa	Adult/egg	Larva in fruit	Adult/egg	Larva in fruit
Leafroller complex	Larva	Pupa	Adult/egg		Up to four generations	Larva
European red mite	Eggs on bark		Juvenile stages/adult/ egg (five generations) on leaves		Eggs on bark	Eggs on bark
Powdery mildew	Mycelium in buds	Mycelium/ conidia	(Many generations) blossoms, leaves, buds, fruit			Mycelium in buds
Scab (black spot)	Perithecium in leaves on soil	Ascospore aerial	Mycelium/ conidia on blossoms, leaves, fruitlet, fruit (conidia splash-dispersed)			Perithecium in leaves on soil

pest (or pest complex), the type of feeding injury, the economic tolerance to damage and the life cycle of the organisms concerned. Fortunately, the chemicals presently used in fruit production are better targeted to specific pest problems (i.e. fewer broad-spectrum pesticides) and are less persistent in the orchard environment than was the case in earlier decades.

Furthermore, modern producers use pest monitoring both to minimize the number of sprays required and to apply chemicals at their most effective time. Thus, chemical usage, with the associated health and environmental risks, is declining, even in situations where chemical pesticides are the only economic alternative.

Pesticide resistance

Pesticides are sometimes applied on a calendar schedule (i.e. insurance sprays) to maximize the production of blemish-free fruit. While this is often a successful approach in the short term, long-term reliance on this practice can have major pitfalls. For example, pesticide resistance may develop over

time. In most regions this already exists for some key pests, particularly among insects and mites, and latterly in several diseases.

The development of pesticide resistance can be slowed by using a variety of chemicals, in a rotation strategy, to control a given pest. Reducing the number of applications also helps to reduce selection pressure and thus resistance build-up.

Chemicals and pest management

A preferred approach to pest control, particularly for many insects, is away from the continual protection insurance strategy and towards the use of chemicals only when necessary. With this approach, pests are managed in such a way as to avoid economic injury yet reduce the deleterious effects of chemicals. Controls are applied at the pest density which constitutes an economic, or 'action', threshold, i.e. the population at which controls should be applied to avoid economic injury. Note that the economic threshold level is always below the economic injury level, due to the delay between pesticide application and the subsequent reduction in pest populations. In order to use this strategy, careful monitoring of pest abundance is necessary. Various trapping systems are used, such as pheromone (sex attractant)-baited traps and individual leaf or fruit assessments. Using simple models to couple these results with environmental measurements, particularly temperature, can give predictions of the timing of life-cycle events. Careful timing of spray applications to weak points in life cycles can greatly improve control efficiency and reduce the likelihood of resistance development. This is also seen in apple scab forecasting systems. By monitoring surface wetness of leaves and temperature, it is possible to determine when infections have occurred and then to recommend the use of an eradicant spray. Thus, in drier springs and early summers, it may be possible to reduce the number of sprays required for scab control.

Biological control

Biological control has often been proposed as the most effective alternative to chemicals. However, its potential in fruit production varies widely, depending on the type of fruit, its market and the pest status category of the target organism. Key pests to which the crop has virtually zero damage tolerance, such as leafrollers, do not lend themselves to biological control, at least not on fruiting trees. More likely candidates for biological control are occasional and secondary-induced pests. In the case of the latter, the destruction of natural enemies by pesticides has led to their elevation to pest status. Efforts to recreate their natural controls are likely to be productive.

While the opportunities for biological controls are often limited by the marketplace demand for pest- and damage-free fruit, there are numerous examples where beneficial organisms have developed resistance to insecticides and acaracides. A notable example is the development of resistance in predatory mites to azinphos-methyl, one of the most commonly used chemicals for the control of leafrollers and codling moth.

After 8–10 years of azinphos-methyl use, the predatory mite *Typhlodromus pyri* may become sufficiently resistant to survive normal orchard applications. This predator could then re-create biological control of the European red mite. This is the basis of integrated mite control, 'integrated' meaning that biological and chemical controls are being used together. A combination of using predatory mites and the careful use of miticides selective for plant-feeding mites has significantly reduced miticide use in apple orchards in many places, thus reducing the risk of resistance to miticides in pest mites. The use of insecticide-resistant predatory mites is also widespread in many parts of the world and in many crops.

In spite of its limitations in the control of key pests within the orchard, biological control can be most useful in reducing the pest-insect populations living outside the orchard. Any reduction in these populations is likely to reduce the later migration of pests into the orchard. In addition to predators and parasites, a third component of biological control is the use of pathogens. Insects are attacked by a wide range of microorganisms and some are now being used for direct pest control. The bacterium *Bacillus thuringiensis* is receiving widespread attention for selective control of caterpillars. Insect viruses may also become useful in the future.

Cultural control

Changing the crop environment to make it less acceptable to insect and plant disease development constitutes cultural control. A wide range of activities can be considered in this light.

Establishing the crop

Most fruit crops are produced on perennial trees and vines that are propagated vegetatively; new plants are produced by obtaining runners, by rooting cuttings and by various budding and grafting operations. All these plant materials used in propagation must be disease free; virus diseases being the principal concern. Where clonal rootstocks are involved, it is essential that they come from a virus-free source. Once plants become infected with viruses, they remain infected. Furthermore, the infected trees may spread the virus throughout the planting.

Thus, it is essential to ensure that only high-health plants are obtained to establish new orchards. For example, this is vital to ensure stone fruit orchards remain free of plum pox virus. There are often private or government schemes to ensure that growers and nurseries have access to disease-free propagation materials. Most schemes begin with high-health material selected or produced by research stations. These plants are then multiplied under supervision and the progeny regularly tested to ensure freedom from known pathogens. This testing is known as 'indexing' and involves the use of reliable, accurate methods to ensure freedom from diseases up to the time of sale. Such schemes imply that there is continuous production and testing of nucleus plants and their propagules. Often the plants arising from such schemes are known as 'pathogen-tested' (PT), 'free of known viruses' (FKV) or simply as 'high health'.

Orchard sanitation

This approach to reducing pest problems can involve a wide range of practices. Examples include the removal of weeds before they produce seed; pruning away diseased limbs and removing dormant prunings, which may harbour pests and diseases; and removing cover for mice and voles. Harvesting and removing diseased or infested fruit unfit for market is an important practice which is often neglected. With powdery mildew (Fig. 10.3), pruning out infected apple buds can reduce the amount of overwintering mycelium. Similarly, mummified fruit remaining overwinter on stone fruit trees are a source of brown rot spores and should be removed. The rotting of leaves infected with black spot can be accelerated with the use of a pre-leaf-fall urea spray, a practice that greatly reduces the number of overwintering perithecia. An eradicant strategy is to spray the leaves with eradicant fungicide prior to leaf fall and so destroy the black spot fungus in the leaves.

With fireblight, infected parts should be pruned out and infected alternative hosts in the near vicinity removed. General hygiene is also important. Maintaining clean packing facilities can reduce overwintering sites for the codling moth and leaf-rollers, and removal of weeds around the crop and packing areas can reduce alternative hosts for pests and diseases.

Soil drainage

Soil-borne fungal pathogens are often more severe in areas where waterlogged or near-waterlogged conditions may occur. Water can assist entry as well as aid movement of pathogens through the soil. With fruit crops, good drainage is essential.

Host plant resistance

The use of natural resistance as the first line of defence against insects and diseases has much to offer. Unfortunately, however, compared to the situation with grain and field crops, there are relatively few good examples of pest and disease resistance in fruit crop horticulture. This probably relates to the length of time required to combine these traits with high fruit quality, the trait that continues to be of greatest importance to most fruit breeders.

Still, there are some important examples of success. Fruit crop rootstocks are available with resistance to woolly apple aphid, grape phylloxera, and phytophthora root and crown rot. There are stone fruit rootstocks with resistance to nematodes. Furthermore, virtually all fruit crops have cultivars with resistance, or some degree of tolerance, to one or more important pests or diseases. There are now apple cultivars resistant to apple scab and pear cultivars resistant to fireblight. Considerable success has been achieved in developing strawberry cultivars with resistance to several important diseases. Other cultivars are sufficiently tolerant of important pests to allow the production of blemish-free fruit using only biological control programmes for that pest or disease.

Integrated Pest Management

Integrated pest management (IPM) involves a combination of all of the above strategies and considerations to achieve economic levels of pests and diseases without undue environmental or social costs. The philosophy embodies a move away from reliance on a single tactic. It is a steadily evolving concept and is variously defined by researchers working in different countries. The further reading list gives examples of this diversity.

In relation to fruit production, full implementation of the IPM philosophy is often constrained by the economic need to produce top-quality produce for many markets. Within that limitation, IPM is based on sound knowledge of crop growth, the processes of disease infection and the development of pest populations. The development of time profiles is one such way of viewing the whole system (Table 10.2). This limited table for apples is an example of the type of time profile which can be constructed for other crops. If this is linked to the crop growth cycle, weak points in the life cycles of pests can be identified.

Control aimed at these points is likely to be more efficient than calendar applications of pesticides.

If this knowledge is used in conjunction with monitoring pest abundance by various methods and measuring environmental factors, decision making in relation to pest control will be even more effective and responsible. Computer-based models are now being used to aid in the prediction of pest outbreaks.

Monitoring, prediction and selection of a range of control strategies is likely to change approaches to future pest control significantly. IPM is an approach requiring considerably more expertise and sophistication than sole reliance on calendar spray applications. However, the benefits in reduced costs, reduced environmental problems and conservation of increasingly scarce pesticides makes the adoption of IPM imperative.

We have outlined here the processes involved in developing an approach to pest control. More specific and detailed information relevant to the management of insects and diseases is available in most regions from government or private advisory services.

Weed Control in Fruit Production

Weeds are unwanted vegetation, usually annual or perennial herbs and grasses that interfere with farming operations or provide unwanted competition with the crop plant. Suckers arising from the tree crown or rootstock also fall into the category of unwanted orchard vegetation and may prove particularly difficult to control.

Controlling weeds is one of the most important tasks facing the fruit producer. It can be relatively simple and very rewarding when accomplished with the right methods applied at the right time. If done incorrectly or at the wrong time, loss of production and considerable heartbreak can ensue.

Weeds in an orchard or plantation have the following undesirable effects.

- They compete with the crop for water and nutrients, and possibly even for light.
- They can make movement on foot or by machines difficult and even dangerous.
- They can harbour insects, mites, diseases and rodents.
- They may cool the orchard environment by reducing heat absorption during the day and interfering with air drainage.
- They make a property appear neglected and reflect on the competence of the grower.

However, some positive effects of weeds should also be acknowledged.

- They can add organic matter and improve soil structure.
- They can slow vegetative growth and promote the hardening off of trees in the autumn.
- They may have a place in biological insect and disease control strategies by harbouring beneficial organisms.

The grower will need to balance the advantages and disadvantages for the particular circumstances.

Before discussing specific approaches in weed control, it is necessary to introduce a few terms in weed biology and to provide a brief introduction to herbicides and factors influencing their effectiveness.

Perennial weeds

Unlike annual weeds that arise each year from seed, perennial weeds are those that survive the winter and have some underground structure that sends up new shoots even if the top is killed. There are several mechanisms by which this is accomplished. *Rhizomes* are underground stems on which axillary buds produce new stems and roots, which give rise to new plants. Examples are: yarrow (*Achillea millefolium*), twitch (couch) (*Elytrigia repens*) and convolvulus (*Calystegia silvatica*). *Stolons* are stems which are mostly above the ground but which can root at nodes. Creeping buttercup (*Ranunculus repens*) is a good example. *Regenerating roots* can regenerate new stems. Examples include dock (*Rumex obtusifolius*), Canada (Californian) thistle (*Cirsium arvense*) and dandelion (*Taraxacum officinale*). *Bulbils* are the regenerating organ for oxalis (*Oxalis latifolia*) and underground *nutlets* explain the persistence of yellow nutsedge (*Cyperus esculentus*).

Herbicides

The herbicides used in fruit production can be divided into two groups: post-emergence and pre-emergence. A full list of herbicides is not appropriate for a book like this, since new products will soon make such a list out of date. In addition, local regulations may prohibit the use of some products which are accepted elsewhere. The following classification may, however, be helpful.

D. Penman and B. Chapman

1. *Post-emergence*: kill existing weeds by contact or by translocation to the roots.
   - Not appreciably translocated, therefore of little value for perennials unless repeated applications are made. Examples: paraquat, diquat.
   - Translocated and can be effective against perennial weeds. Examples: (i) general, most weeds killed (e.g. glyphosate, amitrole); (ii) specific, control some but not all (e.g. asulam – kills docks, clopyralid – kills daisies and legumes, fluazifop-butyl – kills grasses).

2. *Pre-emergence*: kill germinating seeds, little or no effect on above-ground vegetation or perennial weeds, often applied with post-emergence herbicides. Some examples are simazine, terbacil, terbumeton/terbuthylazine and oxyfluorfen.

Quantities

Chemicals are usually sold with instructions as to how much to apply per hectare and it is sometimes a little difficult to calculate the amount needed per tank. The best way is for the grower to make a trial run with a full tank of water and determine the area covered. The amount needed per tank can then be calculated. This procedure may still be adopted, even if using small quantities, such as in a knapsack sprayer.

Weather

The pre-emergence herbicides that kill germinating seeds are best applied to moist ground or just before a light rain. Materials applied to existing weeds are most effective when the spray does not dry too quickly, as on still, humid days or in the morning or late evening. Heavy rain after spraying will reduce the effectiveness of glyphosate and amitrole; paraquat/diquat needs about 20 min before rain for its action to be achieved. Paraquat/diquat needs light for its effect, thus, if weeds are covered with light-excluding material after spraying, they will not be affected.

Some Vegetation Management Practices

Weed control before planting

Land which has not been used for fruit growing may be infested with a range of perennial weeds, especially if it has previously been in pasture. Such weeds should be controlled before planting; if not, subsequent control will be very difficult and such pre-planting control is especially important in the case of berry fruit plantings.

Land in pasture

One full season should be devoted to preparing the land for planting. There are two main ways of eliminating perennial weeds. The first is to plough in winter and harrow regularly as weeds appear during the following season. The underground portions will be brought to the surface, where they wilt and die. The parts which remain under the soil may send out new shoots and roots, but these will be exposed by later cultivations. Cultivation should cease a month before the first winter frost, and any perennial weeds which appear thereafter should be spot-sprayed with glyphosate.

The second method is to spray in the spring, after the first flush of growth, with a herbicide such as glyphosate or amitrole. This will kill all the vegetation. The grower should wait for 1–2 months to observe regrowth, which should be spot-sprayed with whichever chemical was not used previously: amitrole if glyphosate was used, and vice versa. Annual weeds may be left. Two weeks later the land is ploughed and the soil prepared for tree planting the following autumn or winter. Additional grubbing, harrowing or rotary hoeing may be required in the autumn and a little spot spraying may continue, if required, before winter. Amitrole should not be used within 6 weeks of planting; glyphosate has no such waiting period.

Land that has been used for annual crops

This land would normally be free of perennial weeds. However, if a heavy infestation of perennial weeds is present, the approach should be similar to that used for land out of pasture. If no weeds are present, then normal land preparation in autumn will be adequate. If perennial weeds are abundant, the land should be allowed to lie undisturbed from late summer. If it is dry, water may be applied to stimulate weed growth. After 10–20 cm of growth, amitrole or glyphosate is sprayed over the block and regrowth is spot-sprayed before winter frosts. One week after the second spraying, land preparation may begin. Amitrole and glyphosate will not have killed the plant after 1 week, but these materials are absorbed into the plant and continue to act on all parts, even if the top is separated from the root.

Weed control subsequent to planting

Some fruit crop species are established through black polythene strips as a way to control weeds in the young planting. For example, polythene is used with currants, strawberries and grapevines, and those shelter belts where cuttings are planted, such as willow and poplar. Black polythene is not now recommended for blueberries as the soil tends to compact beneath the mulch. Blueberries cannot tolerate soil compaction. Polythene should not be used if full control of weeds has not been achieved, since they will be impossible to control when the underground parts are covered by polythene. These strips are established as follows. The land is rotary-hoed and the polythene laid with special equipment, which is often hired or borrowed by the grower. The polythene will normally be 75 cm wide and 0.35 or 0.5 mm thick. The thicker polythene will persist for several years. If cuttings are used, they may be pushed directly through the polythene; for plants, a cross slit is first cut with a knife. Growers not using polythene must find another way to control weeds in young plantings. For the first year it may be necessary to cultivate and hand hoe these plantings. In plants established for at least a year, a combination of contact and pre-emergence herbicides is usually recommended. Contact herbicides, whether translocated to the roots or not, must be used with great caution around young plants. Even pre-emergence herbicides, those chemicals that prevent the growth of germinating seedlings, must be used carefully. This is especially true on sandy soils, where they may leach into the root zone and cause damage to the young tree or vine.

The main orchard floor management practices used in fruit growing are:

- Weed control by chemicals in the rows and by grass between the rows.
- Complete weed control in and between the rows, either with chemicals or by mechanical means.
- A third method, gaining acceptance on a wider range of crops as herbicide usage declines, is using a deep mulch around the plant to control weeds. This is a long-established practice with blueberries on mineral soils, where sawdust or bark is used.

Controlling root suckers

There are several approaches to dealing with this problem. Paraquat can be used on mature trees to burn off very young, tender root suckers. Success has also been achieved with glyphosate in a number of crops. A more lasting treatment is to apply naphthalene acetic acid (NAA), as described in Chapter 8.

Further Reading

Alford, D.V. (2007) *Pests of Fruit Crops: a Colour Handbook*. Manson, London.

Anon. (2010) UC IPM online Statewide Integrated Pest Management Program. www.ipm.ucdavis.edu/PDF/PMG (accessed 9th June 2010).

Cooke, T., Persely, D. and House, S. (2009) *Diseases of Fruit Crops in Australia*. CSIRO Publishing, Collingwood, Victoria, Australia.

DeJong, T.M. (ed.) (2002) VI International Symposium on Computer Modelling in Fruit Research and Orchard Management. *Acta Horticulturae* 584.

Dent, D. (2000) *Insect Pest Management*. CAB International, Wallingford, UK.

Flint, M.L. and Gouveia, P. (2001) *IMP in Practice: Principles and Methods of Integrated Pest Management*. University of California, Publication No. 3418.

Jones, A.L. and Aldwinkle, H.S. (1990) *Compendium of Apple and Pear Diseases*. American Phytopathological Society, St Paul, Minnesota.

Libek, A., Kaufmane, E. and Sasnauskas, A. (eds) (2005) *Proceedings of the International Scientific Conference: Environmentally Friendly Fruit Growing*, Polli, Estonia, 7–9 September, 2005. Tartu University Press, Tartu, Estonia.

Lind, K. (2003) *Organic Fruit Growing*. CAB International, Wallingford, UK.

MacHardy, W.E. (1996) *Apple Scab*. American Phytopathological Society, St Paul, Minnesota.

Martin, N.A. and Harrington, K.C. (eds) (2005) *Pesticide Resistance: Prevention and Management*. New Zealand Plant Protection Society (www.nzpps.org).

Metcalf, R.L. and Luckman, W.H. (1994) *Introduction to Insect Pest Management*, 3rd edn. Wiley-Interscience, New York.

Müller, W., Polesny, F., Verheyden, C. and Webster A.D. (eds) International Conference on Integrated Fruit Production. *Acta Horticulturae* 525.

Ogawa, J.M. (1995) *Compendium of Stonefruit Diseases*. American Phytopathological Society, St Paul, Minnesota.

Paul, R.E. and Armstrong, J.W. (1994) *Insect Pests and Fresh Horticultural Products*. CAB International, Wallingford, UK

Pedigo, L.P. and Rice, M.E. (2009) *Entomology and Pest Management*. Prentice Hall, Upper Saddle River, New Jersey.

Samietz, J. (ed.) (2008) VIII International Symposium on Modelling in Fruit Research and Orchard Management. *Acta Horticulturae* 803.

D. Penman and B. Chapman

Suckling, D.M. and Butcher, M.R. (2001) *Plant Protection Challenges in Plant Protection*. New Zealand Plant Protection Society (www.nzpps.org).

Ware, G.W. (1994) *The Pesticide Book*, 4th edn. Thomson Publications, Fresno, California.

Whitson, T.D. (ed.) (1996) *Weeds of the West*. The Western Society of Weed Science, Newark, California.

Williams, K.M. (1991) *New Directions in Tree Fruit Pest Management*. Good Fruit Grower Publications, Yakima, Washington.

11 Propagation of Fruit Plants

Michael Thomas and David Jackson

The propagation of trees, bushes and vines is a task for specialists and is usually performed by nursery operators. Nevertheless, there are occasions when growers might need or wish to propagate their own material. This chapter discusses some important principles in plant propagation and describes some commonly used techniques. It also describes the techniques of top-working and frame-working, used to convert a tree from one cultivar to another.

It is very desirable to propagate only from true-to-type trees regularly tested for known viruses. Freedom from major viruses and other systemic disease is especially important for long-term crops, but it is also important for short-cycle plants like strawberry. Virus, bacterial and systemic fungal diseases can all be spread by vegetative propagation. In the case of some viruses, this is the predominant means of transmission.

Propagation by Cuttings

Many fruit and nut species can be propagated by cuttings (Table 11.1). In some cases, satisfactory results can be obtained from more than one kind of cutting, depending on the time of year.

Hardwood cuttings

Hardwood cuttings are widely used in deciduous fruit plant propagation. Currants, gooseberries and grapes are all relatively easy to propagate by this method. Hardwood cuttings of shelter trees such as poplar and willow can also be rooted with ease. Cuttings are prepared during the dormant period, preferably using stem sections formed the previous season.

Cuttings from plants which are easy to propagate are usually made from autumn to early winter. They are immediately planted either in a nursery – a protected, well-drained, well-watered piece of ground, where they remain for a year – or directly into the orchard, as sometimes happens with grapes and currants. Land where the soil is stony, heavy or poorly structured, especially where waterlogging may occur in winter, should be avoided. Although it is preferable to plant cuttings as early in winter as possible, planting often continues right through until spring. The cuttings must not be allowed to dry out, especially in early spring, and irrigation may be essential. Hardwood cuttings are commonly 25–40 cm long. They are usually inserted to a depth equal to half their length, but in situations where drying out is possible, two-thirds to three-quarters of their length should be under the soil (Fig. 11.1). If suckers are undesirable, it may be worthwhile to disbud the cuttings, leaving only two buds at the top.

The soil should be thoroughly prepared. If it has been deeply tilled, the cutting can often be inserted without using a spade. Some modern residual herbicides may then be sprayed over the plants to kill germinating weed seeds.

Alternatively, excellent results can be obtained by inserting cuttings through black polythene spread over the ground. Several techniques can be used to improve rooting of hardwood cuttings from plants which are more difficult to propagate (Table 11.1). Hormone application is often beneficial. For example, cuttings can be dipped in 'Seradix', a powder formulation of the auxinic plant hormone indolyl-3-butyric acid. This product is available in three strengths to be used according to the type of cutting. Liquid formulations, containing the same hormone in dilute alcohol, can also be used. 'Double wounding' involves the

Table 11.1. Fruit and nut species propagated from cuttings.

Fruit	Type of cutting[a]	Time of year	Treatment	Hormone[b]	Comments
Apples	Hardwood – thick basal shoots from hard-pruned stock	Early spring (best) or autumn	Double wound	Seradix 3	Need bottom heat. Seldom used, however – stooling and budding is the commercial method
Blackberries, boysenberries and other brambles	Softwood – shoot tips	Spring		Seradix 2	Mist or polytunnel; also grown from suckers and tip layers
Blueberries	Hardwood – thick shoots. Softwood – four leaves and at least 7 cm long	Late winter–spring	Double wound	Seradix 3	
Citrus	Semi-hardwood – shoot tips	Spring		Seradix 2	Mostly for Meyer lemons; other cultivars budded
Currants	Hardwood – use thick basal shoots, approximately 30 cm long	Autumn		Seradix 3	Can be inserted into open ground. Can also be grown by mound layering
Feijoas	Semi-hardwood – three-node cuttings from low on the bush	Early or late winter	Remove soft tips	Seradix 3	Mist or polytunnel with bottom heat
Gooseberries	Hardwood – use thick basal shoots	Autumn		Can use Seradix 3	Can be inserted into open ground or heated bin
Grapes	Softwood – thin stems, near tips	Late spring, early winter	Remove tips	Seradix 1	Mist or polytunnel. Apply captan sprays on leafy cuttings
	Hardwood – thick shoot	Autumn, winter			Can be inserted into a bin or into open ground
Kiwifruit	Semi-hardwood – use current season's wood	Summer	Remove tips	Seradix 3	Preferably under mist; bottom heat preferred
Mulberries	Hardwood – thick shoots from current season's wood	Autumn	Double wound	Seradix 3	Bottom heat desirable
Olives	Hardwood – thick shoots	Autumn	Double wound	Seradix 3	Need bottom heat
	Semi-hardwood – 10–15 cm tip cuttings	Summer	Double wound	Seradix 3	Need mist and bottom heat
Peaches and nectarines	Hardwood – 12–15 cm tip cuttings	Spring	Double wound	Seradix 2	Need bottom heat using a bin. Seldom used, budding preferred

Continued

Table 11.1. Continued.

Fruit	Type of cutting[a]	Time of year	Treatment	Hormone[b]	Comments
Pears	Hardwood – thick basal shoots	Autumn	Double wound	Seradix 3	Need bottom heat using a bin. Seldom used, grafting and budding preferred
Plum rootstocks	Hardwood – thick basal shoots	Autumn	Remove all but top two buds on rootstocks	Seradix 3	Can set cuttings of myrobalan directly into open ground or use heated bin, at 15°C, 3–4 weeks
Quince rootstock for pears	Hardwood – thick basal shoots	Autumn	Remove all but top two buds on rootstock	Seradix 3	Need bottom heat using a bin at 15–21°C for 2–3 weeks
Raspberries	Suckers from non-flowering certified stock.	Winter, spring	Cut into 5 cm lengths		Bottom heat beneficial
	Softwood shoots from root cuttings	Early spring	Shoot tips	Seradix 1	Preferably under mist
Shelter trees					
Poplars	Hardwood – large, thick basal shoots	Autumn to late winter			Can be set directly into open ground
Willows	Hardwood – large, thick basal shoots	Autumn to late winter			Can be set directly into open ground

[a]'Hardwood' refers to dormant deciduous cuttings taken at or just after leaf fall.
[b]Hormones – the commercial preparation 'Seradix' is a powder formulation, which can be replaced by a made-up solution of indolyl-3-butyric acid (IBA) in a 50% solution of water and ethanol. Concentrations of 750, 2500 and 5000 ppm IBA are equivalent to 'Seradix' 1, 2 and 3, respectively. Commercial liquid formulations of IBA are also available. Cuttings should be dipped in the solutions for 5 s.

removal of two thin slivers of wood from the basal end of a cutting. It is claimed that this improves uptake of the hormone.

Another method which improves the rooting of hardwood cuttings of more difficult to root species such as cherry and quince is to raise the temperature of the propagating medium, using bottom heat. Heating cables are placed at the base of a frame (Fig. 11.2). The purpose of the bottom heat is to promote the initiation of roots. Maintaining the soil temperature at 15–21°C for 2–4 weeks is often adequate. When roots begin to emerge, the cuttings are potted-up into potting medium. Non-uniform heating or an excessively long period of warm storage can lead to failure due to carbohydrate exhaustion.

Selection of hardwood cuttings

The optimum times for collecting hardwood cuttings are late autumn to early winter. The best material for rooting comes from shoots which grew the previous season and should be as close to the base of the plant as possible. Thick cuttings survive better than thin cuttings, although thin shoots collected from severely pruned stock plants will give the most rapid root formation. Thin cuttings will survive so long as they do not dehydrate and are not allowed to get too wet. The main danger is in spring, as the new shoots are beginning to emerge. At this stage the roots are small and may not be able to absorb sufficient moisture to replace that lost by the leaves. If thick cuttings are not available, deep planting or pre-rooting in a hot bed will help thinner cuttings to survive harsh conditions.

The wood at the tip of a shoot should not be used if it is not yet sufficiently lignified. Lignification is the formation of secondary cell walls and is characterized by shoots changing from green to brown. A non-lignified shoot is soft and easily bent; it has fewer reserves and may die in winter frost.

M. Thomas and D. Jackson

For most cultivars, 20–30 cm is considered an adequate length, but shorter cuttings may be used if necessary. For material such as blueberries, which tend to be thinner than other plants, shorter cuttings of 15–20 cm are suitable.

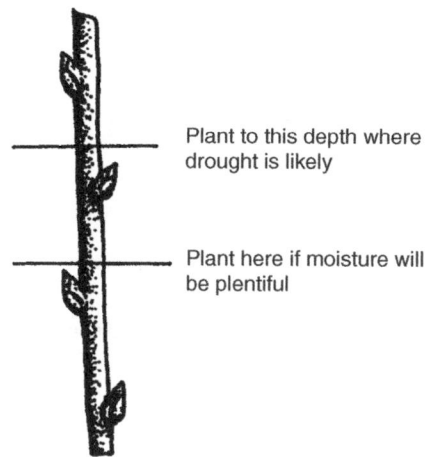

Fig. 11.1. Hardwood cuttings.

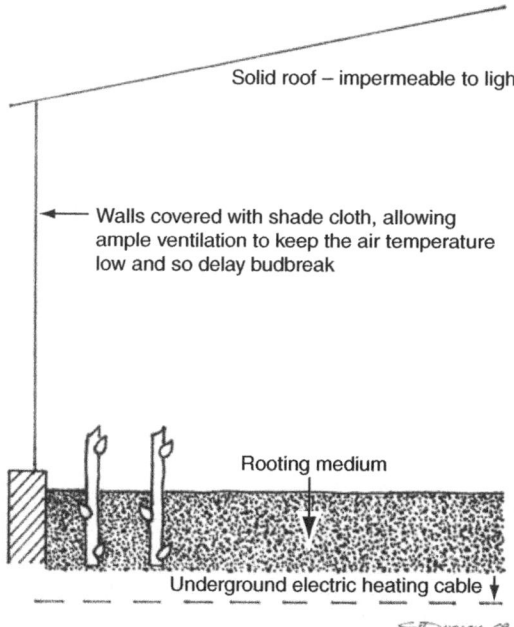

(Note: rooting media may be of several types but a 50–50 mixture of peat and sand would be suitable.)

Fig. 11.2. Outdoor bed (bin) for rooting hardwood cuttings.

Softwood and semi-hardwood cuttings

Softwood cuttings are those prepared from succulent new growth, usually in the spring (Fig. 11.3). Semi-hardwood cuttings are made from partially mature wood, usually in summer. Taking cuttings in the growing season requires additional precautions since moisture loss by leaves cannot readily be replaced until the roots form. To overcome this problem, high humidity is maintained over the bed and bottom heat is commonly employed. Two common ways of maintaining humidity over the tops are shown in (Figs 11.4 and 11.5).

Softwood and semi-hardwood cuttings can be propagated in polythene tents called 'polytunnels' or mist units (Figs 11.4 and 11.5). This method is often used if a rapid multiplication of plants is required. For example, if a new grape cultivar is introduced and only limited planting material is available to the viticulturist, the dormant cuttings can be made one or two nodes long, rooted

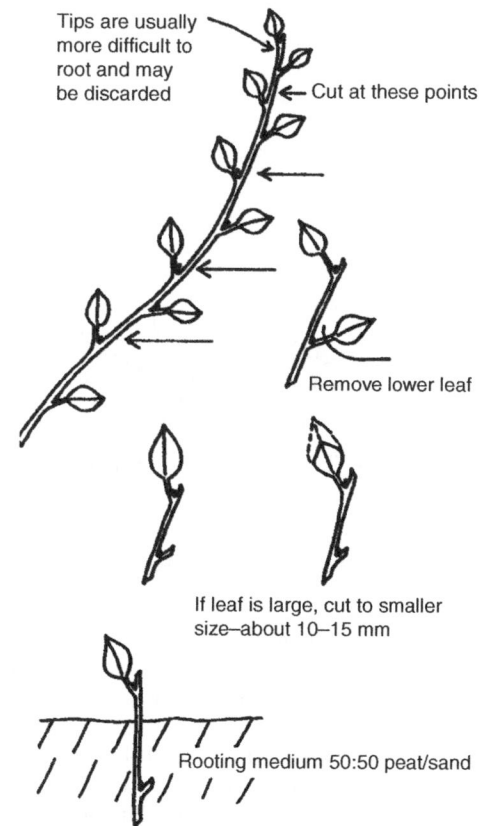

Fig. 11.3. Preparation of softwood cuttings.

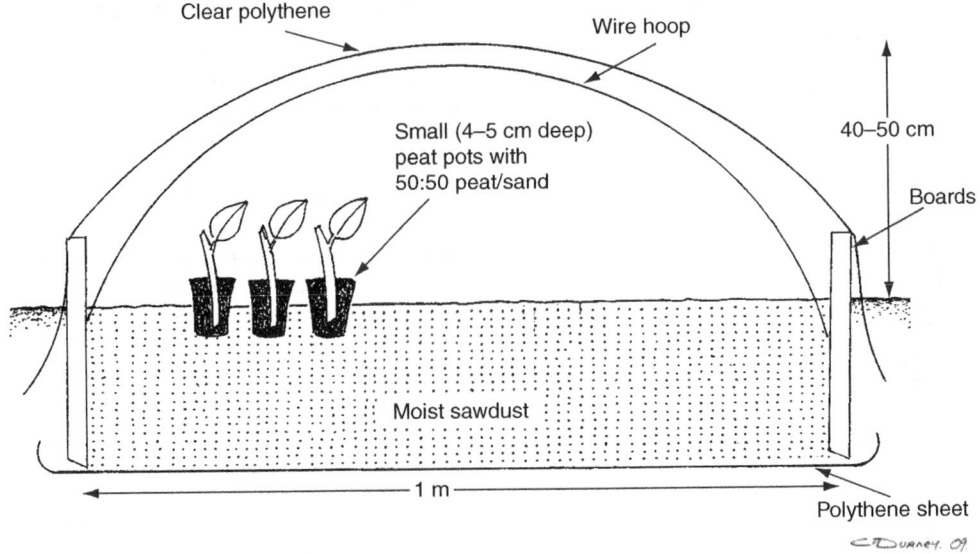

Fig. 11.4. Outdoor structure for softwood and semi-hardwood cuttings.

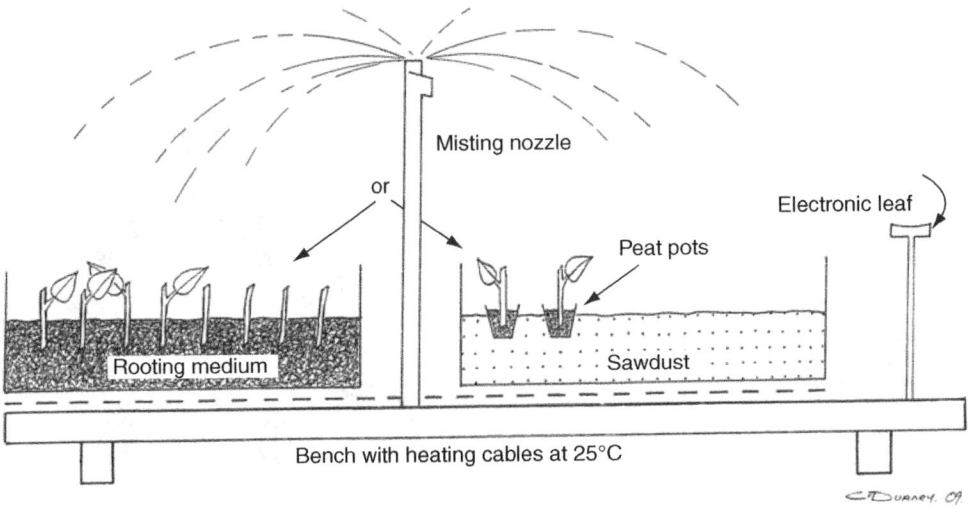

Fig. 11.5. Mist unit. Bench with heating cables at 25°C.

on a hot bed and kept growing over summer in individual pots in a greenhouse. As shoots grow, they can be used to produce new softwood cuttings with five nodes.

The structure shown in Fig. 11.4 is cheap to construct and bottom heat is not required for easy-to-propagate species (Table 11.1). The bed should be in a warm location but shielded from direct sunlight by a structure providing about 50% light transmission, for example a shade house. Beds may be 1–10 m long.

Fungicides are important. As cuttings are made, they are dropped into a bucket containing a fungicide solution (0.1% benomyl is commonly used) before inserting into small peat pots. The pots are placed in the beds as shown in Fig. 11.4 and the whole bed is sprayed with 0.1% (1g/1) captan solution prior to sealing the polythene.

M. Thomas and D. Jackson

The polythene is left undisturbed for about 2 weeks, after which one side is lifted to observe the development of roots. If roots are present, the peat pots are removed and the cuttings replanted in a larger pot, 10–15 cm in diameter. They remain in the shade house and are kept moist. If rooting has not occurred, the polythene is replaced.

Semi-hardwood cuttings may be rooted in a mist unit, normally housed in a greenhouse maintained at or near 25°C. The electronic 'leaf' shown in Fig. 11.5 will switch on the misting unit when its surface dries out. Thus, the foliage is regularly covered by a fine film of moisture, preventing dehydration. When rooted, the plants are potted or re-potted as described above.

Propagation by seeds

Since the fruiting characteristics of plants produced from seeds are nearly always inferior to the parent plant, vegetative propagation is used in almost all instances. However, seedlings are sometimes used for rootstocks and the following notes are intended as a guide to their handling.

Peaches, nectarines, apricots, cherries, loquats, walnuts, pecans, hazelnuts, chestnuts, avocados, citrus and plums

With these species, seeds can be planted directly into the nursery where the plants are to be budded or grafted the following summer. Alternatively, the seeds can be placed outside under sawdust until germination begins (Fig. 11.6). They are then planted in the nursery. This avoids sowing seeds which may fail to germinate and consequently leave gaps in the nursery row. Seedlings should be placed in the nursery as soon as possible after germination.

For peaches and nectarines, seeds from peaches used in the canning industry are often used. Seeds generally do not transmit virus diseases to their offspring. However, prunus necrotic ring spot of peaches and nectarines, and prune dwarf and plum pox virus of plums are transmitted in seeds. Therefore, when these diseases are known to occur it is important to obtain seed only from certified high-health trees.

In the nursery, seedlings are spaced 15–20 cm apart in rows 60–80 cm apart. In practice, the distance is usually determined by the machinery used to cultivate between the rows. This may involve a hand-operated rotary hoe (rotary tiller).

Apples, pears (European and Asian), quinces, medlars, kiwifruit, tamarillos, passionfruit and feijoas

The seeds of these plants are generally small and need more careful handling. After collecting, the seed usually needs 'stratification'. This refers to a period of moist chilling given as a treatment to dormant seeds, without which germination is poor and very irregular. With the larger seeds described previously, stratification is achieved by planting outdoors. Smaller seeds are placed in a mixture of equal parts moist peat and sand, sealed in a plastic bag and stored in a refrigerator for 2 months. The seeds are then planted in seed trays in a glasshouse maintained at approximately 15°C. When the seedlings are about 15 cm high, the trays are moved to a shade house, where the plants are hardened off. They will then be planted in the nursery in early spring.

Use of Stool Beds

Stool beds or mound layering is used to propagate rootstocks of apples, pear and sweet cherry and plants of hazelnut cultivars. The plants are placed

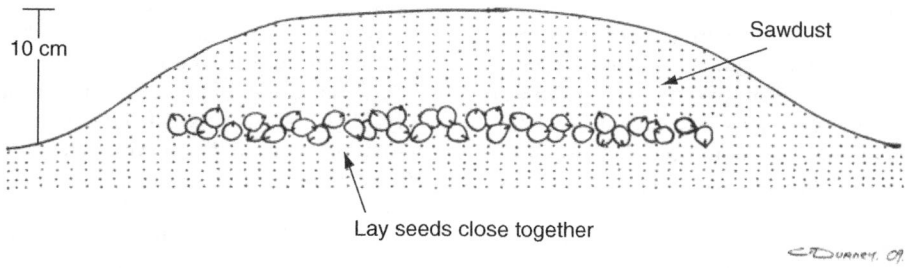

Fig. 11.6. Sawdust seed germination bed.

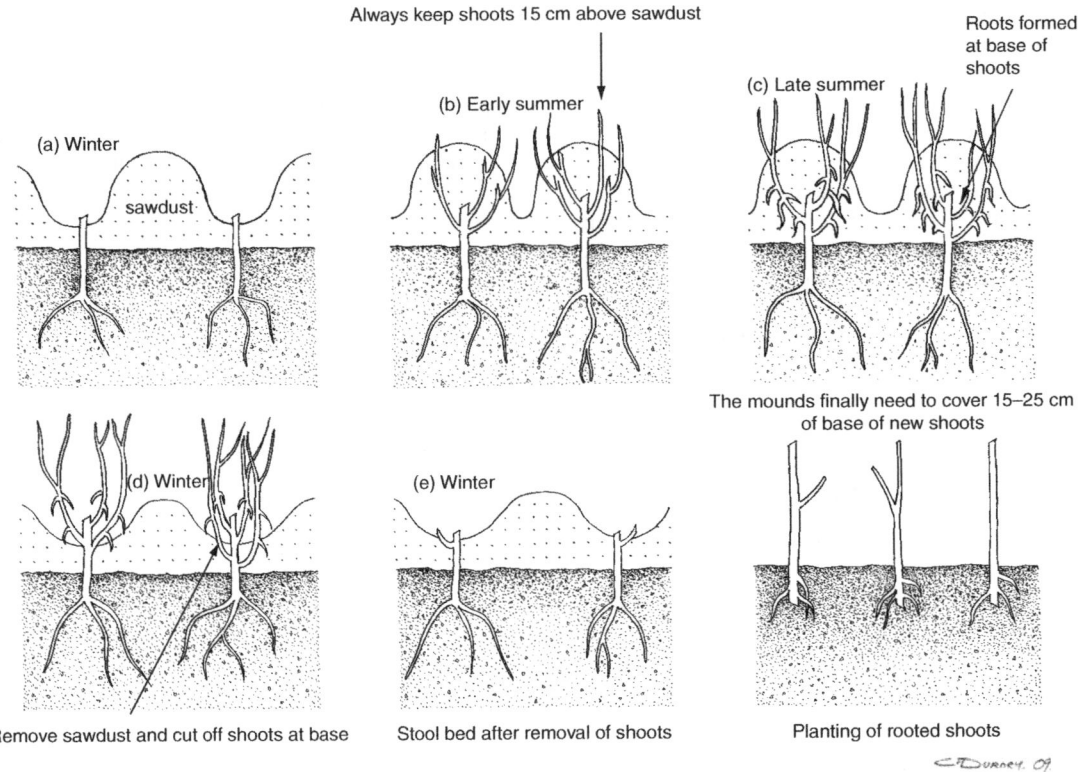

Always keep shoots 15 cm above sawdust

(a) Winter

sawdust

(b) Early summer

(c) Late summer

Roots formed
at base of
shoots

The mounds finally need to cover 15–25 cm
of base of new shoots

(d) Winter

(e) Winter

Remove sawdust and cut off shoots at base

Stool bed after removal of shoots

Planting of rooted shoots

Fig. 11.7. Propagation by stool beds.

30 cm apart in rows about 1 m apart, depending on the machinery used. The surface in and between rows is covered with 15 cm of sawdust. Each winter, plants are cut to ground level and the shoots which emerge are partially covered with soil or sawdust raked from between the rows, as shown in Fig. 11.7a–e.

Budding

The seedlings, rooted cuttings or rootstock layers described above are converted to a finished nursery tree or vine by budding or grafting to a commercial scion cultivar. The bud or graft becomes the cropping part of the tree or vine and the fruit produced is identical to that of the parent tree from which the bud or graft wood was obtained. Apples, pears (Asian and European), quinces, loquats, peaches, nectarines, apricots, plums (Japanese and European), cherries (sweet and sour), almonds, citrus and avocados are commonly propagated by budding.

A number of budding techniques are described in detailed propagation books. Only two will be described here.

Shield or T-budding

Towards late summer the rootstock in the nursery will be growing well and should have achieved the thickness of a pencil. From the time that it was about 30 cm tall the leaves and lateral branches within 15 cm of the ground will have been removed. To be suitable for budding the rootstocks need to be well supplied with moisture so that the bark separates readily from the wood (bark slip).

The buds are obtained from strong-growing, current-year shoots taken from the tree of the scion cultivar of interest. The leaves are cut off, leaving a 10 mm petiole stub at each node. Shoots, or the tips of shoots, less than 4 mm in diameter are unsuitable since the axillary buds will be too poorly developed. To prevent desiccation, the 'bud sticks' are placed in a plastic bag or stood in a bucket of water.

M. Thomas and D. Jackson

Budding should proceed as soon as possible. However, if the budding operation must be delayed, the bud sticks can be stored for a week or more in the refrigerator if wrapped in damp newspaper and sealed in a plastic bag. The process of shield or T-budding is illustrated in Fig. 11.8.

If the petiole falls off or can be pushed off with a touch within 2 weeks of budding, a successful

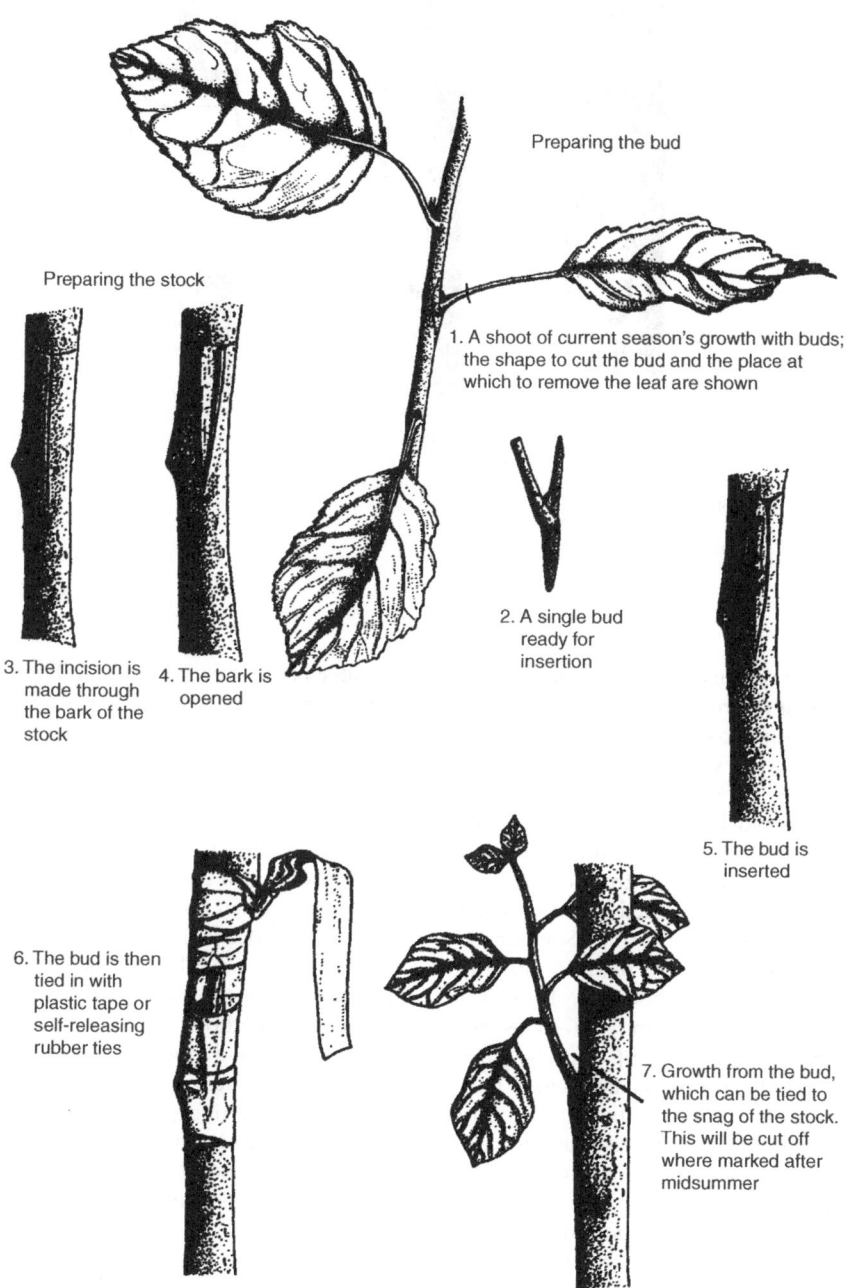

Fig. 11.8. Shield or T-budding.

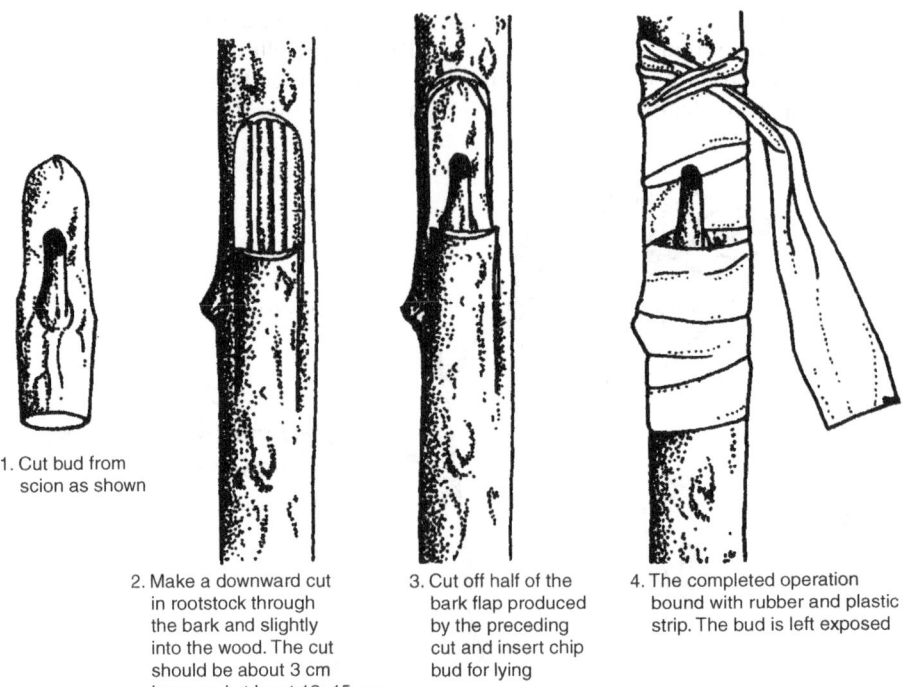

1. Cut bud from scion as shown

2. Make a downward cut in rootstock through the bark and slightly into the wood. The cut should be about 3 cm long, and at least 13–15 cm above ground level

3. Cut off half of the bark flap produced by the preceding cut and insert chip bud for lying

4. The completed operation bound with rubber and plastic strip. The bud is left exposed

Fig.11.9. Chip budding.

union or 'take' has probably been achieved. Six weeks after budding the remaining ties are usually cut so that they do not constrict the rootstock as the girth increases. In late winter the stock is cut off about 4 cm above the inserted bud.

In spring the inserted bud will grow and other buds from the stock that burst should be quickly removed. This plant is normally grown on in the nursery for one season and planted in the orchard the following winter or early spring. Sometimes, however, budded rootstocks are planted into the orchard the first winter after budding. These so-called 'dormant bud' or 'sleeping eye' trees are quite satisfactory and may develop rapidly, since the disturbance to the roots occurs before they become too large. However, the standard of care during the first year in the orchard must be very high, equivalent to that normally given in the nursery, or losses can be excessive.

In cases where the 'sleeping eye' does not 'awake' it may be replaced by re-budding in early summer and the rootstocks cut back 3 weeks after budding. This will force the bud to grow and a shoot 30 cm or more may be produced in the same year. This practice is possible because the rootstock will have accumulated sufficient reserves in the previous growing season.

Chip budding

Chip budding is an alternative method, which does not require the bark to lift easily. It can therefore be used in spring or autumn, or during dry conditions in summer, when normal T-budding is not possible. The tying procedure is slower than for T-budding, since the cuts must be carefully covered to prevent drying and to achieve good contact of stock and scion. This method is illustrated in Fig. 11.9.

While chip budding is a bit slower than T-budding, it can prove highly successful, often resulting in stronger, more upright growth. It is used increasingly for a wide range of fruit and ornamental trees. It may also be used in the process of frameworking described later.

Grafting

In grafting, a short piece of stem with one or more buds is inserted into the rootstock. It is

M. Thomas and D. Jackson

normally done in late winter or early spring and is used for grapes, cherries, kiwifruit, feijoas, tamarillos, passionfruit, walnuts, pecans, chestnuts, macadamias and sometimes hazelnuts. As with budding, the essential ingredient for success is ensuring that the cambial layers of the two parts being joined make good contact. There are many types of grafts, but only two will be described: the whip-and-tongue graft and the cleft graft.

Whip-and-tongue

This is a useful graft to use when the sizes of the rootstock and the grafting stick are similar. It is described in Fig. 11.10.

Cleft graft

This technique is more suitable when the rootstock is larger than the scion. It is described in Fig. 11.11.

Cutting grafts (bench grafting)

Because the cuttings of some grape rootstocks root so easily, grapes may be grafted before the rootstock has roots. Cutting grafts are produced in the nursery by grafting on to an unrooted cutting.

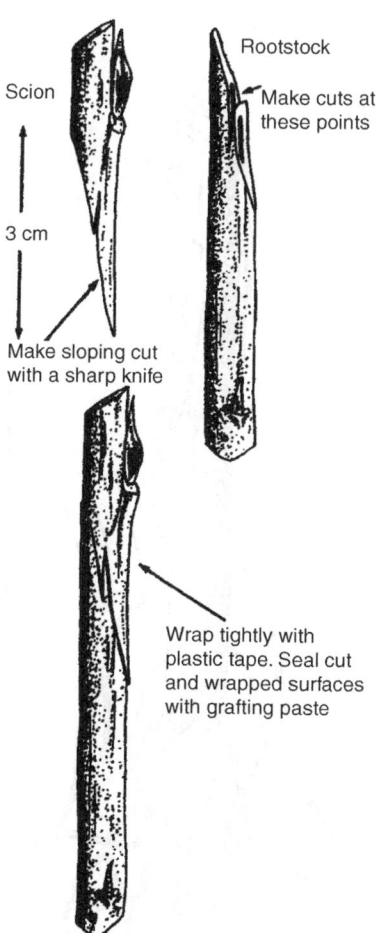

Fig. 11.10. Whip-and-tongue graft.

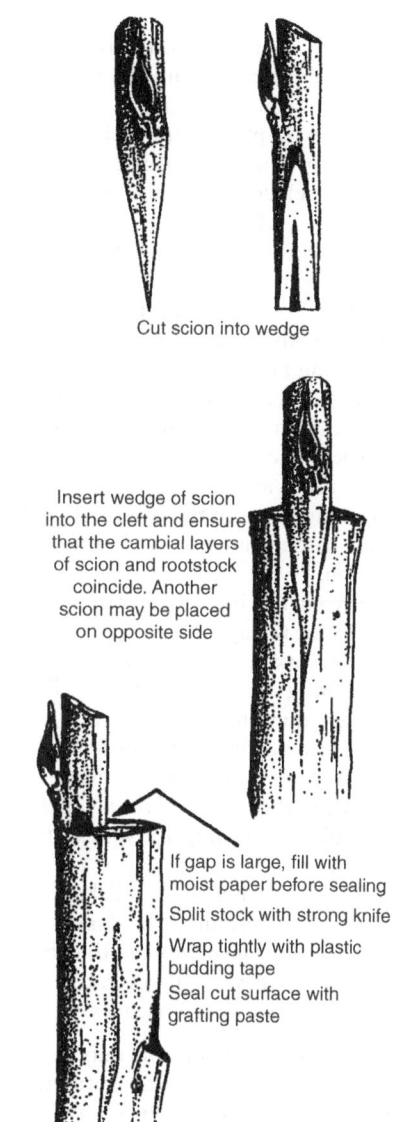

Fig. 11.11. Cleft graft.

This involves simultaneous root development and graft union formation on a single cutting. Bench grafting can be carried out indoors, over a long season, and is cheap and reliable.

Grafting is done in mid- to late winter with grape canes which are between 8 and 12 mm in diameter. After collecting the material, the following steps are taken.

- Make a cutting of the rootstock cultivar approximately 30 cm long; cut off all buds with a cut parallel to the stem.
- Cut a one-bud scion.
- Graft to the rootstock using the whip-and-tongue or a grafting machine (e.g. Fig. 11.12), which does an equivalent graft.
- Tie the graft with plastic tape and dip the whole scion area in melted grafting wax (or use grafting grease). An alternative using staples is shown in Fig. 11.13.
- Place in bucket of water – preferably deep enough to cover the graft.
- When 50 or 100 are done, tie them together, place in a polythene bag and seal with a rubber band. The bag must be airtight. An airtight bag eliminates the need to surround the grapes with moist packing material, but the grafts must not be dry when placed in the bag.
- Place in a room at 25–30°C until callus is seen to be forming between the rootstock and scion. Buds will be beginning to swell and new roots will be seen.
- Remove from the high temperature and from the bag. Insert the apical grafted portion for 1 s in paraffin wax that is at a temperature just above melting point but below 70°C and immediately quench in cold water.
- Place the base in potting mix in a small pot and apply bottom heat (25°C) inside a cool glasshouse. In this position, roots will form and the bud will burst through the wax as it develops (Fig. 11.14).
- When the roots are well established, place the pot in a shade house until the danger of spring frost has passed. The plants can then be planted directly in the vineyard. If vegetative growth is excessive, trim back to two leaves.

On the other hand, grafted plants may be grown in a nursery prior to planting next autumn. This is recommended if vineyard conditions are not good. Rootstocks and scions benefit from soaking in 'Chinosol' prior to grafting. This material reduces

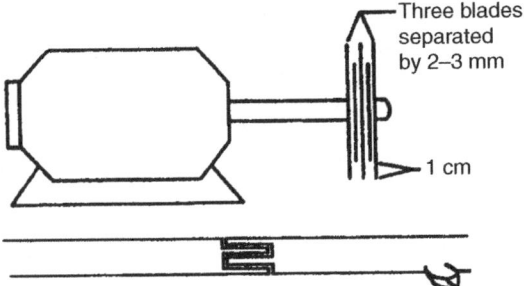

Fig. 11.12. Parallel-saw grafting machine.

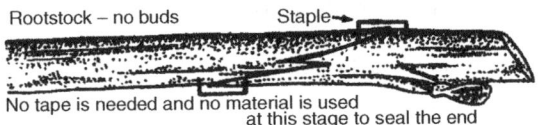

Rootstock – no buds   Staple→

No tape is needed and no material is used at this stage to seal the end

Fig. 11.13. Cutting graft.

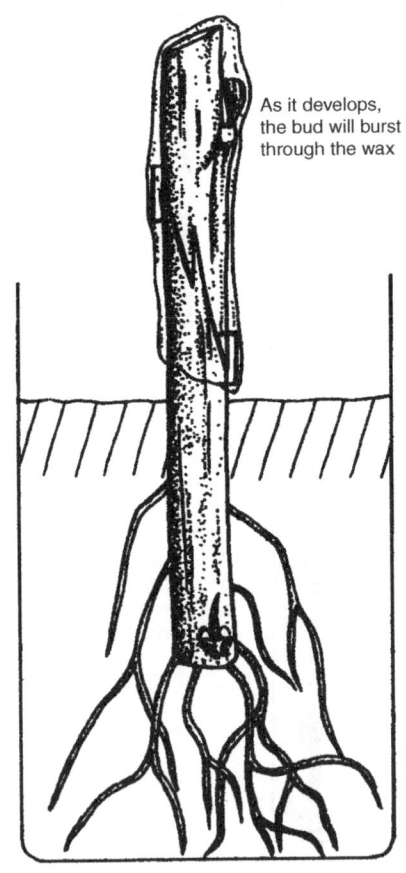

As it develops, the bud will burst through the wax

Fig. 11.14. Root production on a grafted grape cutting.

M. Thomas and D. Jackson

fungal infection in the plastic bag and improves take. Use according to the directions on the packet.

As mentioned above, grafting machines are available for bench grafting, to replace the difficult job of preparing whip-and-tongue grafts. Many brands are available, including the very popular 'Omega'. Another which can be constructed by a technically capable grower uses parallel circular saw blades attached to an electric motor (Fig. 11.12).

The scion and rootstock are separately pushed on to the rotating blades to leave a castellated end. These are then pushed together. They should make a tight fit and may need no tying.

General Comments on Grafting and Budding

It is an essential prerequisite that scion and rootstock are capable of forming a permanent graft union. This means that they are compatible. Plants that are closely related and are free of virus and other disorders have the highest chance of success. However, some closely related species can be incompatible, and it is therefore important to check for compatibility. If there are no records, grafting and assessment of its success may be required over several years.

An expert in grafting or budding will do hundreds a day. The beginner following the instructions given in this chapter will find it may take an hour to do a dozen. Nevertheless, this may be all an orchardist needs to do in a season and, provided that care is taken, there is no reason why success should not be achieved. The following points are important.

- Use very sharp knives. Sharpen on an oiled carborundum stone and, to get an even finer edge, finish with a leather strop.
- T-budding, as illustrated, ensures that the cambial layers adjoin, but, when grafting, the operator must make certain that the cambial layers are together. With whip-and-tongue, it is better if the cambium on both sides is adjacent, but if the stock and scion are not exactly the same size, ensure that at least one side adjoins.
- Strong, healthy material of appropriate size is needed. Pencil thickness is usually the minimum for scions and rootstocks for grafting, as well as rootstocks for budding, while some budsticks

for budding will occasionally be slightly smaller.

- Do not let material dry out, especially after cutting. This is particularly important when budding, as the T-buds are small and dry rapidly in the warm summer temperatures when this job is done.
- Tie tightly – plastic or rubber strips are normally used.
- Material for grafting is collected in midwinter, while still dormant, wrapped in moist newspaper and placed in a sealed plastic bag in the refrigerator (not a freezer). It must be inspected from time to time to ensure that it is not drying out. Grafting is then done in spring, when the buds of the rootstocks are swelling, using the scion material from cool storage, which, of course, is still dormant. Cool-stored scion wood is almost essential for some plants, e.g. cherries.
- Label carefully – never rely on memory; it inevitably fails.

Any special aspects of grafting appropriate to specific crops will be mentioned later when discussing that crop.

Changing Cultivars of Established Trees

Top-working

The tree is cut back to a 1 m stump with four or five main limbs stubbed back. Cleft grafting is used to introduce a compatible, and presumably more desirable, cultivar on to each branch and the main stem. This is best carried out in early spring, using the technique described earlier. Usually several grafts are applied to each branch and the large exposed cut area is sealed with paint of a grafting compound.

Frame-working

In this method of reworking, grafts are inserted throughout the main frame of the tree. Most of the small laterals are replaced, using a large number of scions of the replacement cultivars. Frame-working allows a rapid return to fruiting but is very laborious. Growth from buds of the original tree will need to be removed in subsequent seasons to ensure good establishment of the scions. Whip-and-tongue, cleft and even budding techniques can be used.

Double-working

In some climates where freeze injury to mature trees can be a serious problem it is considered important that the trunk, and perhaps even the base of the primary scaffold branches, possesses greater freeze tolerance than exists either in the rootstock or in the cultivar that will make up the fruiting canopy. Such a tree can be produced by double-working, i.e. budding or grafting the hardy cultivar to the rootstock and, 1 or 2 years later, top-working the framework to the desired commercial cultivar. This can be done in the nursery, but, more commonly, the second operation takes place in the orchard.

Use of Suckers

Raspberries will send up suckers from the roots and these can be dug up and used for propagating. Special rows of any cultivar can be set up specifically for propagation. These will be cut back each winter to ground level to encourage suckering. High-health material must always be used for such an operation.

Tip-layering

Brambles have long been propagated by tip-layering, although they are now most commonly grown from softwood cuttings. Propagation from cuttings (Table 11.1) in the first 2 months of the growing season greatly reduces the spread of dry-berry disease and foliar nematode, especially when high-health stock plants are used.

For tip-layering, the plants are grown in beds, and shoot tips, pegged into the ground, develop roots. Mother plants are grown 40 cm apart in rows 2 m apart. In winter, before shoots emerge, the row is treated with a knock-down herbicide combined with a pre-emergence herbicide. Before pegging down the shoot, remove 5 cm of topsoil in a 15–20 cm circle and peg into this soil. After pegging, cover this area with 10 cm of sawdust. Care must be taken with pre-emergent herbicides in case residues build up.

Further Reading

Bose, T.K., Mitra, S.K. and Sadhu, M.K. (eds) (1991) *Propagation of Tropical and Subtropical Horticultural Crops.* South Asia Books, Columbia, Montana.

Debergh, P. and Zimmerman R.H. (eds) (1991) *Micropropagation: Technology and Application.* Kluwer Academic, Dordrecht, the Netherlands.

Economou, A. and Read, P.E. (eds) (2003) I International Symposium on Acclimatization and Establishment of Micropropagated Plants. *Acta Horticulturae* 616.

Hanke, M.-V., Dunemann, F. and Flachowsky, H. (eds) (2009) I International Symposium on Biotechnology of Fruit Species: BIOTECHFRUIT2008. *Acta Horticulturae* 839.

Hartmann, H.T. and Kester, D.E. (2002) *Plant Propagation Principles and Practices*, 7th edn. Prentice-Hall, Englewood Cliffs, New Jersey.

Lewis, W.J. and Alexander, D.M. (2008) *Grafting and Budding: a Practical Guide for Fruit and Nut Plants and Ornamentals*, 2nd edn. CSIRO Publishing, Australia

McDonald, B. (1986) *Practical Woody Plant Propagation for Nursery Growers.* Timber Press, Portland, Oregon.

Toogood, A.R. (2003) *Propagating Plants.* Dorling Kindersley, London.

M. Thomas and D. Jackson

12 Machinery for Fruit Growing

WILLIAM ATKINSON

The range of machinery used for fruit growing is wide. Each year sees new devices on the market, some revolutionary and some old ideas in a new coat of paint. This means that no one book, let alone a single chapter, can comprehensively cover every type or variation of machine. Thus, this chapter will present an introduction to the essential working principles of the machines most widely used in fruit growing. More details of specialist items, calibration procedures, maintenance procedures and design criteria can be found in the operator's manual of individual machines.

Any machine requires an operator with some degree of skill, and its performance will be greatly influenced by that operator's training, experience and attitudes. Machines themselves will not guarantee problem-free production. Nor should it be assumed that the newest or biggest machine will be the best for the property. Only after a careful appraisal of which jobs need mechanizing, followed, if possible, by trials with selected machines actually working on the property, can assessment be made on how a proposed machine will suit the layout, existing machinery and management skills.

Machinery Requirements

Space restrictions allow discussion here of only the larger machines used outside the packing shed. There are many other items, particularly such hand-held equipment as pruning saws and secateurs, ladders, chainsaws, spades and shovels, which have had to be omitted. One major development that needs to be mentioned is powered, hand-held pruners. These are mostly activated by either hydraulic pressure or rechargable batteries. A list of major items that are needed in the initial stages

of development is given in Table 12.1. Some workshop tools will also be required. Items that are likely to be needed for the harvest of the first significant crop are listed in Table 12.2. There will be variations between crops and the suggestions can only be tentative.

Use of Contractors

Many jobs that need to be done during the initial development stages of any property are infrequent or non-recurring. Such activities may include initial cultivation, drainage, erecting fences for shelter and/or crop support, and sinking irrigation bores. Buying the specialized equipment required is often not justified economically for the small amount of work on one property. Thus it can be good economics to take advantage of available contractor services. It should also be recognized that owning and successfully operating specialized equipment requires experience and skill. Such operators usually have higher rates of working than most beginners. For these reasons, obtaining the services of contractors should be seriously considered for many 'one-off' jobs, both at the beginning and also during later stages of running a property.

However, there are several important points about obtaining the services of contractors. These should be kept in mind before employing, at random, any operator listed in the telephone directory, Yellow Pages or advertised in the local paper or Internet.

- The property manager must have a clear idea of what job is to be done and, if possible, some estimate of the areas or volumes, or times involved, as a background against which contractors' quotes may be checked.

Table 12.1. Machinery required for the initial phase of property development.

Tractor	20–30 power take-off kilowatts (PTO kW) fitted with a three-point linkage, hydraulic lift system and power take-off
Trailer	Heavy car trailer that can be used on and off the property
Carryall transport	Three-point linkage-mounted tray
Rotary hoe (rototiller)	Matched to PTO power output of the tractor. May be hired if inter-row cultivation is not envisaged in later years
Herbicide sprayers	(i) Hand-operated knapsack.
	(ii) Tractor PTO-driven pump. Hand lances and perhaps a small boom
Pesticide sprayer	Motorized knapsack or small tractor-mounted unit
Mower	Three-point linkage-mounted, size matched to PTO power output of tractor

Table 12.2. Machinery required for the season when the first significant crop is harvested.

Large trailer	
Tractor-mounted fork lift	Several forms of three-point, linkage-mounted forks are available. Mounting on the front of a tractor is much more complicated and expensive than mounting on the rear
Fruit bins	Standard fruit bins suitable for handling by fork lift
Pesticide sprayer	A machine with sufficient capacity to cope with the property in full production

- For specialized work, the manager should find out where else the contractor has operated and check how satisfied those clients were.
- At least two quotes should be obtained. If it is a complicated job, then a written list of the work required, including detailed measurements, makes a safer basis for negotiation than unrecorded telephone conversations. A duplicate copy should be kept. As well as prices, information about when a contractor expects to start is a vital part of the quote. A manager has little influence over a contractor's movements and this can cause considerable worry and frustration if the manager expects to be treated as the contractor's only client.
- Careful, clear instructions must be given to contractors and their operators once the machines are on the property ready for work. These people cannot be expected to read the property manager's mind.
- Checks should be made to ensure work is being carried out as expected and agreed. The day the contractor arrives is not a day to take off and go to town because someone else is doing the work.
- Strict attention may need to be paid to the cleanliness of contractors' machines as they arrive at a property. Undesirable weeds or pests, such as phylloxera in grapes, may be readily transported from one property to another on cultivator tines or ditch digger chains. Particularly in newly developing areas, the sensitivity of some fruit crops to hormone spray may not be appreciated. Any herbicide spraying should be done with a machine that has been properly decontaminated from previous hormone spray residues.

Tractors

The central mechanical component of most systems of mechanized fruit growing is some form of 'agricultural' wheel tractor. This one mobile power unit can readily be used to pull, carry and/or drive many machines, thus avoiding a multiple number of engines.

Characteristics of the agricultural wheel tractor that are particularly valuable to fruit growers include:

- Their robust construction and relative reliability, particularly diesel-engine models.
- Their ability to work in difficult traction conditions, from wet mud to soft dry sand.
- The ease with which they can pick up and carry loads or attached machines.
- Their manoeuvrability.
- The good view an operator obtains of where the machine is going and what it is doing.

W. Atkinson

The three-point linkage and power take-off systems (Fig. 12.1)

To enable tractors, which are made by many different manufacturers in many different countries, to be matched to the multitude of machines available, a series of 'standard' features has been agreed to worldwide. The most important standard features for fruit growers are:

- The sizes of the 'lift and carry' parts, known as the three-point linkage system.
- The external drive, known as the power take-off (PTO).

When buying a tractor, it is important to ensure that the three-point linkage system is of a standard size, so that it will fit all standard machines, both one's own and those so often borrowed from neighbours. The common three-point linkage standards are designated Category I and Category II, although there are others, which are not normally used in fruit growing. For practical purposes, these categories refer to the size of the holes in the balls on the end of the tractor lifting links and their corresponding lifting pins on the machines to be attached to the tractor. The diameter of the Category I pin is 22 mm and that of the Category II pin 28 mm. Machines lifted and carried on the three-point linkage are described as 'mounted' machines. Those machines with their own wheels,

which are towed from the drawbar, are described as 'trailed'.

It is also important that the PTO shaft is a standard size; usually on small- to medium-powered tractors it is a six-splined shaft, designed to run at 540 revolutions per minute (rev/min) at a stated engine speed. To indicate this speed, the tractor must be fitted with an engine revolution counter, sometimes called a tachometer. On second-hand tractors, the tachometer may be broken. It is vital to have it working.

Engine speeds, PTO shaft speeds and tractor ground speeds all need to be known for many operations. Guessing the setting will be unreliable and may lead to expensive, ineffective operation, such as poor spraying or expensive repair bills on machines damaged by being driven at the wrong speed. Success in fruit growing depends on attention to detail. Having machines running at the correct speed is one of the more important details.

Tractor power measurement

The tractor most likely to be used by fruit growers would develop PTO power in the range of 10–50 kW (1 kW = 1.34 hp – horse power). Confusion may easily occur here as engine power is often quoted and this is always more than the PTO power available. It is the available PTO power that has to be known to check that a tractor will drive another machine! To operate a PTO-driven machine on undulating and hilly properties, the tractor must provide sufficient power to propel itself and the machine uphill and, in addition, enough PTO power to drive the machine.

Special models for fruit growing

Although standard tractors are often used, manufacturers do provide narrow 'vineyard' and lowered 'orchard' models to accommodate the peculiar requirements of some crops – as is true for some other machinery items, such as low body sprayers (Fig. 12.2).

For all tractors, special shielding mudguards are available to reduce damage to crop plants by the large-lugged tread on their tyres. Track-laying or crawler-type tractors are also available, even in vineyard models. On particularly steep land, the extra traction available from a crawler has advantages. But, in comparison with wheeled tractors, they are more expensive to purchase, have a higher maintenance requirement and are much less

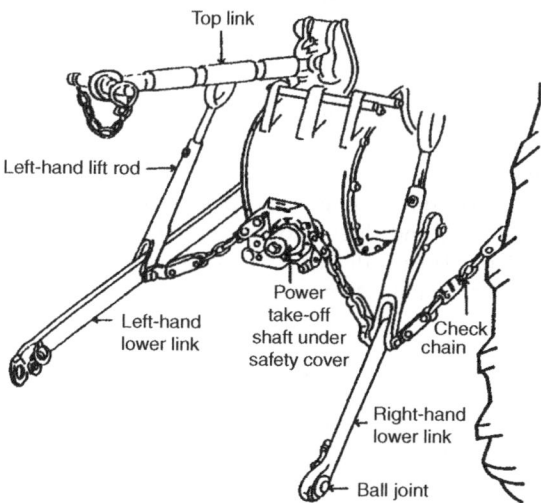

Fig. 12.1. Rear-end view of a standard wheel tractor showing three-point linkage system and power take-off shaft.

Top link

Left-hand lift rod

Left-hand lower link

Power take-off shaft under safety cover

Check chain

Right-hand lower link

Ball joint

Fig. 12.2. Adapted machinery – slimline sprayers that can be towed between narrow orchard and vineyard rows are an example of machinery that can be adapted for specific purposes.

versatile. The increase in availability of many forms of four-wheel-drive tractor has made the crawler even less attractive for many properties.

Two-wheel drive compared with four-wheel drive

On level ground, it can be demonstrated that a four-wheel-drive tractor may marginally outperform the equivalent two-wheel-drive tractor if both are properly ballasted. However, in terms of the initial purchase price and the overall running cost, the four-wheel-drive version will be the more expensive.

For orchard use, there are two main conditions where four-wheel drive is worth considering. One is if the orchard is on hilly country. Here, four-wheel drive will provide additional brake performance and better 'directional control' when travelling across hills. The second is if heavy, front-mounted implements, such as a front-end loader or fork lift, are to be used. The larger front tyres of a four-wheel drive have better load-carrying capacity and provide additional traction.

Tractor ballasting

The amount of pull any particular tractor develops depends largely on the loading of its driving wheels. When using mounted machines this loading can be altered by the correct use of the hydraulic lifting system. When using trailed machines, or where extra loading is required, 'ballast' can be added to the tractor. Adding liquid to the front and rear tyres is the cheapest form of ballast but this is not easy to adjust. Removable cast iron weights are available, and these allow easy adjustment to the ballast. Tractor dealers and consultants can advise on correct ballasting.

Safety frames

Detailed investigations of fatal tractor accidents show that many people are killed in what might be regarded as 'safe' conditions. High speed, steep country, youth or inexperience are not necessary prerequisites for fatal events. Many deaths occur when the tractor rolls over sideways or tips backwards or forwards. This can occur, for example, on

W. Atkinson

a flat, straight road bounded by a ditch. In most cases of rolling or tipping, it has been clearly shown that approved safety frames are likely to prevent death or serious injury. Safety frames are also known as roll-over protection structures (ROPS). Many cabs are designed to act as ROPS.

Any operation using a tractor involves the risk of overturning. In some countries, legal requirements for the fitting of approved safety frames exclude certain horticultural operations. However, such exemptions must not be seen as officialdom saying horticultural tractor work is safe. There are some operations where fitting suitable fully rigid frames is impracticable, such as under low-trellised crops. In such cases it may be possible to fit a folding frame, where the driver-protection hoop can be lowered to lie along the bonnet or back over the rear-wheel mudguards.

Where an approved safety frame can be fitted, it should be. For the saving of a few hundred dollars of tax-deductible expense it is not worth risking a life.

Never fit a 'repaired' ROPS (cab or simple frame) unless it has been certified as reinstated to its original specification by a registered engineer competent and experienced in such designs.

Buying a tractor

In summary, it is essential to get a tractor with enough PTO power for the work to be done (Table 12.1). It should have a standard three-point linkage, a standard PTO drive and an operational tachometer. The choices are between a standard or special model, normal or shielding mudguards, two- or four-wheel drive or crawler, with or without a safety frame or ROPS cab. The ease of operation in terms of driver access, visibility and convenience of control layout should also be carefully appraised.

Having made these decisions, one should then look at the range available from experienced and reputable dealers within a reasonable distance of the property. Regardless of make, all tractors need attention, if not for major repairs then at least for regular maintenance. If essential spare parts and expert servicing knowledge are not easily available, vital crop production jobs may not be completed on time. A cheap, new arrival on the market with no local agent may well cost more in the long run than any savings that may have been made in the initial purchase price.

Fruit harvesting and handling

Despite much research and grower invention only a few fruit crops, such as currants, have fully automated harvesting machines. However, picking and handling aids, such as self-propelled elevating work platforms and mobile picker platforms, are common, especially on larger properties. Most fruit is collected from the orchard in some kind of bin. Common sizes are close to a metre square and up to half a metre deep. How deep depends on the nature of the fruit and how easily it is squashed. Bins are picked up by forklifts for depositing on tractor-pulled trailers or road trucks. Forklifts are available as front- or rear-mounted attachments on tractors or as specialized, off-road, purpose-built, high-capacity machines capable of lifting a tonne at a time. All forklifts have the limitation that they essentially lift and lower vertically, with only small amounts of tilt movement. Tele-porter machines have an extendable arm that allows them to 'reach' to deposit or collect their load. Industrial forklifts require level and sealed surfaces to maintain traction so are of limited application in most on-orchard operations.

Soil cultivation machinery

The prime purpose of cultivation is to kill off and or bury unwanted weed plants or previous crop residues. Further cultivation may then be necessary to provide suitably sized soil particles, 'tilth', for planting the next crop or perhaps to incorporate a fertilizer or herbicide. Specialized deep cultivation may be required for some fruit crops.

Ploughs, tines and discs in various combinations are traditionally used to prepare ground for annual crops. But for most fruit crops nowadays, cultivation is not an annual activity. Usually, soil cultivation is necessary only during the development phase of a property and then most of it can be done with a rotary hoe (rototiller).

The rotary hoe (rototiller)

The rotary hoe is a mounted, PTO-driven machine. It consists, essentially, of a driven horizontal shaft to which soil-cutting blades are rigidly attached: together these form the rotor. This is enclosed by a hood. The hood has an adjustable section, the shield, at the rear. The rotor turns in the same direction as the tractor wheels. As the rotor turns, it slices off blocks of soil and throws them backwards and upwards. Depending on the position of the shield, these sliced-off blocks may hit more or less of the enclosing hood and be further reduced in size, or may just pass out and be deposited in the newly cultivated ground. As a guide, power requirements are in the order of 15–25 PTO kW per metre of cutting width. Rotary hoe dealers should be able to assist in the selection of a suitable size of machine for the tractor available. Normal cultivating widths are from 0.8 m upwards. This type of cultivating machine is very versatile, and this is reflected in its popularity.

Advantages of the rotary hoe

1. Robust.
2. Manoeuvrable.
3. Versatile.

- Will operate under a wide range of soil conditions.
- Will produce a fine tilth (small particles) or any intermediate tilth (coarser particles) from undisturbed ground.
- Will effectively incorporate crop residue, weed growth and fertilizers.
- Depth of working is positively controlled.
- Can be 'offset' from the tractor to work close to or beneath crop plants.

Rotary hoe adjustments and soil tilth

The fineness of tilth obtained depends on a combination of tractor and machine adjustments. These are shown in Table 12.3 and Fig. 12.3.

Soils are easily abused when cultivating with a rotary hoe. Overworking can damage the soil structure and in some conditions lead to erosion.

The PTO shaft should be run at the manufacturer's recommended speed, regardless of the tractor forward speed over the ground.

Table 12.3. Rotary hoeing adjustments and resulting soil conditions.

	Soil condition desired	
Adjustment	Coarse tilth Large blocks	Fine tilth Small particles
Tractor ground speed	Fast	Slow
Adjustable shield position	Raised	Lowered
Rotor speed	Slow	Fast

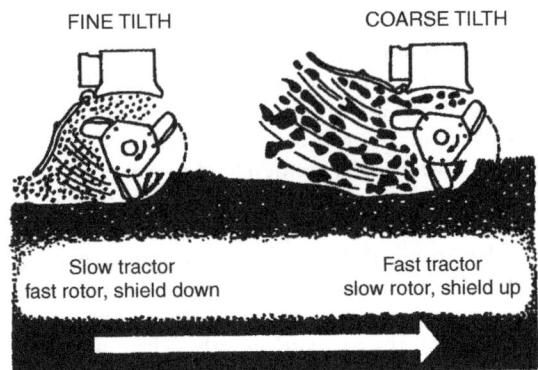

Fig. 12.3. Rotary hoe (rototiller) adjustments.

Rotary hoe maintenance

Because of tough operating conditions, regular maintenance procedures are essential if the working life of a rotary hoe is to be prolonged. These include:

- Daily greasing.
- Checking tightness of nuts and bolts.
- Checking correct setting of the overload clutches.

Deep cultivation

Pan breaking

In some areas hard pans may exist, either naturally occurring at various depths or man-made from previous cultivations and so usually no deeper than 300 mm. Deep subsoiling or ripping may be advocated to overcome shallow pan problems. Where land needs such treatment this should be done before planting. Soils known to have hard pans are less than desirable for establishing permanent crops. The high power requirement of subsoiling demands the use of large tractors. A subsoiler (Fig. 12.4) should penetrate to at least 450 mm and preferably deeper, though the power requirement increases dramatically as depth increases. The depth of operation should be measured while work is in progress. Strong-looking industrial rippers on crawler tractors often have an effective working depth of less than 300 mm.

Root pruning

Deep cultivation may also be recommended to 'root prune' shelter hedges every 2–3 years.

W. Atkinson

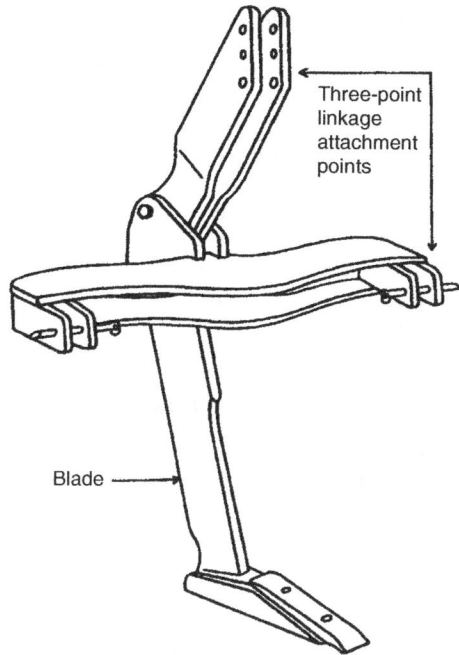

Three-point linkage attachment points

Blade

Fig. 12.4. Agricultural subsoiler.

A subsoiler will do the job using one run, close and parallel to the hedge. If this is not done, spreading roots from the hedge plants may cause severe competition for food and water with the crop plants.

Drainage

On land where the soil below the depth of deep cultivation is poorly drained, a grower should only consider subsoiling where underground drainage can be installed at the same time. Without underground drainage, rain falling on subsoiled land quickly percolates to the depth of the cultivation and accumulates, as the water has no outlet. The land then becomes saturated. Planning and installing such a drainage system requires expert advice and this book will not attempt to cover the technical aspects of drainage. For specific drainage designs, readers are recommended to contact local consultants.

Fertilizer-spreading machinery

The fertilizers used on fruit crops can be roughly divided into two types:

- Those of organic origin, e.g. poultry manure.
- Inorganic materials ('chemical' or 'bag' fertilizers).

Small quantities of liquid or readily soluble nutrients may be applied in solution, either when spraying the crop or when irrigating. Organic mulch material is important for the management of some crops. Often particles are much harder and the sizes are much bigger than those of bag fertilizers. Special equipment is discussed below.

Organic fertilizers can be applied to bare ground using 'muck spreaders'. Once a crop is established, access for such machines is often impossible, and their spreading action, which involves throwing the material indiscriminately up into the air, is highly undesirable. The alternative, hand spreading, may be distasteful and uneconomical.

Most crops, therefore, are fertilized with inorganic materials, for which a range of suitable machines is readily available. Such machines are usually tractor mounted and driven from the PTO. Two spreading mechanisms are in common use: the spinning disc and the oscillating spout, and both throw the fertilizer horizontally (Figs 12.5 and 12.6).

The spreading width of these machines can be varied by machine setting but it is also affected by the density and particle size of the material being spread. For accurate application rates, a machine should be calibrated for each new batch of fertilizer before field application is started.

Spinning disc

Fertilizer is fed on to a spinning disc and flung outwards. The distribution pattern from these machines is greatly affected by where the fertilizer is delivered on to the disc. Ideally the disc is kept near horizontal to prevent the point from changing. For this reason the spinning disc type of spreader is not recommended for use on sloping or rolling country.

Oscillating spout

Fertilizer is discharged through a rear-facing spout that swings rapidly from side to side. This mechanism is not affected by sloping ground but its uniformity of application can be seriously impaired if

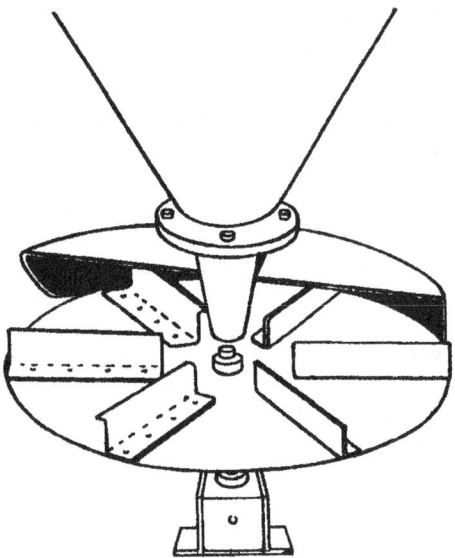

Fig. 12.5. Spinning disc.

Fig. 12.6. Oscillating spout.

the spout is damaged. When not in use, the spout should be removed and stored safely.

By using a shortened spout, application of fertilizers can be confined to a band on each side of the machine. This option is particularly useful for young plantings, where roots have not covered the full ground area. Fertilizer can be banded on top of where the young roots can use it. In established plantings, where the irrigation-wetted pattern may be restricted, as from micro-sprinklers or trickle emitters, banding of fertilizer may also be useful.

Organic mulch spreaders

Some tree crops, such as avocados, benefit from organic mulch spread over their root zones. Often this material comes from 'chippings' of bark or woody plants and trees. Such mulches usually are not as easy to spread as bagged fertilizers, so standard fertilizer machines are not suitable. Mulch can be placed close to the trees by skid–steer loaders (bobcats) or some kinds of animal forage feed-out machinery. Final distribution is often by hand.

Mowing machinery

Most fruit production units will need a machine to cut grass and weeds. It may be for appearances' sake along the road verges near a farm-gate sale entrance or it may be as an essential part of the management of a grassed orchard. Except for areas inaccessible to tractors or on very small properties, mowing machines are usually tractor mounted and driven either from the PTO shaft or from a hydraulic pump. Most mowers used by the fruit grower are of the slasher or the flail type, as these are robust and require little maintenance (Fig. 12.7). Reciprocating-knife, drum or disc hay mowers are not recommended.

The impact cutting process

Both slasher- and flail-type mowers cut by impacting the vegetation at high speed (45–90 m/s or 160–320 km/h). Neither machine therefore needs to have sharp cutting edges. To ensure the correct blade speed, the tractor PTO needs to be set according to the mower manufacturer's recommendation using the tractor tachometer.

Too slow a blade speed will result in ineffective cutting and perhaps blockages, while too fast a speed may result in damage to the mower drive shafts and other components. With flail mowers, too high a speed will require considerably increased power for no improvement in cutting.

Slashers

These machines have one or more horizontally rotating blades, mounted on the lower end of a vertical shaft. Many domestic lawn mowers use the same basic design. Power requirements are typically in the range 8–15 PTO kW per metre of

W. Atkinson

Fig. 12.7. Slasher mower in raised position.

cutting width. They are usually cheaper per metre of cut width than flail types. To provide a measure of protection against striking obstacles, some slashers have the ends of the vertical drive shafts protected by steel saucer-shaped dishes. Replaceable, freely pivoting cutting blades are attached to the circumference of these dishes. The bottom of the dish then provides a skid to ride over bumps or obstacles. The cutting blades are also kept off the ground by this arrangement. Some more heavily built machines may be used as brush cutters.

Where it is desired to cut grass underneath overhanging crop branches, the mower can usually be 'offset' to one side. To reduce the risk of hitting trees, some 'swinging arm' slashers have been developed. These employ springs, or pneumatic or hydraulic actuated arms, to retract an independently pivoted cutting head once a sensor has detected an approaching tree. But such systems are complicated and expensive, and most under-tree growth is now controlled by the use of herbicides.

Slashers are not particularly good at chopping material into short lengths. Thus they are not favoured where orchard prunings are to be chopped up and left as a mulch.

With all slashers there is a very real danger from ejected material. This may be flung in any direction.

Flails

These machines have a single horizontal shaft running across the full width of cut. Attached to this, somewhat like paddles, is a series of flail blades, often pivoted so they will fold inwards on meeting an obstruction. The shaft of a flail mower rotates in the opposite direction to that of the tractor wheels. Flail blades are thus travelling forwards and upwards into the vegetation, so that when it is cut it is carried over the top of the shaft and thrown out to the rear of the machine. Any other material, such as prunings, going through this process will be chopped several times by successive flails and so be reduced to quite short lengths, suitable to be left on the ground as a mulch.

Power requirements are in the order of 25 PTO kW/m cutting width, considerably higher than that of slashers. This high power requirement is caused by the fan action of the flails.

It is important that these machines are not run faster than their manufacturer's recommendations as the power requirements will be increased for no improvement in cutting.

Machinery for Fruit Growing

Because the cut material is carried by the flails in a vertical arc, it can easily be deflected downwards by shields at the rear of the machine. This reduces the risk of the cut material being discharged as a projectile. Although they are not completely safe, flails are preferred to slashers where bystanders are likely to be present, such as on roadside verges.

All mowing machines are inherently dangerous. To ensure safe operation, the following rules should be observed:

- Do not make adjustments or try to clear any blockages on mowers until the PTO drive has been disengaged, the tractor engine stopped and the blades have ceased turning.
- Ensure that slip clutches, if fitted, are adjusted so that when a major obstacle is encountered they will function as designed.

Spraying machinery

Spraying is the application of a chemical product to the ground and plants. The product is normally mixed with a carrier, which is usually water. This 'spray liquid' is projected towards the target area in the form of small droplets by a spraying machine (a sprayer). The coverage of the target depends a great deal on the choice and operation of the sprayer.

Spray chemicals

Most of the chemical products applied in spray form can be classified into two main groups.

- Group 1, herbicides for killing unwanted plants.
- Group 2, fungicides and insecticides for controlling diseases and pests.

The temptation to save money and have only one machine to apply chemicals from both groups should be resisted. The danger of accidentally applying herbicides to crop plants as a result of unsuspected residues cannot be overemphasized. Sprayers should be clearly labelled and never be used for applying the alternative materials.

Nutrients and growth regulators may also be applied by spraying. These products should not be applied with the herbicide sprayer.

It must be recognized that many chemical products can be lethal to humans and all of them should be kept locked away, preferably in a special, separate shed. Whenever spraying, protective equipment should be worn appropriate for the hazard ratings of the chemical product being used.

Chemical mode of action

Spray chemicals may also be classified according to their mode of action as either contact or systemic. This has an important bearing on the method used to apply them.

Contact materials are only effective if the chemical and target pest come into direct contact. A full overall cover of the plant is therefore required.

By contrast, a systemic material is able to enter the sap stream of the plant being sprayed and so be carried to every part of the plant. Application to only part of the plant's surface may be adequate for control. A thorough coverage, although not critical, will nevertheless improve the result.

Chemical formulations

Spray chemical products may be further classified by their physical properties. The following list of physical spray types covers most common formulations:

- *Aqueous concentrate* (AC). Aqueous concentrates are concentrated solutions of the product in water.
- *Emulsifiable concentrate* (EC). Emulsifiable concentrates are oil-like liquids that will disperse uniformly in water.
- *Flowable* (FL). A flowable consists of very finely ground solid particles mixed into a liquid carrier to form a stable suspension.
- *Wettable powder* (WP). Wettable powders are dry chemical powders that readily disperse in a water carrier. To ensure even application of this product, it needs to be kept constantly dispersed throughout the carrier by the agitation system of the sprayer.

The physical properties of the concentrated chemicals have a great influence on the mixing and the subsequent agitation requirements of each product.

Always read the container label thoroughly before attempting to use any spray material.

Spraying volumes

Commonly used spraying volumes and their rates of application are given in Table 12.4 in litres per hectare (l/ha).

Each of these techniques has advantages and disadvantages. No one technique does every spraying job equally well. The density of foliage, the shape and size of the target plant, and the type of chemical all need to be considered.

High-volume spraying

Complete coverage of the target plants should be achieved and 'runoff' drips will be observed. The volumes of spray used are large, so each tankful will cover much less orchard area than low-volume techniques. This means more frequent filling of the tank.

Low-volume spraying

Total wetting is not always necessary for economic control in every situation. Low-volume or concentrate spraying may be quite adequate. Runoff will be negligible, so less of the chemical will be wasted. The operator needs to be more careful in calibrating the machine, in measuring the chemicals and in driving, than with high-volume techniques. Smaller spray droplets may cause drift problems on to adjacent areas. Tall or dense-foliaged trees, such as citrus, or hard-to-wet plants, such as kiwifruit, may not be effectively protected from some diseases and pests by low-volume techniques. For many other crops, both high- and low-volume techniques may be used at different times, depending on the season and the problem to be controlled.

Ultra-low-volume spraying

Ultra-low-volume (ULV) spraying systems depend upon the formation of droplets that are, as much as possible, all the same size. Skilled management and operation are needed to ensure success with ULV systems.

Spray droplet formation

Droplets are formed by:

- Forcing liquid through a small hole in a nozzle under medium to high pressure, giving 'hydraulic' break-up (Figs 12.8–12.10).
- Introducing low-pressure liquid to a high-speed airstream, giving 'pneumatic' breakup.
- Introducing low-pressure liquid to the centre of a rotating disc or cylindrical cage, from where centrifugal force flings the liquid outwards. These rotary atomizers are the basis of many controlled droplet application (CDA) devices.

Hydraulic breakup through nozzles is the most common method of forming droplets and it produces a relatively wide range of droplet sizes. Larger droplets tend to be projected further, be less diverted by wind and evaporate more slowly than smaller droplets. However, smaller droplets may give more effective coverage. The higher the pressure of liquid entering a hydraulic nozzle, the

Table 12.4. Spraying volumes.

	For application to:	
	Low-growing plants	Trees and vines
High volume	More than 600 l/ha	More than 1000 l/ha
Low volume	50–200 l/ha	200–500 l/ha
Ultra low volume	Less than 5 l/ha	Less than 50 l/ha

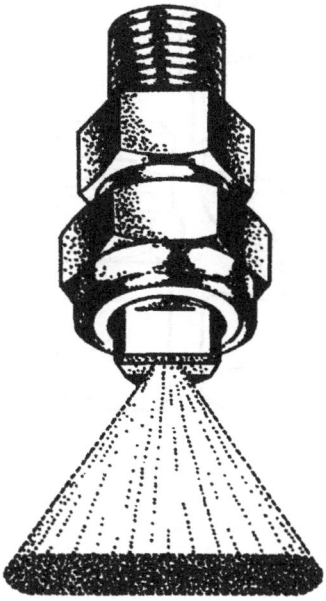

Fig. 12.8. Flat fan nozzle.

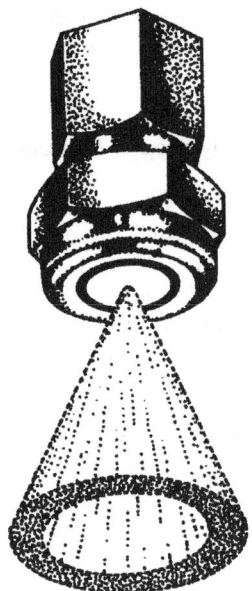

Fig. 12.9. Hollow cone nozzle.

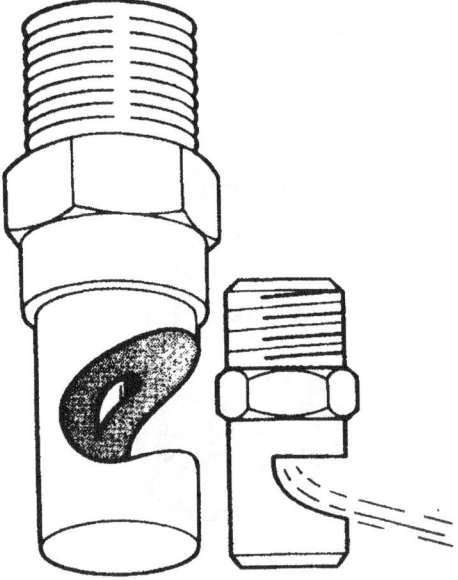

Fig. 12.10. Flood jet nozzle.

greater the proportion of smaller droplets that are formed.

A wide range of interchangeable nozzles is available. Types can be characterized several ways including: the shape of the spray liquid deposit,

e.g. flat fan, hollow cone, solid cone; the mechanism of making the pattern, e.g. orifice, core and disc, anvil or flood jet, air inclusion; the material of the nozzle tip, e.g. brass, stainless steel, plastic, ceramic; and the colour of the nozzle, indicating the output range. The colours have been standardized internationally. Where drift needs to be reduced the use of air-inclusion (AI) nozzles is recommended. They produce a droplet spectrum with few small droplets (driftable) and many larger, air-filled droplets, which are less likely to drift. Much herbicide spraying is now done using AI fan nozzles. Recently AI cone nozzles have become available. They are useful in situations requiring low drift from air-blast sprayers applying high-hazard chemicals to overhead pergola crops, e.g. hydrogen cyanamide to leafless kiwifruit in winter.

Pneumatic breakup, sometimes called air-shear, is commonly employed in machines usually referred to as 'mist-blowers' or 'motorized knapsacks'. Droplet sizes are small, so drift can be a major problem. Low-pressure liquid is introduced into the high-speed airstream without passing through a nozzle. Such mist-blowers are quite different from air-blast sprayers, which are discussed later.

Rotary atomizers are common in ULV systems. It is claimed that the range of droplet size is narrower from these and less spray is needed than when forming droplets in other ways. Currently available CDA mechanisms are more complex and more expensive than conventional hydraulic nozzles.

There are several very important points about operating sprayers.

- Most winds will cause drift, which could be hazardous to crops, people or animals, and operators should be alert to this. However, there may be occasions when a steady, light breeze from an appropriate quarter can be used to advantage to carry droplets away in a safe direction.
- Operating in very low humidity conditions is undesirable. Droplets may evaporate before they reach their target. Avoid dry, hot, midday operation.
- Operating at higher pressures than recommended is highly undesirable because increasing numbers of drift-prone, easily evaporated small droplets will be produced.

W. Atkinson

Spraying machines

Knapsacks

Hand-operated or motorized, these machines are in common use.

HAND-OPERATED These units form droplets by hydraulic breakup and operate at low pressures. A high proportion of large droplets is produced, so drift problems should be minimal. However, when applying herbicides close to sensitive crop plants, such as young shelter belts, shrouding shields over the end of the lance may be added and pressures should be kept as low as possible. Hand-operated knapsacks are normally used for high-volume spraying. A range of hydraulic nozzles is available. Tank sizes are usually in the range 15–20 l, so frequent refilling is necessary, and the area covered per hour is small. Prolonged operation may be tiring.

MOTORIZED Most motorized knapsacks form droplets by pneumatic breakup, and the many small droplets formed have considerable drift potential. These machines should not be used for applying herbicides among fruit crops. They have a place in the low-volume spraying of fungicides and insecticides. However, tank capacities are small, so these machines are best suited for use in young orchards or small plantings.

Boom and 'fixed-head' sprayers

Horizontal booms for spraying weeds with herbicides are usually fitted with fan nozzles, often of the air-inclusion (AI) type. These are designed to be run at relatively low pressures (100–400 kPa; 15–60 psi; 1–4 bar). For fungicide and insecticide spraying of low-growing crops such as strawberries, horizontal booms fitted with cone nozzles are normal. These run at high pressure (500–2000 kPa; 75–300 psi; 5–20 bar). Similarly, for spraying of trellised berry fruits and some grapes, vertical booms fitted with cone nozzles are normal. 'Fixed-head' sprayers have booms formed into an arc with cone nozzles arranged to project spray upwards and outwards for tree, bush and vine crops. These machines offer a cheaper alternative to air-blast sprayers.

The essential components of a typical sprayer using any hydraulic nozzles are set out in Fig. 12.11. Such machines are usually driven from the tractor PTO and may be either mounted or trailed.

Herbicide sprayers are normally fitted with an air-inclusion fan or flood nozzles (Figs 12.8 and 12.10). For sprayers intended for fungicides and insecticides, cone nozzles are fitted (Fig. 12.9).

Air-blast sprayers

Air-blast sprayers are especially designed so that spray droplets penetrate fully inside the tree or crop canopy. Traditionally they have similar basic components to fixed-head orchard sprayers, with the addition of a fan mounted so as to divert large volumes of air past the nozzles. It is these nozzles, usually cone types, that form the spray droplets, which are then conveyed to the crop by the airstream from the fan. A recent development has been to mount smaller, axial-flow fans on booms angled to direct their airstreams into the canopies from above and below. Such booms may be several metres tall to reach the tops of citrus and avocados or as a gantry allowing two sides of a crop row to be sprayed at once.

Fan output in most cases should not be less than 850 m³/min and air velocity at the outlet in the order of 35 m/s. Air-blast sprayers have large power requirements, in the range of 20–35 kW, because of the heavy demands of the fan.

To be effective, air-blast spraying must fulfil two conditions.

- It must replace all the air within the tree with spray-laden air. To do this, the sprayer must not travel too fast.
- The spray liquid must be injected into the airstream so that the droplets are uniformly deposited on the tree. To achieve this, the correct nozzles must be fitted.

Pumps

In general, boom sprayers for herbicide work require low-pressure pumps. Relatively cheap gear, vane or roller pumps are adequate. However, small diaphragm pumps are also now popular. The higher pressures required by cone nozzles usually mean that either diaphragm or piston pumps should be fitted to these sprayers. Their PTO power requirement is usually less than 10 kW. Pumps must be selected on the basis of developing adequate pressure at the required flow rates.

Many modern tractor-driven sprayers rely on a return flow to the tank from the pump to provide hydraulic agitation of the spray mixture. It is a

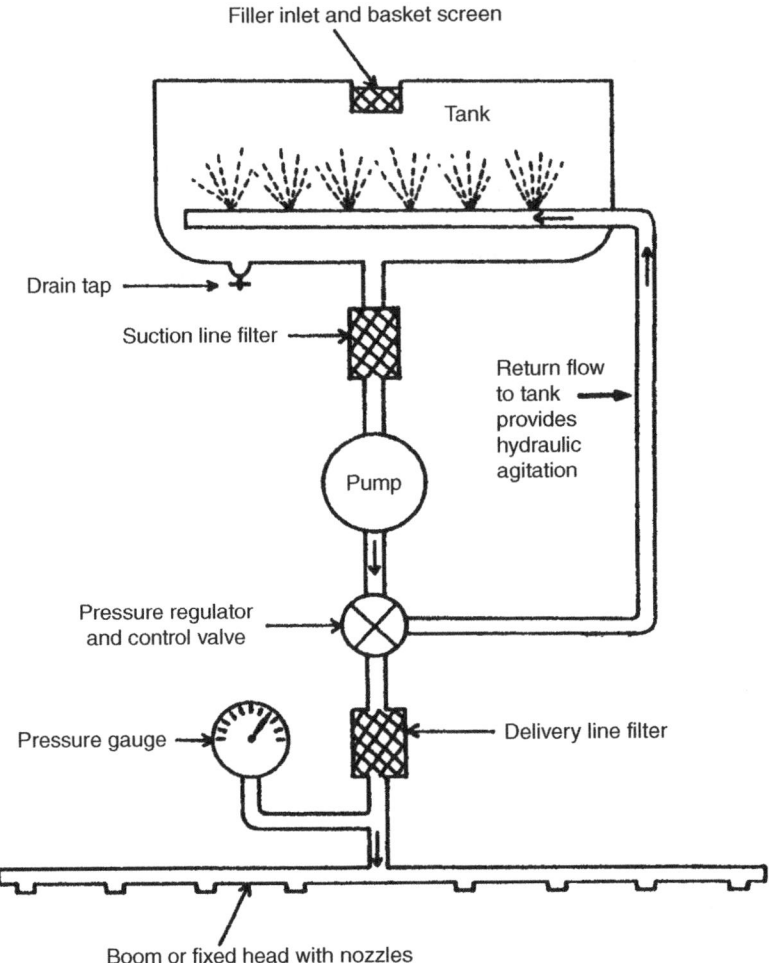

Fig. 12.11. Basic components of a hydraulic nozzle sprayer.

simple method and is now preferred to mechanical agitation by paddles or stirrers. Sprayer pumps should have ample capacity to ensure adequate agitation by returning at least 10% of the volume of the full tank per minute, in addition to that required to supply the nozzles.

Points to consider when choosing a sprayer

- The spray tank should have an unobstructed and rounded base. It is difficult to provide adequate agitation when using wettable powders in square or irregular-shaped tanks.
- The spray tank should be manufactured from corrosion-resistant materials. The most common

are stainless steel, polyethylene and fibreglass, also known as glass-reinforced plastic (GRP).
- The tank filler opening should be more than 300 mm in diameter to allow rapid filling. It should be fitted with a large basket-type screen and a tight-fitting lid equipped with breather. The lowest point of the tank should be fitted with a drain tap rather than a plug. A clearly graduated sight gauge should be fitted where it can be seen from the tractor seat.
- The power required to drive the sprayer, as well as pull it when the tank is full, should not provide a load greater than the power available from the operating tractor.

W. Atkinson

Calibration

All sprayers should be calibrated from time to time to check that their rate of application is correct. Wear rates of nozzle openings can be rapid, particularly with wettable powders. Worn nozzles will cause overspraying. Pressure gauges and tractor tachometers may be reading wrongly. Even machines equipped with computer controllers need to be checked. Growers should consult the operator's manual for the correct procedure for their particular machine. Operator manuals are often lost. It may be that there is information online but, if not, the further reading at the end of the chapter should help. After the recalibration of a machine, check its performance by looking at the spray-deposit pattern, particularly at the centre and top of the crop canopy. Small cards of special water-sensitive paper attached throughout the canopy are particularly helpful for gaining a clear picture of the density and distribution of spray droplets coming from the sprayer under test. For specific chemical recommendations and their rates of application for particular crops, local advisers should be consulted.

Water supply

Clean water is essential for successful spraying, as both physical and chemical contamination can reduce the efficiency of some spray materials. A plentiful supply is required to achieve a quick turnaround when refilling. If town mains or a good bore supply are not available, an elevated storage tank may be useful to accumulate water from a low-volume supply. Such a tank should be covered and its outlet filtered to prevent dust and other airborne debris from entering the sprayer. For fast filling of the sprayer use a large-diameter, flexible hose.

If possible, the water source should be centrally located on the property to reduce travelling time between fills.

Disposal of sprayer washings and chemical containers

The dangerous nature of spray chemicals cannot be overemphasized. This raises the question of how to dispose safely of surplus spray mix as well as the washing water used to flush out the sprayer after use. The contamination of both surface and underground water sources is a real possibility if the nearest ditch or deep pit is used. Many countries have specific regulations for disposal of chemicals and the containers, and growers should first check the local protocol. If there are no local requirements these guidelines are based on first principles.

Small quantities of surplus spray mix and tank washings are best diluted and sprayed back on to the target area. A second method is to spray on to waste ground not used for the production of any food crop or animal feed. There must be no runoff to surface water or percolation to underground aquifers. Livestock must be excluded from this waste ground. The area should be changed from time to time.

The empty chemical product containers must also be disposed of and never used as buckets or drums for other purposes. They should be triple-rinsed and the rinsings poured into the partly filled sprayer tank. The containers should then be punctured. Metal and cardboard items should be buried in an approved, sealed landfill. In some countries plastic containers can be shredded and recycled. Otherwise they also should go to the approved landfill. No containers should be burnt.

Further Reading

Bell, B. and Cousins, S. (1997) *Machinery for Horticulture*, 2nd edn. Old Pond Publishing, Ipswich, UK.

Defra (2006) *Pesticides: Code of Practice for Using Plant Protection Products*. Defra Publications, London.

Furness, G. (2005) *Orchard & Vineyard Spraying Handbook for Australia & New Zealand*. SARDI, Loxton SA, Australia.

Mathews, G.A. and Hislop, E.C. (1993) *Application Technology for Crop Protection*. CAB International, Wallingford, UK.

PART II
Cultivation of Specific Fruits

13 Stone Fruits

Norman Looney and David Jackson

The word stone fruit refers to any one of hundreds of fruit species belonging to the genus *Prunus*. These include apricots, cherries, peaches, plums, almonds and some interesting interspecific hybrids such as the plumcot. Some of these crops are represented by several species; for example, sweet cherries (*Prunus avium*) and sour or tart cherries (*Prunus cerasus*). There are several species of commercial plums, the most important being the European plums and prunes (*Prunus domestica*) and the Asian or Japanese plums (*Prunus salicina*).

Except for almonds and a few apricots where the seed is consumed, the edible portion of stone fruits consists of fleshy exocarp and mesocarp tissues overlying a stony endocarp (the stone or pit). A single seed is contained within this pit, and fruit development seldom continues if this seed aborts. A drupe is the botanical term used to describe such fruits.

Many *Prunus* species are graft compatible, a fact that has proven to be important in commercial fruit production. It is used to adapt some crops to a wider range of soil types and conditions.

Apricots

The apricot species most widely seen in international trade, *Prunus armeniaca*, is native to China and central and western Asia. However, it evolved into the fresh market and processing fruit now popular throughout the western world after it reached the Mediterranean region in about 100 BC. Records suggest that it reached northern Europe before AD 1500. The English called it 'abricock' or 'apricock' (probably arising from the Latin *praecoquus*, meaning early-ripening fruit). It was grown against south-facing walls or in glasshouses but, even after centuries of selection and improvement, there are no commercial apricots well suited to cool maritime climates.

To illustrate the potential of the apricot consider its present-day role in parts of China. On northern China's massive loess plateau, one of the native homes of the apricot, locally adapted cultivars are being planted on steep hillsides, with the mature orchard/forest composed of many different cultivars. The aim is to control soil erosion and provide: (i) fruit for fresh eating, drying, processing and animal feed; (ii) kernels for fresh consumption or as a source of oil for cosmetics and health food products; and (iii) apricot pits and wood for heating.

World production has gradually increased to over 3.0 Mt/year. Turkey, with an average of 528,000 t/year, has been the world leader in recent years, with Pakistan, Iran, Spain, Italy, France, Morocco and the USA being other major apricot-producing countries. Sixty-three countries reported apricot production, but almost half of the total world production comes from countries near the Mediterranean Sea. Apricots are most successfully grown where the summers are warm and dry and the winter-to-spring transition is characteristic of continental climates. Warm temperatures too early in the spring will cause bud break and excessive vulnerability to spring frosts.

Some fruit breeders believe that the great genetic diversity evident in the native apricots of China, north-central Asia, and in countries like Iran, Turkey and Syria will eventually contribute to the development of commercial cultivars suitable for a much wider range of climates. Supporting this belief is the fact that nearly all presently important cultivars belong to the 'European group', just one

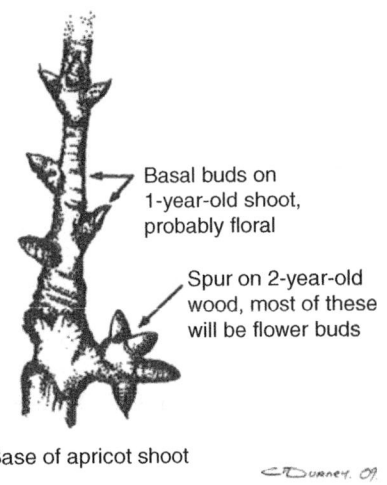

Basal buds on 1-year-old shoot, probably floral

Spur on 2-year-old wood, most of these will be flower buds

Base of apricot shoot

C Durney. 09

Fig. 13.1. Apricots.

of five recognized eco-physiological groups within *P. armeniaca*. This group is believed to have originated from a few introductions into Europe from Armenia and other Middle Eastern countries during the past 2000 years.

Most apricot cultivars are considered to be sensitive to their environment. Thus, important production regions typically rely on a small number of locally adapted varieties. Climatic adaptation is often the overriding factor determining which cultivars are grown, but preferences also relate to tolerance to pests and diseases present in different regions (Table 13.1) or to differing use patterns. Few cultivars are equally suited for fresh sales and processing. Furthermore, some processed products are associated with a specific cultivar. Fresh-market cultivars must have good appearance, medium to large fruit size, good flavour, reasonable tolerance to brown rot, and be capable of maintaining good quality after handling, storage, shipping and marketing.

Prunus mume and *Prunus ansu*, two close relatives of apricot, are also grown as food crops. They perform better in humid regions and are especially important in China, Japan and Korea. *P. mume* was introduced into Japan from China about 2000 years ago. *P. ansu*, Japanese apricot, is also believed to have originated in China.

Mume is valued for its beautiful springtime display of white, pink or red flowers. Only in recent centuries has it become important as a food crop. The astringent fruit is not suitable for fresh eating but 'ume-boshi' pickle and 'umishu' liquor

are considered to be essential contributors to a healthy diet in modern Japan. Mume is cultivated on about 20,000 ha in Japan. The Japanese apricot, locally called anzu, is a minor crop grown side-by-side with apple. It has larger fruit than the mume but high acidity makes them unsuitable for fresh eating. Anzu is used to make jams and syrups.

Key points

Site selection

Even though apricot trees can tolerate quite low winter temperatures, commercial apricot production is most commonly found in climatic zones where the hazard of spring frost is not excessive and where warm, dry summers reduce disease pressure. There are such regions on all continents and across a wide range of latitudes. At the higher latitudes, however, site selection is especially important to minimize spring frost injury. Choosing a sloping site with good air drainage will reduce the risk of frost damage. Spring bud development is delayed on a north-facing slope (in the northern hemisphere) or a site close to a large body of water.

Lack of chill units may be a limitation at low latitudes. Here, sites at higher elevations may be superior. Apricots prefer silt loam or sandy loam soils with good depth and drainage, but with careful selection of rootstocks it is possible to achieve good productivity on heavier and wetter soils.

Cultivars and pollination

While there are hundreds of important cultivars of apricot, only a few are important in more than one country or region. For example, very few of the cultivars grown in Italy are found in France. This indicates that apricot cultivars are very site sensitive and closely tied to specific market demands. One must obtain local advice about suitable cultivars before planting. Relatively few fresh-market cultivars are considered well suited for long-distance shipping, and few of the apricot cultivars used primarily for processing are also suitable for fresh-market sales.

Pollination of apricot is seldom a problem since most cultivars are self-fertile. With these cultivars, even a small population of insects will result in

Table 13.1. Apricots.

Botanical, anatomical and physiological aspects	Climatic, geographic, soil and water requirements	Cultural aspects
Common names Apricot	*Temperature needs* Temperate zone fruit. Apricots very demanding, prefer Mediterranean or continental climate	*Propagation* By budding in late summer. Most rootstocks are grown from seed
Botanical name *Prunus armeniaca*		*Rootstocks* Apricot and peach seedlings commonly used, Myrobalan plum often recommended for heavy or wet soils
Botanical name of related and useful species *Prunus mume* syn *Prunus ansu* (Japanese apricot)	*Frost tolerance* A problem due to early flowering. −4°C will damage swelling buds, −2°C will damage flowers and −1°C small fruit	*Spacing* Often planted on the square at 5–6 m, but rectangular spacing and hedgerow systems are increasingly popular
Type of plant and size Trees 4–5 m high, canopy 4–5 m in diameter at base, round and bushy	*Winter hardiness* Fully dormant flower buds hardy to about −25°C. Leaf buds and wood to about −35°C	*Training and pruning* Trees adapt well to vase or open-centre training. Prune to encourage annual growth, but in areas prone to silver leaf (*Chondrostereum purpureum*) or gummosis (*Botryosphaeria* spp.) prune either after cropping or with minimum cuts in winter, which should be sealed. Prune in early summer where bacterial canker (*Pseudomonas syringae* pv *syringae* or *Pseudomonas syringae* pv. *morsprunorum*) is a problem
Sexuality Hermaphrodite		
Pollination Mostly self-fertile but some important exceptions. Insect pollinated	*Water needs* Requires adequate moisture during growing season to produce a commercial crop	*Thinning* Blossom thinning with desiccants such as ammonium thiosulphate has shown promise but may be risky in frosty areas. Hand thin before pit hardening to two fruits on spurs and to single fruits spaced about 8 cm on 1-year-old wood. Remove small fruit preferentially
Flower buds Produced laterally on 1-year-old shoots or on spurs on older wood. On laterals, two flower buds, each with one flower, normally sit either side of one vegetative bud. On spurs large numbers of flower buds are grouped together. Flowers have five petals, five sepals, numerous stamens and a single carpel. Flowers are initiated in late summer	*Tolerance of wet soils* Peach and apricot rootstocks lack tolerance to waterlogging	*Tillage* Orchards normally managed with grassed alleyways and herbicide or tillage under the trees
	Drought tolerance Moderate to good	*Time to first harvest* 2–3 years
	Humidity tolerance Apricots do best in dry areas, especially with low rainfall at blossom and at maturity. Rain close to harvest may cause severe loss due to fruit cracking and fruit rots. See section on Cherries. *Mume* and *ansu* are more tolerant of high humidity and seasonal rains	*Time to full production* 5–6 years
		Expected yields 2 years: 5 t/ha 4 years: 10 t/ha 6 years: 20 t/ha but depends greatly on cultivar

Continued

Table 13.1. Continued.

Botanical, anatomical and physiological aspects	Climatic, geographic, soil and water requirements	Cultural aspects
Growth of fruit Drupe with double-sigmoid growth curve. Stone hardens in second stage of growth and is not attached to the flesh (freestone)	*Wind tolerance* Moderate *Edaphic features* Since flowering is early, frost-free sites are especially valuable	*Normal productive life* 25–40 years. Less in areas with severe disease problems *Storage* Will store at 0 to –1°C for 1–2 weeks. Cooling prior to and during transport will give longer shelf life. CA storage has not been successful
Time of bud burst Early spring	*Soil needs* Prefer good deep soils. Avoid wet areas	*Method of harvest* Hand harvest. Fruit will last 1–2 weeks if picked when green has faded to pale straw colour. Best flavour when picked mature on tree
Time of flowering Flower buds are first to open. Leaf buds open several days later	*Nutrient requirements* Fairly demanding for nutrients; nitrogen considered very important for promoting good crops. See Chapter 9	*Main diseases* Blossom blast (*Pseudomonas syringae*) can be a major problem in wet springs and this pathogen also causes bacterial canker. Brown rot (*Monilinia* spp.) at harvest (and also at blossom time) can be a major problem. Shot hole (*Wilsonomyces carpophilus*) also known as Coryneum blight is a serious disease in most production areas. The following diseases are regionally important: perennial canker (*Leucostoma cincta*), *Eutypa* canker (*Eutypa lata*), bacterial spot (*Xanthomonas campestris* pv. *pruni*), anthracnose (*Glomerella cingulata*) and sharka (plum pox virus)
Time of fruit maturity Most cultivars ripen in early to midsummer, depending on latitude		*Main pests* Compared with many other fruit crops, apricots have relatively few insect and mite pests. The following can be damaging in some locations and seasons: peach twig borer (*Anarsia lineatella*), shothole borer (*Scolytus rugulosus*), peach tree borer (*Synanthedon exitiosia*) in North America, leafrollers (several species of tortricid moths), spider mites (several species), lecanium scale (*Parthenolecanium corni*) and aphids (several species)

N. Looney and D. Jackson

adequate fruit set. However, there are important exceptions, such as Sundrop, an important cultivar in New Zealand, where provision must be made for pollinizer varieties and pollinating insects.

Pruning and training

Apricot trees are often trained to resemble open-centre peach trees. The mature tree has four to six main scaffold limbs and enough sublateral branches to achieve a full canopy. Annual pruning is required to control tree size, regulate tree shape and maintain fruit yield and quality. However, since apricot bears on 1-year shoots as well as on spurs on older wood, annual pruning differs from that of peaches. Still, a great deal of crop regulation (thinning) can be accomplished by removing or shortening 1-year-old shoots.

Some cultivars, Tilton being a good example, are prone to biennial flowering. This can be avoided by good nutrition, pruning to encourage new wood and by thinning the fruit very early in the growing season.

Pests and diseases

The skin of apricots is easily damaged by pests and diseases. Freckling or spotting on the fruit surface seriously reduces the value of the crop. This may be caused by early attacks of blast, European red mite, shot hole and bacterial spot (Table 13.1). Powdery mildew may also cause fruit marking. Brown rot at harvest causes serious fruit loss in many regions. This pathogen also attacks at blossom time, where the symptoms are blossom blight and twig cankers. Failure to control these bloom-time infections will result in serious losses later in the season. Careful spraying is required to keep such problems under control.

Each producing region of the world has its own array of insect and mite pests of apricot. Many of these pests also attack peach and other stone fruits. The European red mite can be a problem with most fruit trees. The peach twig borer and the shot hole borer are lepidopterous pests of both apricot and peach. Leafrollers and fruit worms attack many fruit crop species and can be a problem early in the growing season. Green peach aphid is a pest common to many regions. Seek local advice for materials and procedures to control these pests.

Regulating cropping

Most apricot cultivars set very heavy crops, which must be thinned by hand early in the growing season if fruit size and quality is to meet market expectations. Thinning often starts during the flowering period, where a proportion of the bloom is removed by desiccating chemicals or by hand. More commonly, hand thinning of young fruits occurs within 30–45 days after bloom.

Harvesting and handling

Apricots develop excellent quality when allowed to ripen on the tree, but for practical reasons this is seldom desirable. Fresh-market apricots, as well as fruit intended for canning, are harvested by hand, and multiple picks select those fruits in the early stages of ripening. Flesh colour and background skin colour are important guides to harvest maturity, as are fruit size and shape (fill). Apricots can be held in cold storage (0°C) for 2–3 weeks to permit orderly marketing and processing.

Apricots for drying are picked much riper than for the fresh fruit trade, and in some regions are removed by shaking. They are split into halves and the stones removed, either by hand or by machine. The halves are spread on trays, which are put on racks in a shed in which sulphur is burned for 6–7h to preserve the fruit and its colour. The trays are then placed on the floor of the plastic houses for 3–5 days or, in hot climates, in the sun.

Cherries

More than 100 species and at least 1500 cultivars of cherries are known. The cherry is indigenous in some form or another in all the countries of the northern hemisphere temperate zone, but the present commercial varieties probably originated in the Caucasus, between the Black and Caspian Seas. Records of cherries are found in the very earliest historical writings and, even today, many important cultivars are discovered as chance seedlings.

Progenitors of most important cherries in modern horticulture are thought to be 'races' referred to in the early English literature as Mazzard or Geans and Bigarreaux (modern sweet cherries; *P. avium*) and Kentish (Amarelles) and Morellos (i.e. two races of tart or sour cherries; *P. cerasus*).

In the Pacific Northwest, USA, large new sweet cherry plantings began to replace less-profitable

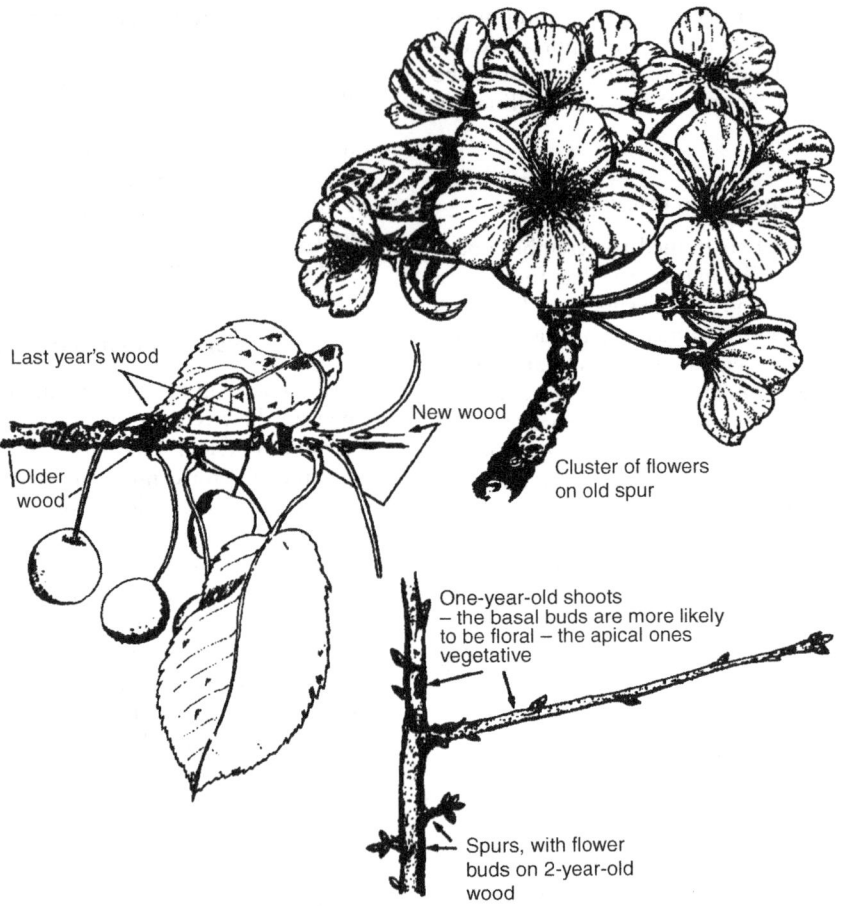

Last year's wood

Older wood

New wood

Cluster of flowers on old spur

One-year-old shoots – the basal buds are more likely to be floral – the apical ones vegetative

Spurs, with flower buds on 2-year-old wood

Fig. 13.2. Cherries.

apple orchards in the 1990s. At that time, in Washington State, it was common to see huge piles of uprooted apple trees mounded together, ready for burning. Soon after, those blocks were planted to cherries, and that expansion has continued to the present.

However, this trend has not been confined solely to the Pacific Northwest or even to the USA. Data from the UN, FAO indicate that harvested acres of cherries have increased from 285,000 to over 400,000 hectares in the last two decades, ending 2005. According to the same source, harvested acreage expanded in the last decade in every region of the world except Eastern Europe. The largest expansion was recorded in Asia (less the Near East), with an increase of nearly 54%, followed by North America at 48%.

Key points

The following points relate mainly to sweet cherries (Table 13.2).

- They are particularly attractive to birds, partly because they are one of the first fruits to ripen.
- Pollination and fruit set can be problematic. Many cultivars are self-incompatible; wet weather at flowering can dramatically reduce set.
- Rain at or near harvest will induce fruit cracking.
- They are susceptible to difficult-to-control diseases, such as bacterial canker and silver leaf.
- Labour costs are high and hand harvest is tedious. Large trees are part of the problem. Harvest must be completed in a short time. Getting the fruit to market in good condition is a major challenge.

N. Looney and D. Jackson

Table 13.2. Cherries.

Botanical, anatomical and physiological aspects	Climatic, geographic, soil and water requirements	Cultural aspects

Common names
Sweet cherry, sour (or tart) cherry

Botanical names
Prunus avium
Prunus cerasus

Related and useful species
Prunus mahaleb (Mahaleb cherry)
Prunus fruticosa (ground cherry)
Prunus tomentosa (Nanking cherry)

Type of plant and size
Bushy tree; unchecked, *P. avium* may grow to 20 m or more. Annual pruning needed to keep trees to a manageable size. Individual shoots branch less than other stone fruit and tree is naturally more open. *P. cerasus* forms a smaller and bushier tree, usually with a pendulous growth habit

Sexuality
Hermaphrodite

Pollination
P. avium mostly self-incompatible, and choice of pollinators is important (see Key points). Poor set can be a problem and contributes to poor yields. *P. cerasus* mainly self-fertile but with important exceptions

Flower buds
Produced laterally and terminally on 1-year-old shoots and on spurs on older wood. Flower buds are separate and contain three or more flowers. Flowers have five sepals, five petals, one carpel and numerous stamens

Growth of fruit
Fruit is drupe, borne in a cluster on a long pedicel. It exhibits a double-sigmoid growth curve with endocarp hardening during stage II

Time of bud burst
After apricot but often ahead of plums and peaches

Time of flowering
Flowers open slightly before leaves

Temperature needs
Temperate but fairly tolerant of a wide temperature range. Not successful in warm, humid regions or areas with warm winters. Cherries require more chilling than peaches, sour cherries more than sweet cherries

Frost tolerance
Bud swell and flowering, –3°C; small fruit, –2°C.

Winter hardiness
Fully acclimatized flower buds of sweet cherry hardy to about –30°C, xylem tissues to about –35°C. Comparable values for sour cherry –35 and –40°C

Water needs
To maintain good fruit size and annual cropping it is important to avoid serious drought stress. Water requirements for both species similar to other deciduous fruit trees, even though the crop matures in midsummer. Fruit size reduction is main cost of water stress

Drought tolerance
Moderate. Probably greater for sweet cherry

Humidity tolerance
Poor for sweet cherry. Rain at blossom reduces set and increases disease. Rain at, or just before, maturity causes splitting of sweet cherry

Wind tolerance
Moderate

Edaphic features
Flat land preferred

Soil needs
Most particular of all stone fruit. Do best in deep, rich loams. Mazzard stock has moderate tolerance of wet soils and performs poorly in light soils; Mahaleb stock better on lighter, better-drained soils

Propagation
Mostly grafted or budded on seedling or clonal rootstocks

Rootstocks
Mazzard selection of sweet cherry is most common and generally used under favourable conditions. Mahaleb is sometimes used and may result in a smaller tree; while more drought tolerant, it is incompatible with some cultivars. Dwarfing rootstocks now becoming available for sweet cherry

Spacing
Row spacing for sweet cherries often 5–6 m. In-row spacing may be as low as 3 m, depending on rootstock and pruning and training strategy. Sour cherries require less space

Training and pruning
Usually grown as open-centre trees; after formation of the desired framework, little detailed pruning is required. Periodic branch renewal helps to maintain good fruit size. Trellis systems such as Tatura trellis and high-density central-leader systems increasingly popular

Thinning
Neither feasible nor sufficiently beneficial

Tillage
Orchards normally grassed, with chemical weed control under trees

Time to first harvest
3–4 years

Time to full production
7–10 years

Expected yields
4 years: 2 t/ha
6 years: 5 t/ha
8 years: 10 t/ha

Normal productive life
25–35 years. Less where diseases such as bacterial canker or pollen-transmitted viruses are prevalent

Continued

Table 13.2. Continued.

Botanical, anatomical and physiological aspects	Climatic, geographic, soil and water requirements	Cultural aspects
Time of fruit maturity Earliest stone fruit, mostly before midsummer but new late-season cultivars have extended season. Sour cherries tend to mature later than sweet cherries	*Nutrient requirements* Needs good nutrition and responds to nitrogen. See Chapter 9	*Method of harvest* Shake-and-catch harvesting common with sour cherries but most sweet cherries are hand picked with stems attached
		Storage Sweet cherries: Rapid removal of field heat is recommended. Will store for up to 3 weeks at 0°C. Sealed polythene box liners will further extend storage life. CA storage has been used with 3–10% O_2 and 10–12% CO_2. Growing interest in modified atmosphere (MA) packaging. Sour cherries: processing should occur immediately after harvest
		Main diseases Cherry leaf spot (*Blumeriella jaapii*) is a serious disease throughout Europe and North America; sour cherry most susceptible. Bacterial canker (*Pseudomonas* spp.) is a worldwide problem and most severe on sweet cherry. Brown rot (*Monilinia fruticola* and *Monilinia laxa*) destroys both flowers and fruit; sour cherries are somewhat resistant. Silver leaf (*Chondrostereum purpureum*) is a particularly serious fungal disease in New Zealand, Chile and France. *Verticillium* spp. wilt is caused by this soil-borne fungi, and soil fumigation is required if cherries are to be planted after other susceptible crops such as tomatoes, strawberries, etc. Cherries are susceptible to a wide array of virus and virus-like diseases spread by pollen or by insect vectors. The use of certified virus-free propagation wood is essential
		Main pests The most important pests of cherry are the various species of cherry fruit fly (*Rhagoletis* spp.), several species of leafrollers (*Archips* spp. and *Pandemis* spp.) and the black cherry aphid (*Myzus cerasi*). Spider mites can build up to damaging levels and the pear sawfly or cherry slug (*Caliroa cerasi*) is emerging as a serious pest in many regions

N. Looney and D. Jackson

Although pruning has not been considered a pivotal operation in successful cherry production, it is increasingly evident that more attention to this practice is warranted. Fruit size and quality decline very significantly as spurs age and main lateral branches lose vigour.

Cultivars and pollinators

SWEET CHERRIES There are hundreds of important cultivars of sweet cherry, but in any given country or region it is common for two or three cultivars to predominate. There are usually good reasons for this, often relating to historical consumer preferences (e.g. the popularity of Bing sweet cherry in North America). But there are other important marketplace forces. Since the earliest cherries often give the highest per unit returns, early-ripening cultivars predominate at lower latitudes. The opposite strategy is used in western Canada and Norway, where late-maturing cultivars are preferred, both for their fruit quality and for the fact that they are free of competition from warmer production regions. Australia, Chile and New Zealand export fruit to the northern hemisphere during the Christmas and New Year celebrations. Here the important considerations are time of ripening, suitability for long-distance shipping and fruit quality attributes appropriate for the market served. For example, white- or yellow-fleshed sweet cherries have traditionally been preferred by Japanese consumers.

Thus, while an extensive list of cultivars could be included in this section, it is much more important to recognize that the cultivar chosen for planting must have some clear marketplace advantage. This kind of information is best obtained from experienced and innovative producers, packers and shippers in a particular country or region.

There are, none the less, serious problems in sweet cherry production and marketing that are clearly related to cultivar. Some cultivars with very attractive, large and high-quality fruit cannot be grown profitably because of poor fruit set and cropping. Others are prone to fruit disorders that appear after storage and shipping. Many cultivars are considered to be too susceptible to rain cracking and others to diseases such as brown rot (*Monilinia fructicola*), leaf spot (*Blumeriella jaapii*) or bacterial canker (*Pseudomonas* spp.). Unfortunately, strong field resistance to any major pest or disease is yet to appear in a commercially attractive sweet cherry cultivar.

While virtually all of the sweet cherry cultivars important in commerce over the past 50–100 years arose as chance seedlings, most modern cultivars have come from formal breeding programmes. Improved fruit size and quality seem to be the main considerations when choosing a new cultivar, but there have also been important advances in tree productivity, precocity, cropping reliability (via self-fertile cultivars) and in extending the marketing season with later cultivars. Some new cultivars are considered to be significantly less prone to fruit cracking and others seem particularly well suited for shipment to distant markets.

With the advent of self-fertile cultivars (i.e. cultivars not requiring cross-pollination), the issue of sweet cherry pollination has become less problematical. These cultivars effectively pollinize all other sweet cherries. However, it is important to understand that among the self-sterile cultivars (and such cultivars still predominate) not all combinations are compatible. One must plant a cultivar known to produce pollen that is compatible with the main crop cultivar.

SOUR CHERRIES One sour cherry cultivar, Montmorency, accounts for virtually all North American production. This ancient French cultivar has proven to be well suited for mechanical harvesting and mechanical pit removal. Furthermore, Montmorency, an Amarelle-type sour cherry with clear juice and distinctive flavour, is well known to North American consumers of cherry pies. Thus, since the North American industry is totally geared to the efficient production of a fruit for use in pie fillings, the predominance of Montmorency is unlikely to change.

By comparison, the sour cherry cultivar picture in Europe is exceedingly complex. *P. cerasus* is believed to have arisen via natural hybridization between *P. avium* (sweet cherry) and *Prunus fruticosa* (the ground cherry), and across Europe one can find sour cherries that clearly favour one or the other of these parent species. Most of the sour cherries important in Europe belong to the group of cultivars classed as Morellos with red to dark red juice. Some of these cultivars are used in the manufacture of a very specific product (e.g. Stevnsbär and Kelleris, used in Denmark for juice and wine production) but others, such as Schattenmorelle, are grown for the manufacture of a wide range of products. A few sour cherry cultivars have sufficiently large, sweet, attractive and durable fruit for fresh eating. The several clones of the high-quality Hungarian sour cherry, Pándy, are good examples.

Pruning and training

SWEET CHERRIES Bearing sweet cherry trees have a tendency to produce flower buds near the base of new shoots (usually the lower four to five nodes) and vegetative buds towards the tip. Spurs are formed on older wood. Thus, heading cuts on 1-year-old shoots do not remove many flower buds. Apical dominance is quite strong for sweet cherry and such heading cuts are often needed to stimulate lateral branching as well as to control tree height.

Sweet cherry trees have traditionally been planted on the square, spaced 6 m or more apart, to accommodate a spreading, open-centre tree. However, hedgerow orchards with central-leader or modified central-leader trees can be successful, especially if a size-controlling rootstock is employed.

To establish the fruiting structure of a central-leader tree, the leader of a strong nursery-grown tree, preferably a feathered tree, is headed back after planting. About ten lateral branches are selected over the next 3 years and each is headed back annually to encourage the development of spurs and sublateral branches, thus increasing the potential fruiting area. However, excessive pruning will delay fruiting. Branch positioning with spreaders or a trellis system can sometimes be used to achieve the desired tree shape with less pruning.

Once the framework has been formed, pruning should be reduced to the minimum essential to maintain a balance between fruiting and growth, while ensuring that the tree is not overcrowded and that all parts are accessible to pickers. Spurs remain productive for many years and renewal of older branches is only needed if a decline in productivity or fruit size is noted.

The time of pruning is critical in areas subject to bacterial canker or silver leaf infections. Just before or after bud break or during the summer growth period are pruning 'windows' usually considered best for avoiding these diseases. Summer pruning over several seasons, however, can lead to reduced tree vigour. Winter pruning, on the other hand, tends to invigorate the tree and may explain the larger average fruit size achieved in regions where this practice predominates.

SOUR CHERRIES Sour cherry exhibits considerably less apical dominance than sweet cherry and, without annual pruning, will develop a bushy tree with a weeping growth habit. There may also be an excessive amount of blind wood and many branches with poor attachment angles. Poor light penetration results in poor fruit quality and such trees are not ideal for shake-and-catch harvesting. Fruit removal is especially difficult from long, pendulous branches.

A tree shape better suited to mechanical harvesting and producing good-quality fruit is the modified central-leader tree. Such a tree has six to eight scaffold branches arising from a central trunk. These branches are selected during the first 2 or 3 years in the orchard, during which time it is important to retain the dominance of the central leader. This involves removing lateral branches below the tip that threaten to overgrow the leader. As the tree matures and reaches full production, the central leader, and perhaps one or two scaffold branches that shade the interior and lower portions of the tree, can be removed.

Some detailed annual pruning will be required throughout the life of the tree to prevent blind wood and to ensure that there is enough vigorous new growth to maintain an adequate supply of flower buds.

Harvesting and handling

SWEET CHERRIES Cherries do not ripen after picking, they only undergo senescence. Thus, they are picked when fruit maturity (often judged by surface colour) matches the marketplace requirement. Sweet cherries for fresh-market sales (and those used for brining – the manufacture of Maraschino or glace cherries) are harvested with stems and ahead of full ripe colour development. Interestingly the condition of the stem (colour, turgidity) is used in marketing channels to judge condition of the fruit, even though this relationship is not particularly strong. Fruit intended for canning, freezing or juice and jam making is harvested, sometimes mechanically, without stems. Sweet cherries for processing (other than brining) are harvested when fully mature. Pre-harvest fungicide sprays are a normal requirement for quality sweet cherry production.

A variety of containers are used for picking, from small buckets with a hook that can be attached to the tree, to boxes which can be attached by straps to the picker's shoulders. Depending on the quality of the fruit on the tree, sorting can occur as the fruit is picked and transferred to boxes (i.e. field packing) or it takes place later in a packing shed. In addition to sorting, cluster cutting and size grading are important operations in most commercial packing sheds. The rapid removal of field heat and continuing cold storage are considered essential if fruit is to have any significant storage or shelf life (Chapter 7).

Cherries are packed in small wooden or cardboard cases, according to the preference of the market. Fruit is normally jumble packed in boxes, but for special lines, specified packs will be needed (Chapter 7). Hand positioning of fruit (facing) is increasingly rare.

Peaches and Nectarines

Despite its botanical name (*Prunus persica*) the peach originated in China and not Persia. Nevertheless, it was cultivated in the Middle East long before being introduced to Europe, probably reaching modern-day Iran before 100 BC. In China, it was mentioned in poems of the 10th century BC.

Nectarines are peaches without pubescence, a characteristic defined by a single recessive gene but associated with other traits, such as increased pest

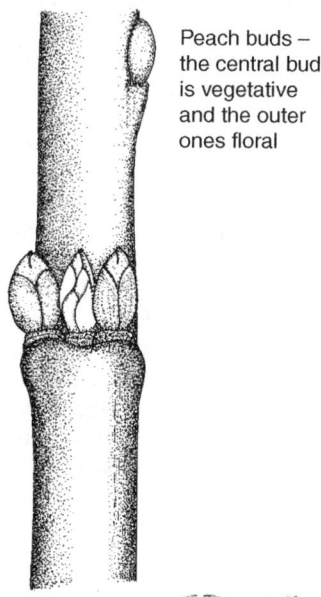

Peach buds – the central bud is vegetative and the outer ones floral

CDurney. 09.

Fig. 13.3. Peaches.

susceptibility, important to the commercial producer. Most peaches and nectarines have yellow flesh, but white-fleshed cultivars have always been known and valued. Again, the yellow flesh and aromatic white flesh characteristics are defined by a single dominant/recessive gene pair. Red-fleshed peaches are also known.

Another characteristic of great importance in commerce is the extent to which the pit remains attached or separates from the flesh during fruit ripening (clingstone versus freestone). The most important of the processing peaches are of the clingstone type and have firm flesh, even when ripe. On the other hand, most fresh-market peaches and nectarines are classed as freestone and are more likely to have tender, juicy (melting) flesh when fully ripe. Some freestone peaches are also processed for pie filling and other products.

World production of peaches and nectarines (FAO statistics do not separate these two crops) stands at about 17 Mt. While more than 70 countries report annual production statistics, two regions stand out as being the most important: the countries of the Mediterranean basin produce about 40% of the total, and China, Japan and Korea combined produce another 30%. Considering individual countries, China is by far the largest producer, with annual production approaching 8 Mt; Italy is second at 1.7 Mt and the USA third (1.3 Mt). Spain, France and Greece are major European producers. Chile, South Africa and Australia are important southern hemisphere producers.

While peaches are an important processing crop in California, Greece, Australia and South Africa, the bulk of the industry is devoted to producing fruit for the fresh market. In fact, increased production in recent years is largely explained by the growing international trade in fresh-market peaches and nectarines. Improved cultivars that extend the season and are specifically adapted to certain regions and marketing channels have contributed importantly to this growth.

The peaches grown for processing peach halves, slices or fruit cocktail in bottles and cans are primarily firm-fleshed clingstone cultivars. These fruits are mechanically harvested and subsequently mechanically peeled and pitted in very efficient processing facilities. Peaches intended for drying or freezing are freestone cultivars harvested when firm-ripe to maximize sugar content.

Peaches and nectarines grown for fresh consumption are usually classed as freestone (although there seem to be considerable differences among cultivars) and nearly always have tender, juicy flesh when fully ripe. They are harvested while still firm but with the expectation that they will ripen quickly once removed from cold storage. With a range of cultivars with different ripening dates, California and other countries with a Mediterranean climate harvest fruit over a period of more than 4 months and deliver fresh fruit to the market for at least 5 months.

Key points

Cultivars and pollination

There are many active peach and nectarine breeding programmes around the world and improved cultivars intended for the fresh market are introduced every year. This availability of new plant materials, added to the fact that peaches and nectarines have a relatively short tree life, has led to a constant renewal of the peach and nectarine variety picture. In California, for example, more than 100 nectarine cultivars are represented in that state's annual production of about 225,000 t. None of these cultivars were popular 30 years earlier. A similar picture exists elsewhere. Obviously, it is very important for producers to stay up to date with this rapidly changing cultivar scene.

Most of the recent progress in breeding has been with fresh-market cultivars, where productivity, fruit appearance and suitability for handling and shipping have improved substantially in recent decades. Families of cultivars are now available where individual cultivars differ little except for season of ripening. This permits the shipping of a highly uniform product over several months.

Less attention has been devoted to breeding new peaches for processing. The ideal canning peach would be a highly productive tree with firm, fine-textured and non-browning fruit flesh, attractive flesh colour and small, round pits not prone to splitting. Mechanical harvesting technology demands a uniform population of mature fruit on a given day. One can also be quite sure that the introduction of important pest and disease resistance would be welcomed by this industry. One or two varieties dominate production in each of the important peach-processing countries. Examples are Golden Queen in Australia and Halford in California. It appears that existing cultivars are considered largely satisfactory.

Most new peach and nectarine cultivars have proven to be adapted to a rather narrow geographic and climatic range, which may explain the large number of cultivars in use around the world. Furthermore, little progress has been made in introducing pest and disease resistance to a widely adapted peach or nectarine scion cultivar. For example, while bacterial leaf spot (*Xanthomonas campestris* pv. *pruni*) resistance has been achieved with some peach cultivars, there are many strains of the disease, and broad-spectrum resistance has yet to be discovered. Important differences in susceptibility to peach leaf curl, powdery mildew and brown rot have been reported but, to date, there are no fully resistant cultivars.

With very rare exceptions, peach trees are self-fertile and can be planted in solid blocks. While they are normally insect pollinated, the addition of bee colonies is not usually required to get a full crop.

Pests and diseases

In addition to the many pests mentioned in Table 13.3, plant parasitic nematodes can be a very important problem in some parts of the world. They transmit serious virus diseases, reduce root efficiency and generally debilitate the tree. Pre-plant fumigation is often advised. The use of nematode-resistant rootstocks is commonplace in some regions.

There are many important diseases of peach and nectarine trees and fruit. Fortunately, not all are present in all producing regions. Some of these diseases can cause tree death. Silver leaf (*Chondrostereum purpureum*), bacterial canker (*Pseudomonas syringae*) and armillaria and phytophthora root and crown rots are good examples. Important diseases of flowers, foliage and fruit include blossom blight in the spring and brown rot at maturity, both caused by *Monilinia* spp. fungi. This pathogen is difficult to control, especially in wet climates. A similar disease at blossom time, blossom blast, is caused by the bacterial canker organism. Peach leaf curl (*Taphrina deformans*), while relatively easy to control, can damage both foliage and fruit. Shot hole disease (*Wilsonomyces carpophilus*), also known as Coryneum blight, can damage fruit and kill twigs and buds.

Various virus or virus-like diseases reduce tree productivity and fruit quality and some can even result in tree death. The most important peach tree

N. Looney and D. Jackson

Table 13.3. Peaches and nectarines.

Botanical, anatomical and physiological aspects	Climatic, geographic, soil and water requirements	Cultural aspects
Common name Peaches and nectarines *Botanical name* Prunus persica *Type of plant and size* Round bushy tree 4–6 m high, 3–5 m in diameter *Sexuality* Hermaphrodite *Pollination* All important cultivars are self-fertile and insect or wind pollinated. Introducing bees usually considered unnecessary. Dry weather at blossom time preferred *Flower buds* Produced laterally on 1-year-old shoots. Normally two flower buds surround a leaf bud. Flower has five sepals, five petals, a single carpel, numerous stamens and is sessile *Growth and type of fruit* Natural fruit set usually excessive, requiring careful hand thinning. Peach fruits are a drupe exhibiting a double-sigmoid growth curve. Stone hardens during the last phase of fruit growth (stage II). Peaches have a fuzzy skin and nectarines a smooth skin. Flesh colour white or yellow, some red-fleshed cultivars known. Peaches are either freestone, semi-freestone or clingstone	*Temperature needs* A temperate zone fruit but grows more satisfactorily and develops better fruit quality in regions with warm summers *Frost tolerance* Can be a problem because of early flowering: −4°C will damage bursting buds, −3°C blossom and −1°C small fruit *Cold hardiness* Among the least hardy of the stone fruits but fully dormant buds will withstand about −20°C and bark about −25°C *Water needs* Need plentiful water supply, especially in final fruit swell (stage III). Reducing irrigation after harvest can help control vigour *Tolerance of wet soils* Do not select areas where water fails to drain away; peach rootstocks have no tolerance to waterlogging *Drought tolerance* Moderate	*Propagation* Budding is the normal method. Some success with cuttings has been obtained but this practice is not in common use *Rootstocks* Peach seedlings most common, but source of seed considered important with respect to graft compatibility and cold hardiness. Some seed lines (e.g. Nemaguard) resistant to nematodes; others may impart some tree vigour control *Spacing* Traditionally planted on the square at 6 m. Now commonly planted much closer in hedgerows –see below *Training and pruning* Trees are naturally bushy and adapt well to open-centre systems. However, central-leader trees can be developed and maintained in hedgerow plantations. The most popular technique is to space trees about 2–3 m in rows spaced 4.5–5.0 m apart and train as central leaders. Annual pruning practices aim to encourage the development of new shoots along the full length of scaffold branches. One-year-old shoots are thinned out and often shortened back before bud-break and removed after cropping. Where silver leaf and bacterial canker occur, prune after harvest but before leaf fall. Summer pruning to expose the fruit to direct sunlight and to remove water sprouts is often considered desirable *Thinning* Flower and fruit thinning is largely done by hand and is often the most expensive single operation in peach production. Early thinning maximizes fruit size *Tillage* Grassed alleyways and a weed-free tree row, often slightly raised, is the most common orchard floor management system. However, spring tillage of the alleyway to incorporate prunings and brown rot mummies' is sometimes deemed an essential orchard sanitation practice. This practice is combined with the use of an annual cover crop

Continued

Table 13.3. Continued.

Botanical, anatomical and physiological aspects	Climatic, geographic, soil and water requirements	Cultural aspects
Time of bud burst Early spring – 3 weeks before apples	*Humidity tolerance* Blast and brown rot cause serious problems in humid regions. Nectarines, in particular, are prone to cracking if heavy rain falls close to harvest – see notes under cherries	*Time to first harvest* 1–2 years for close planting, otherwise 2–3 years
Time of flowering First buds to open are flower buds; leaf buds are several days later		*Time to full production* 4–6 years – earlier with close planting
Time of fruit maturity Earliest cultivars ripen early to mid-summer, latest in early autumn	*Wind tolerance* Moderate	*Expected yield* Varies greatly from region to region, cultivar and management. With wide spacing expect: 2 years, 5 t/ha; 4 years, 20 t/ha; and 6 years, 40 t/ha. Narrow spacing: 1 year, 5 t/ha; 3 years, 20 t/ha; and 4–5 years, 40 t/ha
	Edaphic features Flat land or mild slopes preferred. Sunny slopes advance flowering and fruit ripening; trees on shaded slopes less prone to spring frost	*Normal productive life* 10–20 years depending on endemic disease susceptibility
	Soil needs Prefer light, well-drained soils, provided adequate moisture can be maintained. Avoid heavy clay soils	*Method of harvest* Hand pick for fresh market. Mechanical harvesting is possible and acceptable for canning peaches
	Nutrient requirements Foliar applications of minor elements common practice. Tree vigour must be maintained to achieve good fruit size and adequate lateral shoot growth. See notes in Chapter 9	*Storage* Reduction of fruit temperature after picking will assist transport and marketing; cool storage at 0°C for up to 5 weeks is possible. CA storage at 1% O_2 and 5% CO_2 extends storage life
		Main diseases There are many diseases of peach and nectarine, and knowledge about the epidemiology and control of the diseases important in each region is a key ingredient of successful production. Some of these diseases are mentioned in the text. Refer to locally produced bulletins for details about control
		Main pests Direct damage to peach and nectarine fruit and shoot tips is caused by the larvae of the peach twig borer (*Anarsia lineatella*) in most peach-growing regions. The Oriental fruit moth (*Grapholita molesta*) causes similar damage. Other moth larvae that damage peach leaves and fruit include several species of fruitworms and leafrollers, and the larvae of the peach tree borer (*Sysanthedon exitiosa*) can girdle young trees near the soil surface. Green peach aphid (*Myzus persicae*) can be especially damaging to nectarine fruit, as can bloom and post-bloom infestations of the western flower thrip (*Frankliniella occidentalis*). Several of the true bugs such as *Lygus hesperus* and *Lygus lineolaris* can deform young fruit. Several species of spider mites feed on peach trees

viruses are yellow bud mosaic, Prunus ringspot and peach yellow leaf roll. These diseases are vectored by a range of organisms and their biology can be very complex. Purchasing disease-free trees and removing infected trees are the most important ways to avoid and control virus diseases. Other diseases of peach include bacterial leaf spot (*X. campestris* pv. *pruni*), scab (*Cladosporium carpophilum*) and powdery mildew (*Sphaerotheca pannosa* and *Podosphaera tridactyla*). Mildew is especially problematic with nectarine.

Training and pruning

A wide variety of plantation systems and tree forms are found in peach production around the world. Where tree life expectancy is 20 years or more it is common to find vase-shaped trees planted at spacings as wide as 7 × 7 m. However, hedgerow plantations with much closer tree spacing are increasingly popular when centre-leader-trained trees are used (Table 13.3 and Fig. 13.4).

Where diseases such as bacterial canker or silver leaf are known to reduce the expected tree life, it is common to see trees planted at higher densities and trained to encourage early cropping. In these situations minimal pruning is performed and, when absolutely required, carefully timed to minimize disease entry.

It is important to understand that peaches crop on new wood, often flower profusely and set enormous crops that must be thinned by hand. To avoid the situation where the tree gets larger and larger and only crops at the extremities of the canopy, tree management must ensure that new fruiting wood is produced along the whole length of the scaffold branches every year. Pruning then becomes the important tool for positioning the fruit along these major branches and for reducing the amount of fruit thinning that will be required. Two basic approaches are common: (i) reducing the number of 1-year-old shoots to the minimum needed for a full crop; or (ii) leaving more shoots but shortening each to about half its original length. There are advantages and disadvantages to each approach.

Harvesting and handling

A growing proportion of total peach and nectarine production is destined for distant markets. Therefore, all of the concerns about producing a high-quality, blemish-free product that reaches the consumer in good condition are applicable to this crop. Most of the comments made about the handling and marketing of sweet cherries apply to peaches and nectarines intended for the fresh market.

Careful harvesting, at the correct stage of fruit maturity, is very important. Pickers should be well trained and asked to wear cloth gloves. Fully ripe fruit cannot be handled without bruising and will deteriorate before it reaches the consumer. On the other hand, excessively immature fruit will not have good colour, will fail to develop good internal quality when ripened and harvesting such fruit will greatly reduce yield since fruit weight increases rapidly in the days preceding harvest.

The approach is to pick at a stage of maturity consistent with the marketing plan: firm-ripe fruit for the short-haul local markets, a few days earlier for fruit intended for more distant markets. Even fruit destined for overseas markets should show appreciable development of yellow (or white) background colour and the flesh should be just starting to soften.

Fig. 13.4. A centre-leader-trained peach tree (shown here) can be planted more intensively, at closer spacings, than vase-shaped or open-centre-trained trees.

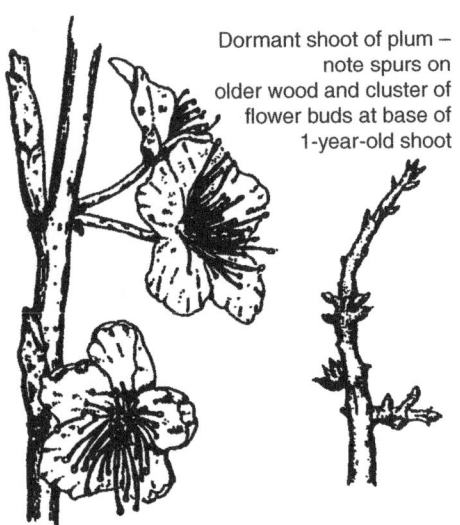

Dormant shoot of plum – note spurs on older wood and cluster of flower buds at base of 1-year-old shoot

Flowers of plum – flowers are normally in clusters

Fig. 13.5. Plums.

The fruit should be pre-cooled as soon as possible after picking and maintained at high humidity and temperatures just above freezing until it reaches the consumer. Low temperature is the best insurance against postharvest diseases and other disorders. However, there are some cultivars that develop a classic chilling disorder, sometimes referred to as 'woolliness', if held too long at temperatures between 2 and 10°C. This disorder seems to be more common with late-season cultivars. Cold storage of peaches for more than 3 weeks is seldom advised.

Grading to market requirements usually occurs in a packing shed using equipment appropriate for handling tender fruits. Much of the bruising occurs during grading.

Plums

There are two important species of plum in commercial horticulture: those classed as European plums and prunes (*P. domestica*), and the Asian or Japanese plums (*P. salicina*). The overriding importance of *P. salicina* plums in China (see below) and their growing presence as a high value item in international trade suggests that they may now be as important as the European plums in world commerce.

The European plum has been cultivated in Europe for at least 2000 years but there is no single known wild progenitor – *P. domestica* appears to be a complex hybrid. The Japanese plum was introduced into the west from Japan but probably originated in China. It requires warmer summers and is not cultivated in the cooler parts of Europe, where a large proportion of the world's plums are grown.

There are many other species of plums native to America, Europe and Asia, but only a few have achieved commercial significance. Damsons and bullaces, *P. domestica* subspecies *insititia*, are acid and are used only for pies and jams. Myrobalan or cherry plum, *Prunus cerasifera*, is most important as a rootstock.

United Nations (FAO) statistics indicate that world production of plums (all species combined) has increased and now stands at 9.7 Mt. The number of countries reporting plum production statistics now stands at 78. China is by far the largest producer, with about 4.8 Mt, followed by the USA and Serbia about equal at 0.7 Mt.

Key points

Cultivars and pollination

Both European and Japanese plums exhibit a wide range of shapes, colours, fruit sizes and internal qualities. If one adds the many less widely grown cultivars arising from the several species of American plums, it is clear that plums are the most diverse of all stone fruit crops. It seems likely that this diversity could be profitably exploited, especially in those affluent countries where consumers are increasingly interested in novelty fruits and fruit products. Furthermore, interspecific crosses, such as the plumcot (*P. salicina* × *P. armeniaca*), may grow in number and commercial importance.

The oval-shaped purple plums used for drying, canning or for making brandy are usually referred to as prunes (e.g. D'Agen, or French prune) but some *P. domestica* cultivars are called prunes even though they seldom achieve the sugar content needed for efficient drying without removing the pit (the generally accepted definition of a prune). Italian prune is a good example. Other classes of European plums used for fresh eating or for processing into canned fruit, jams and jellies are:

- Greengage-type plums of high fresh eating and canning quality – usually round, sweet and yellow to green in colour.
- Yellow egg-type plums, round, with yellow or red skin and flesh colour, most often used for canned fruit.

Table 13.4. Plums.

Botanical, anatomical and physiological aspects	Climatic, geographic, soil and water requirements	Cultural aspects
Common name Plums and prunes	*Temperature needs* Generally, European plums are grown in cooler countries and regions, but there are exceptions. Prunes for drying are mostly grown in warm temperate areas like California and southern France	*Propagation* Budded on to *P. cerasifera* (Myrobalan) seedlings or Marianna 2624 (a cherry plum hybrid) clonal rootstock. Japanese plums often budded on to peach seedlings when intended for well-drained soils
Botanical names *Prunus salicina* (Japanese plums) *Prunus domestica* (European plums and prunes)		*Rootstocks* See above. Some North American *Prunus* species used by nursery trades to reduce tree size for home garden sales
Botanical name of related and useful species *Prunus domestica insititia* (damsons) *Prunus cerasifera* (Myrobalan or cherry plum)	*Frost tolerance* Damaged by −4 to −5°C at bud swell; −2 to −3°C at blossom; and 0.9°C at small fruit stage. Because most European plums flower later than peaches, they often sustain less damage than the Japanese plums	*Spacing* Planted on square, normally at 6m. Closer planting system can be used provided tree vigour is controlled, and hedgerow plantings are becoming more popular, especially with cultivars of Japanese plum with a very upright growth habit
Type of plant and size Vigorous. especially *P. domestica*, bushy tree 4–8m high, 4–5m in diameter		*Training and pruning* Often grown as vase-shaped trees. Detailed and extensive pruning is considered less essential than in other stone fruit. Some *P. salicina* cultivars may prove adaptable to trellis systems such as the Tatura trellis. Some very strong suckers or water sprouts may develop and require removal each year
Sexuality Hermaphrodite	*Winter hardiness* Cultivars vary widely but in general the Japanese plums are similar in hardiness to peaches whereas the European plums rank with sour cherries	*Thinning* Some plums suffer more from lack of set than excessive set but others require thinning to achieve good size and return cropping
Pollination Many plums require cross-pollination so seek advice before planting. Careful consideration of bloom time of pollinators and their cross-compatibility is essential. Nectar is often low in sugar and other flowers may out-compete plums for bees	*Water needs* Provide adequate moisture, especially during the final growth stage	*Tillage* Grown in grassed orchards with herbicide used to control grass and weeds under the trees
	Tolerance of wet soils Moderate to good. European plums are more tolerant of 'wet feet' than other stone fruit	*Time to first harvest* 3–4 years
Flower buds Produced both on 1-year-old shoots and on spurs on older wood. Flower and vegetative buds are separate. Flower buds of European plum contain one to three flowers, Japanese plums two to five flowers on long pedicels. Flowers have five sepals, five petals, a single carpel and numerous stamens	*Drought tolerance* Moderate	*Time to full production* 7–9 years *Expected yield* 3 years: 3 t/ha 6 years: 9 t/ha 9 years: 18 t/ha

Continued

Table 13.4. Continued.

Botanical, anatomical and physiological aspects	Climatic, geographic, soil and water requirements	Cultural aspects
Growth of fruit Drupe, with double-sigmoid growth curve. Stone hardens during lag phase of fruit growth. In most Japanese plums stone is firmly attached to the flesh. European plums often freestone. Flesh is yellow to dark red in both species	*Humidity tolerance* Grown in both desert and humid climates but diseases and fruit set problems are less severe in drier regions	*Normal productive life* 25–35 years. Less where diseases like bacterial canker or silver leaf are a problem
Time of bud burst Most Japanese plums flower early spring; others, including most European plums, flower 1–2 weeks later	*Wind tolerance* Moderate *Edaphic features* Flat land preferred	*Method of harvest* Hand harvested for fresh market. Mechanical shakers often used for prunes for drying
Time of flowering Flower buds open first, leaf buds several days later	*Soil needs* European plums can tolerate heavier clay soils than most tree fruit. Japanese plums are less tolerant of such soils. Nevertheless, deep, well-drained loams are the best for both types	*Storage* Most are sold locally for the fresh market; some can be stored for 2–4 weeks at 0°C. CA storage has been used with 2% O_2 and 2–8% CO_2
Time of fruit maturity Variable, some are among earliest fruit to be seen in any one season	*Nutrient requirements* Need less nitrogen than other stone fruit but fruit quality and yield are enhanced by good nutrition. Use guide in Chapter 9	*Main diseases* Silver leaf (*C. purpureum*) and bacterial canker (*P. syringae*) are serious diseases in some countries but not others. Blossom blast in spring and brown rot at maturity, both caused by *Monilinia* spp. fungi, are difficult to control, especially in wet climates. Bacterial spot is a major disease of some of the newer Japanese plums bred in California but planted in more humid regions. Plum leaf scald (*Xylella fastidiosa*) is the most damaging disease of Japanese plums in the south-east USA. Sharka or plum pox is a serious virus disease of plums in Europe
		Main pests European plums and prunes in general are attacked by fewer pests than some other fruit crops. None the less, several species of leafrollers and cutworms can damage both foliage and fruit. The peach twig borer (*Anarsia lineatella*) damages plums in California and the peach tree borer (*Synanthedon exitiosa*) is damaging in various other regions of North America. Several species of Eriophyid mites are common on plums, but the effect on the tree and crop is usually minor. Aphids, pear slug and scale infestations can occur in poorly sprayed orchards

- Large, oval blue or red plums of the Lombard or Victoria type. Victoria is the most widely grown plum cultivar in northern Europe. President is more popular in North America. They are used solely for fresh eating.

The Japanese plums, *P. salicina*, are even more diverse in colour and shape. Many cultivars are heart shaped with a pointed calyx end, but some others are round or even flat in appearance. Santa Rosa, a bright red, round plum with yellow flesh, developed in the early 1900s by Luther Burbank, is a long-time favourite of the California industry. However, it is being overtaken by a collection of cultivars intended to extend the season for shipping fresh fruit to distant markets. Some of these are flatter and dark red to black when mature. Friar has this kind of appearance.

As warmer regions are being planted to plum orchards, the chilling requirement can be an important selection criterion. A number of new cultivars are considered low-chill and suitable for regions with warm winters.

The pollination requirements for plum trees are variable and often quite complicated. Some cultivars are self-fruitful but others require a rather specific pollinizer. Fruit tree nurseries should be able to provide these details for the cultivars that they sell. Where cross-pollination is essential for achieving a full crop it may be necessary to plant alternate rows of paired cultivars.

The provision of suitable pollinators is also important. Bees are not easily attracted to plum blossom, so a high number of hives may be required (three to five per hectare). In planning a plum orchard it would be wise to ensure that there are no competing flower sources in the vicinity at plum blossom time.

Pruning and training

To date plums have not been successfully grown as central leaders in intensive plantings such as those used for peaches and nectarines. The open-centre tree is therefore still recommended.

The basic method of growing vase trees described in Chapter 5 should be adopted. European plums bear most of their crop on lateral spurs 5–10 cm long, and renewal of branches carrying these spurs is recommended every 5–6 years. Japanese plums fruit on 1-year-old laterals, so heavy pruning (one-third to a half) of new wood is required each year.

If heavy thinning is needed in the following season, the pruning level is possibly too low.

The trees of some Japanese plum cultivars are rather weak and, for these, a two- or three-wire trellis for support is sometimes used. Some orchardists are considering growing a wider range of cultivars on such a trellis, especially for high-quality export plums. The disadvantage of this method is that it has sometimes been found to increase the incidence of silver leaf, by allowing spore entry through small wounds where rubbing occurs.

Pests and diseases

As indicated by the information in Table 13.4, plums and prunes are hosts for a large number of serious diseases. Fortunately, cultivars have been selected that have some resistance to one or more of these diseases, and over the years the cultivars grown in each region reflect these differences. Plum pox virus (sharka disease) is an important limiting factor in plum production in Europe and considerable effort is expended to prevent its movement to other regions of the world. However, even this serious disease is less damaging to certain cultivars, permitting the European industry to continue.

The pests of plum are often also the pests of other stone fruits. However, there are some pests of plums that cause great damage in some regions of the world but not others. For example, the citrus cutworm (*Xylomyges curialis*) causes serious damage to Japanese plums in California but is not mentioned as a pest of plums in other parts of the world. Modern control strategies focus on pest monitoring and the use of biological control agents and other 'soft' pest control options when necessary. It is imperative that the producer be fully aware of the pest control options available.

Harvesting and handling

The remarks under peaches and nectarines are appropriate for plums. Plums for fresh sales are hand picked with multiple harvests. Only French prunes for drying are harvested with tree shakers.

Japanese plums are often stored at 0°C for up to 5 weeks as they pass through the marketing channels. European plums are less tolerant of long storage. Italian prune, for example, should not be stored for more than 2–3 weeks.

Further Reading

Audergon, J.M. (ed.) (2006) XII International Symposium on Apricot Culture and Decline. *Acta Horticulturae* 701.

Eris, A., Lang, G.A., Gulen, H. and Ipek, A. (eds) (2008) V International Cherry Symposium. *Acta Horticulturae* 795.

Infante, R. (ed.) (2006) VI International Peach Symposium. *Acta Horticulturae* 713.

Lang, G.A. (ed.) (2005) IV International Cherry Symposium. *Acta Horticulturae* 667.

Romojaro, F., Dicenta, F. and Martínez-Gómez, P. (eds) (2006) XIII International Symposium on Apricot Breeding and Culture. *Acta Horticulturae* 717.

Scott-Johnson, R. and Chrisosto, C.H. (eds) (2002) V International Peach Symposium. *Acta Horticulturae* 592.

Vangdal, E. and Sekse, L. (eds) (2007) VIII International Symposium on Plum and Prune Genetics, Breeding and Pomology. *Acta Horticulturae* 734.

N. Looney and D. Jackson

14 Pome Fruits

DAVID JACKSON AND JOHN PALMER

The term 'pome fruits' describes the fruit morphology of a select number of commercial fruit species. Pome fruits (Family *Rosaceae*; subfamily *Pomoideae*) are formed by a fusion of the ovary and receptacle. Multiple seeds are borne in five *carpels* made up of mesocarp and exocarp tissues. The fleshy part of the fruit is considered, botanically, as pith inside the core line and cortical tissue outside the core line. Botanists consider the pome fruit as a modified stem.

Apples (*Malus* spp.), pears (*Pyrus* spp.) and quince (*Cydonia* spp.) are species important in fruit production but a number of species with berry-like fruits also fall into this group. These include service berry (*Amelanchier* spp.), hawthorn (*Crataegus* spp.) and mountain ash or Rowan (*Sorbus* spp.). Only apples and pears will be dealt with in this chapter.

Apples

Wild apples are found over much of Europe, in the Caucasus region, in middle Asia (e.g. Kazakhstan) and China. It is generally considered that the domesticated apple derived from complex hybrids of several wild species of *Malus*. The botanical name of the common apple has been variously referred to as *Malus pumila*, *Malus domestica* or *Malus sylvestris*, although it is now generally referred to as *Malus domestica* Borkh.

The apple has long been associated with western civilisation – it was probably cultivated in Greece as early as 600 BC – but remains of apples have been found in excavated sites of prehistoric lake dwellers in northern Italy and Switzerland.

Apples are the most important world fruit crop after oranges, bananas and grapes. World production has averaged slightly over 60 million tonnes over the period 2000–2007, with the leading producers being China, the USA, Iran, Turkey, Poland and Italy. China now produces over one-third of the world's apple production, following large-scale plantings in the 1980s and 1990s. Generally apples are grown under cool or warm temperate conditions at latitudes of 35–55°, where they can receive sufficient winter chilling. The regions suitable for apple growing have been extended by the use of low-chilling-requirement cultivars and dormancy-breaking chemicals, e.g. hydrogen cyanamide (Dormex®).

Most apples are consumed as a fresh product, although the primary utilization in some regions is processing into juice and cider, pie fillings and other products. The storage life of apples after picking can be extended by cool storage and controlled atmosphere conditions. The long storage life of some cultivars has resulted in some areas of the world being able to supply apples over a 12-month period.

Key points

Cultivars

Although there are thousands of cultivars of apple documented throughout the world, commercial production has concentrated on a smaller number of apple cultivars with widely accepted texture and flavour and good storage life. Chinese production is dominated by Fuji. In the rest of the world the dominant cultivars are Golden Delicious (19%), Delicious (19%), Royal Gala (14%), Fuji (7%) and Granny Smith (7%). Many parts of the world are experiencing a rapid replacement of older cultivars. In 1993/4, almost 60% of the apple production in the USA was from two cultivars (Delicious and Golden Delicious). In 2005/6 60% of the apple production in the USA

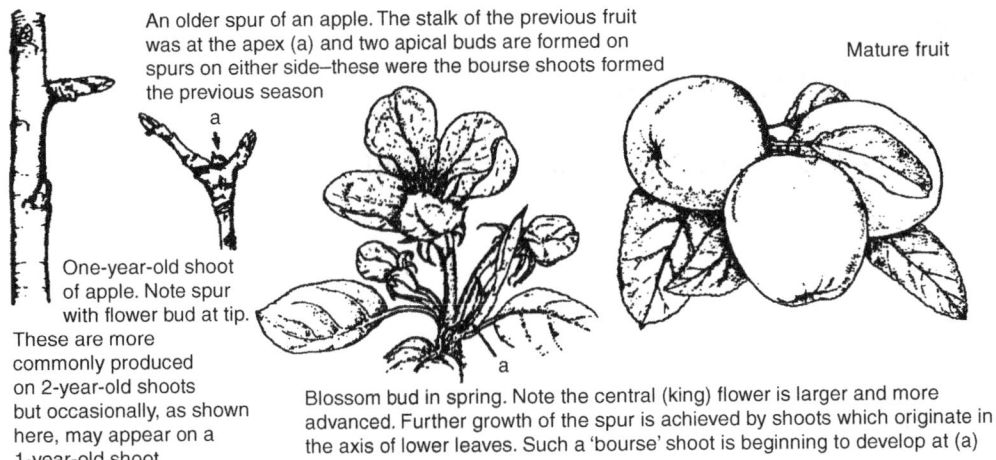

An older spur of an apple. The stalk of the previous fruit was at the apex (a) and two apical buds are formed on spurs on either side–these were the bourse shoots formed the previous season

Mature fruit

One-year-old shoot of apple. Note spur with flower bud at tip. These are more commonly produced on 2-year-old shoots but occasionally, as shown here, may appear on a 1-year-old shoot

Blossom bud in spring. Note the central (king) flower is larger and more advanced. Further growth of the spur is achieved by shoots which originate in the axis of lower leaves. Such a 'bourse' shoot is beginning to develop at (a)

Fig. 14.1. Apples.

was from four cultivars (Delicious, Golden Delicious, Gala and Fuji). In New Zealand, for example, Braeburn, Royal Gala and Fuji made up only 17% of the harvested crop in 1988 but by 1995 they accounted for 65% of total production. Growers worldwide are looking for new high-value cultivars, and introductions bred or selected in one country are often adopted enthusiastically elsewhere, e.g. Jonagold was released in New York State in 1968 but was extensively planted in north-west Europe in the 1980s. There is a continual turnover of cultivars – a selection of important mainstream and newer cultivars is given in Table 14.2. No early maturing cultivars are listed in Table 14.2 as most early cultivars do not store well and therefore the choice of early cultivars is strongly market driven. Length of growing season (flowering to fruit maturity) can vary from 90 (Beauty of Bath) to 230 days (Sundowner™, syn. Cripps Red).

Although nearly all new cultivars attract a tree royalty, payable to the owner of the cultivar via the licensed propagator, there is a growing trend of 'club varieties' e.g. Pink Lady®, Tentation®, Jazz™ and the Kiku® strain of Fuji. 'Club varieties' have arisen in an attempt to control both the quantity and quality of production to try and maintain more stable high prices for these premium cultivars. Many of these cultivars also carry a production-based royalty as well as a tree royalty. Although most of the common apple cultivars today are smooth-skinned green, red or red-striped fruit, russetted or partially russetted cultivars still have a place in some markets, e.g. Boskoop in the

Netherlands. It is important to point out, however, that unless surface russet is characteristic of the cultivar, the presence of russet on the skin of apples normally results in downgrading.

IMPROVED STRAINS Growers and scientists are continually looking for better strains of existing cultivars. In the USA, 'spur-type' strains are popular for some cultivars. These are strains of common apples with shorter internodes and a tendency to form spurs rather than strong lateral branches. Since spur-type strains can revert to the standard growth habit, it is important to select bud wood from mature trees clearly demonstrating the desired growth habit. Interest in spur types has waned of late as non-spur type cultivars on dwarfing rootstocks have been more widely planted in the USA.

With many red or striped red cultivars there is interest in strains with improved colour intensity or colour pattern. For example, there are already many strains of Jonagold, Fuji and Gala that have superior fruit colour. Like the spur-type habit, red strains can revert to the original colour. It is therefore essential to ensure that bud wood comes from trees still holding the improved characteristics, particularly with Royal Gala. With smooth-skinned cultivars that are prone to russet, e.g. Golden Delicious, a number of clones have been selected which are less susceptible to this disorder.

REGIONAL PREFERENCES It is important to remember that not all apples do well in all growing regions. Granny Smith and Fuji, for example, are basically

D. Jackson and J. Palmer

Table 14.1. Apples.

Botanical, anatomical and physiological aspects	Climatic, geographic, soil and water requirements	Cultural aspects
Common Name Apple	*Temperature needs* Needs cool winter. Trees in regions where winters are mild, experience delayed and extended flowering. Winter freeze damage can occur at temperatures below −20°C. Some cultivars, like Granny Smith, do better in warmer areas; others like Red Delicious and Cox's Orange Pippin may have better shape, flavour and colour in cooler climates. There is a range of low-chill cultivars suitable for warmer regions	*Propagation* Budded or grafted on a range of rootstocks, which are usually propagated in stoolbeds or less frequently using hardwood cuttings. Rapid propagation of rootstocks through tissue culture has been successful, although this technique should only be used prior to establishment of new stoolbeds, rather than propagating directly on to tissue-cultured stocks
Botanical names *Malus domestica* Borkh		
Botanical name of related and useful species *Malus baccata, Malus × floribunda, Malus micromalus, Malus robusta, Malus prunifolia*		*Rootstocks* A very wide range of rootstocks are used. Some are: Very dwarf – M.27, P22 Dwarf – M.9, B9, Mark, M.26 Semi-dwarf – MM.106, M.7 Standard – Northern Spy, M.793 Large – MM.115, M.25, Robusta 5
Tree size and shape Tree between 1.5 and 7 m in height, depending on rootstock. Under natural conditions, varies from round-headed to pyramidal in shape. Basal diameter 1–4.5 m		*Spacing* For semi-intensive orchards, plant at 4.5–5.0 m between rows, 2.8–4.3 m between trees, depending on vigour of scion, rootstock and site. Intensive apples on M9 are commonly at a spacing of 1–1.5 × 3–3.5 m
Sexuality Hermaphrodite	*Frost tolerance* Flowers and young fruit will be damaged or killed by temperatures below −2°C after open cluster	*Training and pruning* The semi-intensive, centre-leader system at the above spacing is common in warmer climates. Many of the intensive systems pioneered in Europe are being planted in other parts of the world where the dwarfing stocks do not succumb to cold damage
Pollination Some cultivars partly self-fertile, but cross-pollination will generally improve set. Pollen trees should be no more than two rows from the cultivar to be insect pollinated.	*Water needs* Lack of water reduces yield but excessive watering or rain may produce fruit with poor keeping ability. Maintain adequate water supply with irrigation in dry districts	*Thinning* Hand thinned at about 2 cm diameter; chemicals also used – mainly carbaryl and NAA (Chapter 8)
Flower buds Produced on tips of shoots (tip bearers) or on spurs formed on 2-year-old or older wood. Lateral buds on 1-year-old shoots can also be fruitful in some cultivars. Buds are mixed, with five to eight flowers and a similar number of leaves. Central flower is frequently larger and earlier and is called the *king flower*. Flowers contain 5 sepals, 5 petals, 5 pistils and about 20 stamens	*Tolerance to wet soils* Moderate, depends on rootstock *Drought tolerance* Moderate *Humidity tolerance* Moderate. Diseases such as scab and Nectria canker are worse in higher humidity	*Tillage* Mostly sprayed with herbicides in the rows under the trees and grass mown in the alleyways *Time to first harvest* Intensive: 1–2 years Semi-intensive: 2–3 years

Continued

Table 14.1. Continued.

Botanical, anatomical and physiological aspects	Climatic, geographic, soil and water requirements	Cultural aspects
Time of bud burst Early spring, 1–2 weeks after most stone fruit	*Wind tolerance* Young orchards benefit from shelter in windy sites. This may be removed in some districts as apple trees grow to 3m or more, when they become self-sheltering	*Time to full production* Intensive: 5–8 years Semi-intensive: 8–10 years
Time of flowering Spring		*Expected yield* Very dependent on cultivar/rootstock combination and site but some indication is given below. Intensive: 3 years – 20t/ha 5 years – 40t/ha 8 years – 50–70t/ha Semi-intensive: 4 years – 10t/ha 6 years – 30t/ha 10 years – 50–70t/ha
Growth of fruit and shoot Fruit growth follows a single sigmoidal growth curve. Shoot growth is from either a terminal or lateral vegetative bud. New growth of the flowering spur is continued by a bourse shoot in the axil of one of the basal spur leaves	*Soil needs* Tolerate wide range of soil types. Avoid heavy, wet clays unless structure and drainage can be improved	*Normal productive life* Up to 40 years, although orchards are often replaced with higher-paying cultivars much sooner than this
Time of fruit maturity Variable according to cultivar. Generally after stone fruit but before grapes or kiwifruit	*Nutrient requirements* The best pH range is 5.5–6.5, but in some circumstances 4.5–8.0 is acceptable. Create good nutrient regime but avoid excessive nitrogen fertilizers, which will encourage excessive vegetative growth, and reduce storage quality and colour on red or partly red cultivars. Fertilizer recommendations in Chapter 9 may be used after any soil deficiencies have been corrected	*Method of harvest* Hand harvest into bulk bins. Mechanical harvesting suitable only for processing (e.g. for cider)
		Storage Keeps well in cool and in controlled atmosphere storage, but big variation between cultivars
		Main pathogens and pests Apple scab, powdery mildew, codling moth, red spider mite

Table 14.2. A selection of apple cultivars.

Name	Maturity period	Tree vigour	Comments
Royal Gala	Mid to early	Moderately vigorous	Good quality, small to medium-sized, bicoloured apple. A high-yielding cultivar from New Zealand
Golden Delicious	Mid to late	Weak	Susceptible to russet (physiological) in some districts and russet ring (virus)
Red Delicious	Mid to late	Many strains are vigorous; Oregon Red is weak and Red Chief is very weak	Red Delicious strains have totally replaced the standard Delicious, which is unattractive in comparison. Long, crowned, bright red fruit are especially appreciated in North America and these are easier to produce in areas with cool autumns, e.g. Washington State
Jonagold	Mid to late	Vigorous	A large-fruited, triploid, bicoloured cultivar with good fruit quality. Many red sports are available in Europe
Braeburn	Late	Weak	A New Zealand cultivar with large fruit size and good sugar/acid balance. Tree is subject to biennial bearing, powdery mildew and bitter pit. Some red strains are available
Fuji	Late	Moderately vigorous	A large, red, sweet, low-acid Japanese apple, proving popular in China, Europe and North America. Prone to watercore and fruit russet
Granny Smith	Very late	Vigorous	Full green apple from Australia, suitable for eating or cooking. A high-yielding cultivar. Fruits are susceptible to scald in storage if picked too early and tree pit if not sprayed with calcium nitrate. Also susceptible to powdery mildew
Cripps Pink (Pink Lady®)	Very late	Vigorous	Long-season cultivar from Australia. Medium to large apple with attractive pink blush. Can easily be bruised during picking and handling despite the very firm texture. Flesh browning can be a problem

long-season, warm-climate apples and cannot be grown successfully in cool climates or in regions with a short growing season. Delicious is not suited to wet climates, due to its susceptibility to mouldy core and European canker, whereas Cox's Orange Pippin has better flavour and skin texture in cool maritime climates. Local knowledge should be requested before choosing a cultivar mix.

VIRUS STATUS OF PLANT MATERIAL Most of the common cultivars can now be obtained free of known viruses (FKV). Such trees produce higher yields or better fruit quality but they also tend to be more vigorous. For that reason they may need to be planted further apart or grafted on to more dwarfing rootstocks.

MAKING THE CHOICE The importance of choosing the right cultivar for successful commercial production cannot be overemphasized. It can make the difference between economic success and failure, so it is important to:

- *Know the target market.* Ensure that the cultivars chosen are currently receiving good prices but also be aware of any current and future trends in prices and planting. The choice of cultivars may differ according to the preferred market, e.g. farm sales, local market or export.
- *Spread the risk.* A range of cultivars will give an adequate spread to the harvesting season to spread both the labour requirement and the financial risk.
- *Select the best strains.* Be familiar with the new strains of standard cultivars which are available and plant the best.
- *Choose a reliable nursery.* Trees produced by a good professional fruit tree nursery will usually prove to be superior to home-grown trees.
- *Seek advice when making a selection.* In particular take note of local knowledge as to the suitability of a cultivar or strain for the area.

Rootstocks and growing systems

The common methods of growing and training have been described in Chapter 5 and further details for apples are given in Table 14.1.

Trees on clonal rootstocks have many advantages over trees on seedling rootstocks. These include uniformity, predictable vigour, precocity and resistance to soil-borne pests and diseases. The clonal apple rootstocks mentioned in Table 14.1 are well known to the nursery trade, but keep in mind that new rootstocks continue to appear. Furthermore, as with new cultivars, there are new strains of existing rootstocks being selected, particularly with M.9. Not all apple rootstocks are available all over the world. This is particularly the case for newer rootstocks, which may be covered by plant variety rights or restricted propagation agreements.

The most widely known apple rootstocks are the Malling (with prefix M) and Malling Merton (prefix MM) series. The original Malling series were collected within Europe and classified at East Malling into 16 different types. Later crossing produced the Merton and Malling Merton series, with woolly apple aphid resistance. There were also some additions to the Malling series that were not woolly apple aphid resistant, e.g. M.26, M.27. Important apple rootstocks have also emerged from breeding programmes in Russia (the Budagovsky series, usually prefixed with B), in Poland (with prefix P) and more recently from a breeding programme at Cornell University (the Cornell Geneva series, usually prefixed by CG or more recently by G). The objectives of the Cornell apple rootstock breeding programme have been to produce dwarfing rootstocks resistant to fireblight, *Phytophthora* and woolly apple aphid, and tolerant of low winter temperatures. Rootstock breeding has continued at East Malling and new rootstocks from this source are now prefixed by AR.

Pollination, fruit set and thinning

In orcharding districts, which tend to plant a range of cultivars, and where the climate is generally favourable, fruit set is seldom a major problem. For such conditions, blocks of a single cultivar, up to four rows wide, can be managed in most intensive or semi-intensive systems. In cooler areas, where pollination and fertilization may be more of a problem, specific pollinizer cultivars are often planted within the row. Pollinizer cultivars must have compatible pollen and an overlapping flowering period with the main cultivar. Ornamental crabapple cultivars that flower annually and profusely, even after hard pruning, are popular in some regions. Triploid apples, such as Gravenstein and Jonagold, have an absolute demand for cross-pollination since their own pollen has very low viability. Furthermore, they are of no value as pollinizers for other cultivars.

Golden Delicious, Gala and Braeburn have good pollen and are very effective pollinizers for other cultivars. Growers who have experienced pollination problems should, in addition to planting pollinizer cultivars, introduce two bee hives per hectare of apples planted.

Fruit size is of paramount importance and the grower should aim to achieve maximum pack-out of the preferred sizes to improve financial returns. The methods used to achieve this include pruning to renew fruiting wood continually, regulating crop size and distribution by flower or fruitlet thinning, providing adequate, balanced nutrients, avoiding competition from weeds and other vegetation, and providing enough water to avoid moisture stress.

In reality, since it is normal practice to control weeds, provide optimum water and nutrients and prune adequately, the critical additional tool for managing fruit size is thinning. The level is difficult to prescribe in advance for any new location, because it depends heavily on the cultivar, its cropping habit, the growing environment and the fruit size objective. A number of specific approaches and chemicals are described and discussed in Chapter 8. Keep in mind that some cultivars are partially self-thinning and less requiring of intervention. Granny Smith is such a cultivar.

A further reason for thinning is to reduce the risk of biennial bearing, where a high crop one year (the on year) is followed by a low crop the next (the off year). Susceptibility to biennial bearing varies widely between cultivars. James Grieve shows very little tendency to biennial bearing, whereas cultivars such as Fuji and Honeycrisp show strong biennial tendencies. As has been discussed in Chapter 8, chemical thinning reduces the extent of biennial bearing on these apples.

Training systems

A number of training systems are available. Several are described in Chapter 5 and include centre-leader, semi-intensive systems; centre-leader intensive systems; and various trellis systems, such as the Y and V trellises. Multi-leader, vase-shaped trees, planted at very low tree densities, are now rare in new commercial plantations.

As discussed earlier (Chapter 5), the centre-leader approach can be adapted to suit the size of tree and plantation density desired. Larger trees at wider spacing are cheaper to establish but come into production later, as time is needed to grow the structure on which the future crop will be carried. Intensive systems are more expensive to establish, but earlier in cropping and easier to spray, prune and pick. Thus, there is ongoing debate as to which tree size and planting density is most appropriate. Profitability depends not only on the choice of system but also on the skill of the grower in applying good and appropriate orchard management.

Even modern intensive orchards are evolving with changes to tree management as growers respond to the needs for greater efficiency and as they seek to adapt systems to new cultivars.

Harvesting and storing

Harvesting has been discussed in Chapter 7. Table 14.3 provides information about appropriate temperatures and atmospheres for apples in air and controlled atmosphere storage. Note that these conditions do change appreciably from region to region so it is important to seek local knowledge. For example, McIntosh apples grown in western Canada benefit from oxygen levels as low as 1.5% in CA storage. Those grown in Ontario suffer low oxygen injury if the level is lower than 2.5%.

It is worth reiterating that a key factor for successful storage is the speed with which apples are moved from the field to the cold room. A day's delay can cause several days reduction in effective storage life.

There are many disorders which occur in stored apples and which limit their potential for long storage. It is not possible to mention them all here, but some of the most significant are introduced below.

BITTER PIT Small, brown spots appear in the flesh, most commonly just below the skin at the calyx end of the fruit, and may show through the skin as circular, brown or greenish-brown depressions. If pitting develops while apples are still on the tree it is called 'tree pit', but it may not be evident until after a period of cold storage. Pit is most common in Cox's Orange Pippin but can occur in most other cultivars, with Braeburn and Granny Smith being among the more susceptible. Pitting is most severe in large apples on light-crop trees, on young trees or on trees making vigorous late vegetative growth. Pit is also generally a more serious problem under hot, dry conditions. Pit is related to localized deficiencies of the very immobile element calcium. It can be reduced by multiple sprays of calcium nitrate or calcium chloride during the growing season and by postharvest drenches containing a calcium salt (usually calcium chloride).

CORE FLUSH (BROWN CORE) This disorder affects the flesh of the core area between the carpels and

Table 14.3. Storage temperatures and controlled atmospheres for apples.

Cultivar	Storage period	Recommended temperature (°C)	Controlled atmospheres	
			Oxygen (%)	Carbon dioxide (%)
Royal Gala	3 months	0.5		
	CA 4–6 months	0.5	1.5–2	2
Golden Delicious	3–4 months	0.5	–	–
	CA 5–9 months	0.5	1–2	2
Red Delicious	3–4 months	0.5	–	–
	CA 5–9 months	0.5	0.7–2	2
Jonagold	4–5 months	0.5	–	–
	CA 5–9 months	0.5	1–2	2
Braeburn	4–5 months	0.5	–	–
	CA 6–8 months	0.5	2–3	1
Fuji	4–5 months	0.5	–	–
	CA 5–9 months	0.5	1–2	2
Granny Smith	4–5 months	0.5	–	–
	CA 6–8 months	0.5	1–2	2
Cripps Pink	3 months	0.5	–	–
	CA 6–8 months	0.5	1–2	1

later extends throughout the apple. It is more common in fruit harvested before it is fully mature. The cause of core flush is not known but it seems to be related to the presence of seeds. If the seeds are destroyed by irradiation, core flush does not develop. (This method of control is not commercially successful.)

INTERNAL BREAKDOWN The first symptom of this disorder is a mealiness and brown coloration of the flesh. Later, the fruit becomes soft, is easily bruised and shows a dull, waterlogged appearance. Most varieties will eventually develop breakdown, although some, like Red Delicious, are very resistant. Low-temperature storage, harvesting over-mature fruit or excessive nitrogen fertilizers will all increase breakdown. Delay between harvesting and storage increases breakdown susceptibility, and large apples are particularly prone to this disorder. Internal breakdown and core flush are generally indistinguishable once they are well developed, but in the initial stages the two are quite distinct. Core flush develops from the centre and progresses outwards while internal breakdown does the reverse. Breakdown is perhaps the most serious disorder limiting long storage of apples. Calcium sprays or drenches used to reduce bitter pit also help to reduce internal breakdown. Although in many cultivars internal breakdown is senescence related, in others it can be aggravated by high CO_2 concentrations during cold storage.

SUPERFICIAL SCALD The skin on affected apples after removal from storage becomes brown, grey-green or almost black. The surface may become slightly depressed or wrinkled but the underlying flesh is not normally affected. Scald tends to be more serious on early-harvest fruit, fruit with high nitrogen content, fruit exposed to warm pre-harvest temperatures and in stores with high relative humidity. Granny Smith and Red Delicious are very susceptible. Oiled wraps were formerly used to reduce scald on Granny Smith but diphenylamine (DPA) can be used on wraps or as a postharvest dip, and ethoxyquin ('Stop-scald') can be used as a dip, although not in paper wraps. Dips are more successful than oiled wraps.

SHRIVEL This, of course, can occur in any apple, but certain cultivars, such as Golden Delicious and Braeburn, are very susceptible. It is accentuated by early picking and low storage humidity.

FUNGAL DISORDERS These can be seen on the surface of the fruit, usually as grey-brown spots, which become soft as they develop. Fungi will often invade fruit tissues developing the physiological

D. Jackson and J. Palmer

disorders mentioned above or damaged areas of skin arising from poor handling, e.g. stem punctures. Poor orchard hygiene and inadequate field disease control will accentuate the problems observed in storage.

Pests and diseases

There are scores of pests and diseases that can affect apple production around the world. A few of the major pests and diseases are discussed below. With some exceptions, these same pests and diseases are important in pear growing.

The basic principles of pest management are outlined in Chapter 10, and for specific pests and diseases the reader is referred to the more detailed texts listed at the end of the chapter and to local information on the availability and efficacy of insecticides, fungicides and antibiotics. One clear trend in all fruit-producing regions is reduced usage of pesticides considered risky to the orchard environment, to applicators and to consumers. This is accomplished by better pest monitoring, by carefully targeting specific disease infection periods or egg laying or hatching periods, by using reduced-risk chemicals and by relying increasingly on biological control strategies.

LEPIDOPTERA INSECTS A number of moths lay their eggs on apple flowers, leaves and fruit, resulting in feeding damage to the fruit from the caterpillars or larvae. Codling moth (*Laspeyresia pomonella*) is one of the most serious as the larvae burrow into the fruit to feed on the core and the seeds, rendering the fruit useless for sale.

HEMIPTERA INSECTS Aphids frequently attack the young growing shoots of apple, feeding on the phloem sap and reducing vegetative vigour. Woolly apple aphid (*Eriosoma lanigerum*) is characteristically covered by sticky, white, waxy strands. In some parts of the world, e.g. Australasia and South Africa, woolly apple aphid can migrate down on to the root system of the trees. The use of resistant rootstocks, such as the Malling Merton and Merton series, can prevent this latter type of damage, but they cannot confer resistance to the aerial part of the tree. Rosy leaf curling apple aphid (*Dysaphis plantaginea*) can result in shortened and twisted growth of new shoots and deformed fruit. Other sap sucking Hemiptera causing serious damage include mealybug and

scale insects. If not controlled, populations of these insects can increase very rapidly, causing serious damage to the trees and downgrading of the fruit.

EUROPEAN RED SPIDER MITE (*PANONYCHUS ULMI*) These small mites suck the cell contents from apple leaves, resulting in a mottled appearance to the leaf. In serious cases they can lead to bronzing of the foliage and premature leaf fall. There are several generations a year, the first arising from overwintering eggs. Current control relies heavily upon the predacious mite (*Typhlodromas* spp.) and chemicals where needed. This is but one of several spider mites that can afflict apple and pear trees around the world.

APPLE SCAB (*VENTURIA INAEQUALIS*) Apple scab or black spot (as it is known in Australasia) is a disease attacking leaves, shoots, sepals and fruit. Lesions on young leaves and fruit occur as small brown or green patches. As the infected fruit grows the lesions become brown and corky, resulting in considerable commercial loss. Apple scab is serious in areas that have cool, humid weather during the spring. The disease overwinters on infected leaves and fruit on the orchard floor, and ascospores are released in the spring. Since the time for infection is related to temperature and leaf wetness, sprays can be targeted to infection periods.

APPLE POWDERY MILDEW (*PODOSPHAERA LEUCOTRICHA*) Apple leaves, blossoms and fruit can be infected with this organism. On leaves the disease begins as small white patches of spores and mycelium, which extend to cover the whole leaf lamina, seriously debilitating the growth of shoots. Infections on the fruit can result in fruit russetting. The disease overwinters on the tree in infected buds, which open in the spring to produce white, infected shoots.

FIREBLIGHT (*ERWINIA AMYLOVORA*) This bacterial disease originated in North America and has spread to many pome fruit-growing regions of the world, frequently via infected trees or grafting wood. Those countries that do not have this disease now have strict quarantine regulations to prevent its entry. The disease can affect flowers and young growing shoots and spread rapidly down into the woody tissues of the tree. Diseased tissue can looked scorched – hence the name – and under humid conditions exude droplets of bacterial ooze. Although European pears are normally much more sensitive to this disease, it can cause serious tree losses on apples. Warm, moist conditions favour the development of the disease.

COLLAR AND CROWN ROT (*PHYTOPHTHORA* SPP.) This is a major soil-borne disease that can result in tree death of apples. It is associated with wet conditions around the roots, particularly during the autumn and spring. The aerial symptoms are poor leafing-out in the spring, with small yellowish leaves. Choice of rootstock can decrease tree losses. For example, the M.9 and Mark rootstocks are less susceptible than MM.106. Other management treatments to reduce the disease include the avoidance of poorly draining soils, the provision of field drains and the planting of trees on ridges.

SPECIFIC APPLE REPLANT DISEASE This disease appears when apples are replanted into land which previously grew apples or pears. The newly planted trees grow very poorly, in some cases for the first few years, in other cases for much longer periods of time. Various soil-borne diseases and nematodes have been implicated in causing this disorder, but the causal organism(s) are not known for certain. This problem can be avoided by using land that has not grown apple or pear trees in recent years, or controlled by a pre-planting fumigation with a broad-spectrum biocide such as chloropicrin. In some cases this problem can be dealt with by using potting compost in the planting hole, coupled with a high-phosphorous fertilizer and subsequent fertigation. Fruit tree nurseries frequently use fresh land for each crop of trees to avoid specific apple replant disease.

Finally, it is necessary to comment once more about viruses and mycoplasmas. Some are known to adversely affect apple tree yields and fruit quality. The use of certified virus-free propagation wood is the best way to avoid these disease entities.

Organic production

There is a growing demand for apples and pears grown under certified organic labels. Each country normally has one or more certified standards for organic production, as there is no universally applicable standard. The growing of apples and pears under organic production impacts on two major areas of management – pest and disease control and nutrition. Under organics, disease control relies heavily on sulphur and copper sprays and is generally easier under dry summer conditions than in areas with wet springs. Undoubtedly the widespread future availability of disease-resistant cultivars with good fruit texture and storage potential will make disease control under organic production much easier. Pest control relies on BT sprays, derived from *Bacillus thuringiensis*, for lepidopterous pests and pyrethrum and oil for sap-sucking insects. Biocontrol is a major component of organic production, although biocontrol is also a key component of many conventional, non-organic, integrated fruit production (IFP) systems. Mineral nutrition under organic production is easier on fertile soils, which can release close to adequate nutrients for the trees' requirements. Potassium and nitrogen supply are often the challenge on less fertile soils. It must be remembered that apple fruits are high in potassium, so each 10 t/ha of fruit requires 12 kg/ha potassium, and this is a net loss from the orchard. Nitrogen can be supplied under organics from compost, mulches or legumes. Organic production suffers from not having the wider range of agrichemicals available to the conventional grower, particularly in the case of flower and fruitlet thinners, so hand thinning becomes a major cost in organic production.

European and Asian Pears

The European pear is derived from *Pyrus communis*, a species native to temperate Europe and western Asia. It seems that the pear has been selected and improved since prehistoric times. There were 39 varieties known to the Romans, and it has always been a favourite fruit in Italy. Many of our present cultivars originated in Europe over 100 years ago, particularly in France and Belgium. Williams' Bon Chrétien, syn. Bartlett, the main canning variety, was grown in England before 1800. World production of European pears has been fairly stable at about 7.7 million tonnes between 2000 and 2007, with the major producers being Italy, the USA and Spain. Most European pears require a period of cold storage to develop their characteristic melting, juicy texture.

On a world scale, production of Asian pears exceeds that of European pears and has been steadily rising over the last 10 years to *c.*13 million

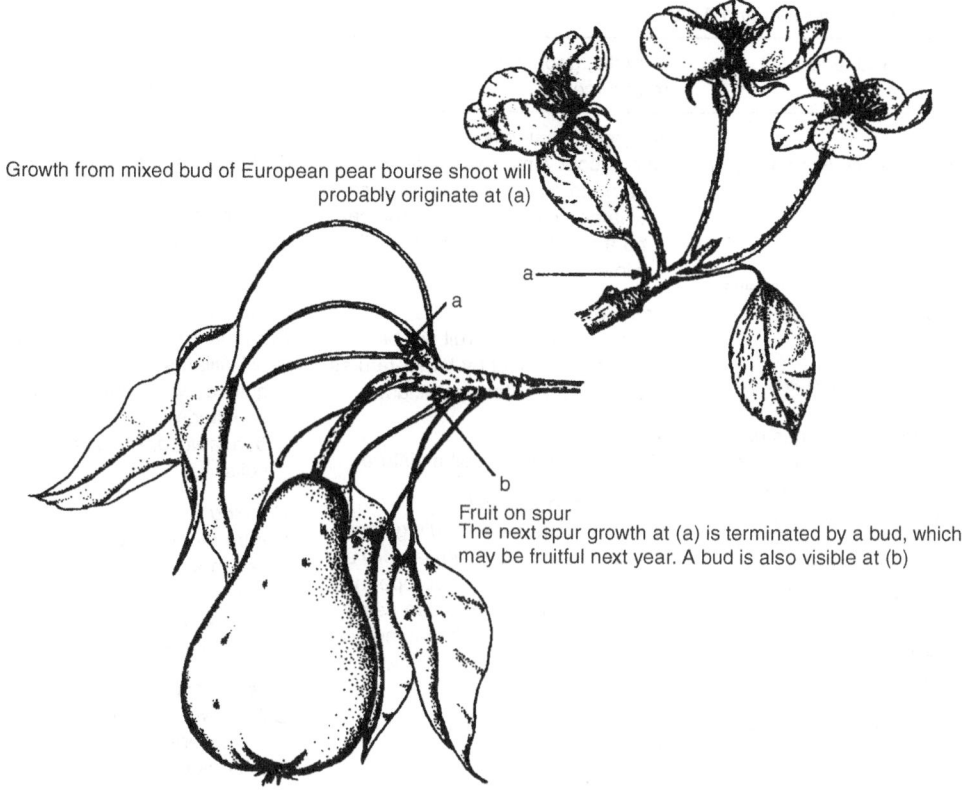

Growth from mixed bud of European pear bourse shoot will probably originate at (a)

a

b

Fruit on spur
The next spur growth at (a) is terminated by a bud, which may be fruitful next year. A bud is also visible at (b)

Fig. 14.2. Pears.

tonnes in 2007, with China dominating production. Asian pears are derived from several wild species. *Pyrus pyrifolia* is endemic to Japan and the southern parts of China and Korea. These pears are typically round with a crisp, juicy flesh. In contrast, Chinese Ya pears are derived from *Pyrus bretschneideri* and show a more European shape but with the characteristic Asian crisp, juicy flesh. Compared with European pears, Asian pears tend to be sweeter, with a lower acid content but with less characteristic pear flavour. Once mature, they also have a longer shelf life.

Pears have always tended to be the poor cousin to apples. Despite the fact that a good, well-matured European pear is a delight to eat, it is much more difficult to present them attractively to the consumer at the correct maturity. Asian pears have yet to achieve the same wide popularity of European pears in western markets. The opposite is true in Japan, where European pears comprise a small proportion of the pear production.

Key points

Reduced plantings of European pears, other than in South America, have led to a general decline in the availability of many formerly popular cultivars. European pears are generally slow to come into full production, and there has not been widespread use of dwarfing rootstocks and intensive management systems to deal with this problem, other than in parts of Europe. The problem associated with presenting the consumer with a ready-to-eat fruit continues to limit the popularity of European pears in some parts of the world. Western interest in crisp Asian pears has been slow to develop, although Asian immigrants to western countries are increasing demand in these markets.

Cultivars and pollination

Table 14.5 provides details of cultivars and suitable pollinizers for European and Asian pears. Although

Table 14.4. Pears – European and Asian.

Botanical, anatomical and physiological aspects	Climatic, geographic, soil and water requirements	Cultural aspects
Common name Pear (European pear); Asian pear or nashi *Botanical name* *Pyrus communis* (European) *Pyrus pyrifolia* – sometimes labelled *Pyrus serotinia* (Asian pear or nashi) *Botanical name of related and useful species* *Pyrus betulaefolia* sometimes used as rootstock for Asian pears. *Pyrus calleryana* – the Callery pear – has been used as an ornamental and rootstock tree *Pyrus ussuriensis* – Ussuri pear, also known as the Harbin, Siberian or Manchurian pear, used for fruit, rootstock and ornamental purposes *Pyrus × bretschneideri* – Chinese white pear or Ya pear (possibly a hybrid between *P. pyrifolia* (syn *P. serotina*) and *P. ussuriensis*) *Tree size and shape* Generally larger than apples, 5–8 m high and 3–5 m in diameter at base. Cylindrical or pyramid in shape under natural conditions *Sexuality* Hermaphrodite *Pollination* Mostly self-incompatible, choice of pollinizers critical. Pollinated by insects, especially bees *Flower buds* Formed on tips of shoots (tip bearers) or on spurs formed on 2-year-old or older wood; occasionally, lateral buds of 1-year-old shoots are fruitful. Buds are mixed, with five to eight flowers and a similar number of leaves. Flower contains 5 sepals, 5 petals, 5 pistils and about 20 stamens	*Temperature needs* Needs cool winter but winter freeze damage can occur with very cold weather. Extended flowering can be a problem in areas with little winter chilling *Frost tolerance* Flowers and young fruit will be damaged or lost by temperatures below –2°C after open cluster *Water needs* Maintain adequate soil moisture *Tolerance to wet soils* European pears on seedlings and Asian pears on *P. betulaefolia* are tolerant of wet soils. European pears on quince or Asian pears on *P. pyrifolia* need well-drained soils *Drought tolerance* Moderate *Humidity tolerance* Moderate *Wind tolerance* European pears – moderate and similar to apples, except for cultivars whose fruit mark easily, e.g. Comice, Cascade. Asian pears need slightly more protection *Edaphic features* Flat land preferred; do not choose frost pockets or undrained sites *Soil needs* European pears are tolerant of a wide range of soil types if planted on seedlings. They are less tolerant on quince, where they prefer non-calcareous soils. Asian pears budded on *P. betulaefolia* are tolerant of heavy soils; for lighter, more fertile soils, *P. pyrifolia* is used	*Propagation* Budded in mid to late summer on to seedlings or quince rootstocks (propagated by stoolbeds) or clonal *Pyrus* stocks propagated by cuttings *Rootstocks* European pears – usually seedlings of Williams' Bon Chrétien or clonal quince. Asian pears – mainly seedlings of *P. pyrifolia*, also *P. betulaefolia*, *P. calleryana* in poor soils *Spacing* Centre-leader trees on seedling at 5–6 m × 3–4 m or on quince, 4–5 m × 2–4 m. Asian pears may suit structures such as Lincoln canopy or Tatura trellis *Training and pruning* Still grown as open-centre trees or as centre-leaders, particularly in conjunction with quince. Renewal pruning most commonly used. Asian pears will be pruned similarly, but sometimes a little harder to encourage large-sized fruit *Thinning* Chemical thinning less effective than with apples and is seldom used. Hand thinning is adopted when good size is required. Generally, only the small-fruited European pears, such as W. Nelis need thinning. Asian pears do need thinning – see Key points *Tillage* Normally, orchards are grassed in the alleyways and a strip sprayed with herbicides under trees *Time to first harvest* Seedling rootstocks: 5–6 years Quince: 3–4 years *Time to full production* Seedling rootstocks: 10+ years Quince: 6–8 years *Continued*

D. Jackson and J. Palmer

Table 14.4. Continued.

Botanical, anatomical and physiological aspects	Climatic, geographic, soil and water requirements	Cultural aspects

Time of bud burst
Spring, usually after stone fruit and before apples
 	
Time of flower
Slightly before apples, after most stone fruit. Asian pears are a few days earlier than European pears
 	
Growth of fruit and shoot
Fruit grows with single sigmoid growth curve. New shoots arise from terminal or lateral vegetative buds. Growth of spur continued by bourse shoot in axil of one or more of the basal spur leaves
 	
Time of maturity
Similar range as apples
	Nutrient requirements
	The best pH range is 5.5–6.5, but they grow, in some circumstances, from 4.5 to 8.0. Create good nutrient regime but avoid excessive nitrogen fertilizers, which will encourage excessive vegetative vigour, reduce storage quality and sometimes reduce flower-bud initiation
		Expected yield
		Seedling rootstocks:
		    5 years – 8t/ha
		    7 years – 20t/ha
		    10 years – 50–80t/ha
		
		Normal productive life
		Pear trees are often very long lived, frequently 40–80 years. There is far less turnover of cultivars with pears compared to apples
		
		Method of harvest
		Hand harvest into bulk bins. European pears are frequently strip picked while Asian pears are more commonly select picked
		
		Storage
		Most late cultivars keep well in cool store at –0.5°C or in controlled atmosphere store. Some, such as Comice, need a period of cool storage to trigger the ripening process
		
		Main pathogens and pests
		European pears: pear scab, fireblight, codling moth, leafrollers, mites, scale insects, pear psilla. Asian pears are susceptible to the same disorders and they are especially attractive to birds

most European pears are green or brown, there are a number of red-skinned cultivars available. Some of these are sports of existing cultivars, e.g. Red Clapp's Favourite, Crimson Gem Comice and Red Bartlett, although some, such as Cascade, are original red-skinned cultivars. For reasons not yet understood, red pears tend to be less vigorous than the green types, and blemishes on the fruit show up more clearly on the red background. There are also a number of pears with a red blush, e.g. Florelle and two releases from South Africa, Flamingo and Rosemarie.

Thinning

While some European pears do not require thinning, others require careful attention to this practice. For example, Bartlett pears grown at high latitudes will not meet market standards unless the crop is reduced substantially. Asian pears set very heavily and definitely require thinning to achieve large fruit size. The level of thinning is difficult to predict, but based on limited experience and on observations in Japan, it is suggested that for trees planted 2 × 5 m about 250 fruit per tree should be left. Japanese nashi growers thin twice: once at 1.3 cm diameter and again 6–8 weeks later. Normally the fruit in the centre of the cluster is retained. Nashi are normally thinned by hand.

Rootstocks

Outside of Europe, European pears have been traditionally grown on seedling rootstocks which

Table 14.5. Cultivars and pollinizers for European and Asian pears.

Cultivar	Suitable pollinizers	Comments
European pears – listed in order of ripening		
Clapps Favourite (CF)	B. Bosc, Conference, WBC, W. Nelis	Very early – useful for local marketing only
Williams' Bon Chrétien (WBC) – also called Bartlett	W. Nelis, B. Bosc, PT, CF. Will set parthenocarpically if temperatures are high enough during blossom	Main canning pear. Short storage life gives it limited value as fresh fruit but very good fruit quality
Conference	WBC, B.Bosc, Comice	Medium size, partly russetted fruit with good storage potential, widely grown in Europe
Packham's Triumph (PT)	WBC, B. Bosc	High-yielding cultivar but in cooler regions the fruit can be rather uneven or knobbly in shape. Stores well
Beurré Bosc (B. Bosc)	WBC, W. Nelis, CF	A long, good-quality, full-russet pear. A rather weak, ungainly growing tree
Doyenné du Comice	B. Bosc, WBC, W. Nelis	Excellent flavour but difficult to grow well. Fruit easily marked on the tree and in handling. There is a fully russetted sport, Taylor's Gold, which ripens a little later than its parent
Winter Nelis (W. Nelis)	B. Bosc, Conference, WBC	Winter Nelis (and Winter Cole) are largely unknown, except in Australia and New Zealand
Asian pears – listed in order of ripening		
Shinseiki	Self-fertile	Similar season to Kosui; main value is as pollinator. Fruit is yellow, medium size, skin thin, firm fresh; it is sweet with medium eating quality
Kosui	Shinseiki	Golden brown russet on a pale green to yellow background. Fruit is sweet, good quality, moderate in size, stores quite well but skin susceptible to damage
Hosui	Kosui, Shinseiki	A fully russet type. It is the largest of the four cultivars. It has good flavour, is sweet, tender, crisp and juicy, and generally of excellent quality and stores well.
Nijisseiki (Twentieth Century)	Self-fertile, use Kosui or Shinseiki if set is not good	Small to medium size with a clear, yellow skin. It needs heavy thinning due to high fruit set. It stores well and is of fair to good eating quality

D. Jackson and J. Palmer

promote vigorous growth; trees eventually reaching 6–8 m or more. Full cropping may not be reached until year ten. This tradition has developed because the quince (*Cydonia oblonga*) clonal rootstocks used in Europe lack hardiness in non-maritime climates. However, quince rootstocks do offer improved precocity and reduced tree vigour compared to seedling *Pyrus* rootstocks. However, graft incompatibility is a problem with many pear cultivars on quince rootstocks. The use of an interstock of a compatible cultivar such as Beurré Hardy can overcome this problem but adds to the cost of the tree.

Unfortunately, although there is dwarfing material within the *Pyrus* germplasm, it has not been possible to date to achieve comparable levels of ease of propagation, dwarfing and precocity that quince rootstocks offer. There is growing interest in North and South America in a series of clonal rootstocks resistant to fireblight and pear decline that were derived from crosses between Old Home and Farmingdale (OH×F). Some of these clones also provide a moderate degree of vigour control and a precocious and productive tree; OH×F 87 is considered one of the most promising in this series. A recent pear rootstock candidate from Europe, marketed as Pyrodwarf, arose from an Old Home × Bonne Louise Jersey cross made at Geisenheim, Germany. It is dwarfing and productive but has not been shown to be as dwarfing as trees on quince QC. The BP series from South Africa (particularly BP 1 and BP 3) are widely planted in that country but not used widely elsewhere.

Asian pears are propagated on *Pyrus pyrifolia* or *Pyrus betulaefolia* seedlings, which give trees a little smaller or somewhat larger, respectively, than European pears on seedling rootstocks. Asian pears are not directly graft compatible with quince, but again an interstock of Beurré Hardy can be used.

Training

There is a good case to be made for growing some pear cultivars as a trellised canopy as this can reduce the marking of the skin common to both European and Asian pears in windy regions. European pears on seedling rootstocks do well on the Lincoln canopy (Chapter 5), a method of training very similar to the Japanese pergola. Other trellised canopies include V-shaped cross-section canopies such as the mini-Tatura. Pears for export need to be well presented and of finest quality and the support given by these methods will help to achieve this result.

For free-standing trees, the centre-leader form is widely used. Using seedling trees for both pear types, distances of 2–2.5 × 5–5.5 m are recommended. For European pears on quince rootstocks, planting distances of 1.5–2.0 × 4.5–5.0 m would be suitable, although with more intensive systems of tree management, plantings in rows as close as 3.5–4.0 m can be successful.

Storage

European pears can be stored at +0.5°C, but storage life can be extended by storing at −0.5°C. Asian pears, by contrast, are stored at +0.5°C.

Further Reading

Barritt, B.H. and Kappel, F. (eds) (1997) VI International Symposium on Integrated Canopy, Rootstock, Environmental Physiology in Orchard Systems. *Acta Horticulturae* 451.

Corelli-Grappadelli, L., Janick, J., Sansavini, S., Tagliavini, M., Sugar, D. and Webster, A.D. (eds) (2002) VIII International Symposium on Pear. *Acta Horticulturae* 596.

Hrotkó, K. (ed) (2007) VIII International Symposium on Canopy, Rootstocks and Environmental Physiology in Orchard Systems. *Acta Horticulturae* 732.

Huang, X.M. and Janick, J. (eds) (2007) I International Symposium on Organic Apple and Pear. *Acta Horticulturae* 737.

Iwahori, S., Gemma, H., Tanabe, K., Webster, A.D. and White, A.G. (eds) (2002) International Symposium on Asian Pears, Commemorating the 100th Anniversary of Nijisseiki Pear. *Acta Horticulturae* 587.

Palmer, J.W. and Wunsche, J.N. (eds) (2001) Orchard systems for apple and pear: conditions for success. *Acta Horticulturae* 557.

Theron, K.I. (ed.) (2005) IX International Pear Symposium. *Acta Horticulturae* 671.

Webster, A.D. (2004) XXVI International Horticultural Congress: Key Processes in the Growth and Cropping of Deciduous Fruit and Nut Trees. *Acta Horticulturae* 636.

Webster, A.D. and Oliveira, C.M. (eds) (2008) X International Pear Symposium. *Acta Horticulturae* 800.

15 Grapes

DAVID JACKSON

The European grape, *Vitis vinifera*, probably originated in the region of the Caucasus Mountains between the Black and Caspian Seas. This one species accounts for the major production of grapes used for making wine, for drying or for fresh fruit. There are also several grape species native to North America; some are used for juicing, some as rootstocks and a limited number (often as interspecific hybrids) as dessert and wine grapes.

Wine, the end product of most grapes, has a long history. It was probably enjoyed by the Assyrians and Egyptians 5000 or more years ago. The first merchant navigators, the Phoenicians, exchanged wines for other goods around 1000 years ago, and wine has been drunk, traded and enjoyed in Europe ever since. As Europeans moved into the New World, grapes and wine inevitably moved with them. In the Americas vines were planted by Cortez in 1524; in South Africa van Riebeeck established grapes in 1652; and Busby introduced grapevines to Australia and New Zealand in the early part of the 19th century.

The settlement of North America was a significant step in the history of grapes. The settlers were surprised to find such a variety of native grapes in their new home, but when some of these were introduced into Europe they brought with them a number of then unknown pests and diseases, of which the most significant was phylloxera. This root aphid devastated large areas of *vinifera* grapes, which carried no resistance. It is for this reason that, in most areas, grapes are grafted on to resistant rootstocks of American origin.

Grapes, one of the world's most widely grown fruit crops, are most common in relatively warm temperate zone climates and are not well adapted to subtropical and tropical areas. Nevertheless, there are special management techniques whereby these areas can be utilized, and two to four crops of dessert grapes per annum are produced in places as hot as Thailand and Indonesia.

Key points

Climate requirements

To a major extent heat accumulation determines which grapes can be successfully grown in any district. Below 700–900°C degree days few grapes will ripen satisfactorily; from 950 to 1500°C degree days table wines are predominant; above 1500°C degree days table and fortified wines are produced; and above 1950°C degree days table (fresh) grapes and dried fruit are dominant.

Some areas which have adequate summer heat cannot grow grapes because of cold damage in the winter – many parts of central (continental) USA have such conditions. Generally if the mean temperature of the coldest month is 0°C or below, freeze damage can be expected since such climates will often have conditions where the temperature can drop to –15 to –20°C.

Most of the more prestigious European wine-growing areas are cool rather than warm. Thus Bordeaux, Burgundy and Champagne are more renowned for table wines than parts of southern France or Spain. Nevertheless, it is believed that, with appropriate technology, warmer districts can produce more distinctive wines than previously was believed possible. Australia and California are showing that warm areas can produce such wines. Fortified wines such as sherry and port are grown in warmer districts such as Spain, Portugal and parts of Asia.

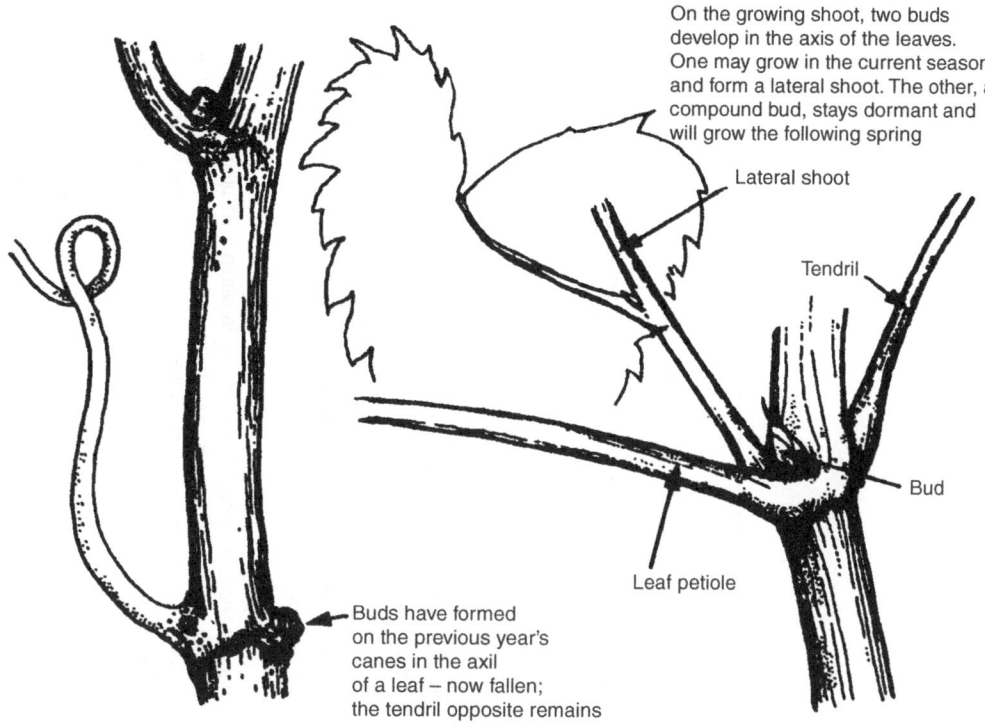

On the growing shoot, two buds develop in the axis of the leaves. One may grow in the current season and form a lateral shoot. The other, a compound bud, stays dormant and will grow the following spring

Lateral shoot

Tendril

Bud

Leaf petiole

Buds have formed on the previous year's canes in the axil of a leaf – now fallen; the tendril opposite remains

Fig. 15.1. Left, part of a dormant cane; right, formation of buds in growing shoot.

Soils and fertilizers

Grapes are very tolerant of a wide range of soils, although they do not like wet and cold conditions. Grapes grown for wine or juice are probably better on less fertile soils, since high fertility induces excessive vegetative growth. This leads to disease problems, additional summer trimming and possibly poorer wine quality. Fertilizers, particularly nitrogen, are used sparingly.

Water needs

The deep roots of grapevines enable them to survive in very dry conditions, and it is sometimes believed that irrigation adversely affects fruit and wine quality. Under certain conditions this may be correct, since, like fertilizers, irrigation may induce excessive vegetative growth. Thus, mild water stress in grapevines is often the preferred state since it helps in the management of vine vigour. However, like most generalizations, there are numerous exceptions and many districts with low rainfall could not produce grapes without irrigation.

Rootstocks

There are many selections which have *Vitis rupestris*, *Vitis riparia* and *Vitis berlandieri* in their parentage. Because local preferences and needs are so diverse, it is very important to seek local advice about the most suitable rootstock for the region.

Pests and diseases

The main pests are phylloxera and nematodes, although the choice of an appropriate rootstock can virtually eliminate the consequences of infestation. Many insects can attack grapes and the incidence varies from district to district. Caterpillars of several moth species, scale insects, mites of various sorts, weevils and mealybugs are some of the many that attack grapes.

A number of fungal diseases can seriously limit grapevine productivity. In wet conditions botrytis needs special attention. Other diseases such as downy mildew, anthracnose and phomopsis can be serious during wet weather and often do not

Table 15.1. Grapes.

Botanical, anatomical and physiological aspects	Climatic, geographic, soil and water requirements	Cultural aspects
Common name Grape	*Temperature needs* Warm, temperate climates required, usually 900 or more degree days (base 10°C) are required	*Propagation* Where phylloxera or nematodes are not a problem, cuttings are used: normally hardwood, or softwood for rapid propagation. Otherwise, field or bench graft (cutting graft) on to resistant rootstocks
Botanical name *Vitis vinifera* (European grape)		
Botanical name of related and useful species *Vitis labrusca* *Vitis aestivalis* *Vitis vulpina* *Vitis rupestris* *Vitis riparia* *Vitis berlandieri*	*Frost tolerance* −20°C dormant, −1°C early bud burst, −0.5°C bloom and fruit *Water needs* Has deep root system and moderate water needs	*Rootstocks* There are many selections which have mostly been bred from combinations of *V. rupestris*, *V. riparia* and *V. berlandieri*. They can impart some or all of the following: resistance to phylloxera and/or nematodes; some measure of size control, although less than, say, apple rootstocks; precocity in terms of earliness of fruit maturity; tolerance to drought, wet, acid or alkaline soils and various mineral deficiencies. Despite these advantages cutting-grown plants are generally used when phylloxera and nematodes are not present. Local advice is the best way to choose an appropriate rootstock
Type of plant and size A vine capable of vigorous growth if left unchecked	*Tolerance of wet soils* Low in summer, moderate in winter *Drought tolerance* Good, especially if established in deep soils	*Spacing* Rows are normally between 2.7 and 3.0 m apart, while within rows plants are spaced from 1 to 2 m apart, depending on soils, rootstocks and climate. Many traditional European vineyards are planted as close as 1 m apart between and within rows. Special machinery is needed for such planting. Some quality improvement is claimed
Sexuality *Vinifera* is hermaphrodite, many native American species are monoecious	*Humidity tolerance* Low; regular spraying for disease control needed in wet climates	*Training and pruning* There are very many different training systems, but possibly the most preferred is the vertical, shoot-positioned (VSP) canopy, described in Chapter 6. Variations depend on local customs and preferences, current knowledge or fashion, climate and the end use of the grape. Spur and cane pruning are the alternative approaches to pruning, with the former being more common in warmer climates
Pollination Wind pollinated and insects are not required; warm, dry conditions improve pollination and fruit set		

Flower buds
The grape has a multiple bud produced laterally on the previous season's cane. The overwintering bud (Fig. 15.1) is, in fact, three buds enclosed by common scales. The central, primary bud normally grows in spring and produces a shoot with two lateral inflorescences. The other two buds usually remain dormant

Growth of fruit
Double-sigmoid growth curve. Veraison (colour change) occurs at beginning of final growth spurt (stage III)

Time of bud burst
Usually later than stone fruit, similar time to apple

Time of flowering
7–9 weeks later

Time of fruit maturity
5–7 months after bud burst, normally autumn

Wind tolerance
Moderate to good when on trellis

Edaphic features
Slopes facing sun sometimes used in cooler districts to promote heat accumulation and reduce frosts. Otherwise flat land preferred

Soil needs
Tolerant of many soil types, provided deep and well drained. Generally, light soils are preferred and those with lower fertility can make canopy management easier

Nutrient requirements
Heavy manuring, especially with nitrogen, causes vigour problems, particularly in humid districts

Bunch thinning
Not needed for wine grapes unless overcropping occurs; detailed thinning needed for glasshouse table grapes and moderate thinning may be adopted for outdoor table grapes

Tillage
Commonly clean cultivated. Herbicides often used between plants in rows; some growers now use chemicals between and within rows

Time to first harvest
2nd or 3rd season after planting

Time to full production
Three to four seasons

Expected yield
Very dependent on cultivars and district. Normally (for wine)
2 years: 4 t/ha
3 years: 8–15 t/ha
4 years: 10–20 t/ha

Normal productive life
40–60 years possible, fashion or disease may shorten life

Method of harvest
Harvesting machines common for wine and juice; otherwise by hand

Storage
Bunches can be cool stored for 3–6 months at 1°C. Fumigation with sulphur dioxide sometimes used

Main pathogens
Problems are caused by insects, nematodes. fungal and bacterial diseases, birds and physiological disorders – see Key points

appear in dry climates. Powdery mildew (*Oidium* spp.) appears in both wet and dry areas. Crown gall is the most serious bacterial disease and usually occurs as a consequence of physical damage to the lower trunk, sometimes due to winter freezes. Many virus diseases, such as fan leaf, leaf roll, grapevine yellow speckle and stem pitting, attack grapes and can only be eliminated by heat treatment at the propagation source. The best insurance is to buy plants of known minimum virus status.

In many districts birds are the most serious problem. Bird damage can be reduced by employing a range of tactics, such as shooting, scare-devices such as bangers, hawk replicas and distress call emitters, or, in the last resort, exclusion netting.

A number of physiological disorders can cause problems. These include early bunch–stem necrosis, which causes sections of the bunch to wither and die before flowering. It is a response to vine stress, such as heavy shade, low water or nutrient deficiencies. Bunch–stem necrosis causes death of stem sections after veraison (grapeberry colour change). Its cause is not fully understood. Millerandage or poor set is due to cool and wet conditions close to flowering, which reduce pollination and fertilization.

Harvesting

The timing of wine grape harvest is very critical. Picked too early, the grapes are too acid; if picked too late, they may contain insufficient acidity or suffer reduced yield because of bird damage or bunch rots. Choice of the most appropriate picking date ensures the grape contains a high level and the correct balance of flavour and aromatic compounds.

Wine grapes generally mature with between 16 and 24% sugar and have acid levels from 0.6 to 1%. The most appropriate levels will vary according to the district, the cultivar and the wine style required. Juice from grapes picked outside their optimal sugar and acid range may need some adjustments by the winemaker.

Many grapes sold for winemaking and juicing are mechanically harvested. This substantially reduces harvest labour costs. Furthermore, because the harvest operation can be completed in a shorter time period, more of the grapes reach the winery or juice plant at optimum maturity.

The production of quality juice and wine is determined by using the best varieties for the climate, combined with good husbandry, determina-

tion of optimum maturity, good strains of yeast, proper care during fermentation and adequate maturation of the wine.

Dried grapes (raisins) are harvested and treated in several ways. They may be hand picked and dried, either directly in the sun or in racks. To assist drying and give a special light-coloured appearance to the fruit, grapes may be treated with alkaline oil emulsions prior to drying. Machine harvesting causes too much damage to fruit for effective drying in the sun or on trays, but by pruning canes with bunches still attached and allowing them to hang on the vine to dry, harvesting machines can be used.

Main cultivars

WINE GRAPES The main requirement for wine grapes is to have an appropriate juice composition that will produce good wine; the appearance of the fruit is of little consequence. Fewer than 30 cultivars are used to make the world's classic quality wines; all of these are cultivars of *Vitis vinifera*. Hundreds more are used to some degree for wine production; some of these are American species and others are hybrids between American species and *V. vinifera*. Some important wine grapes are:

- Red: Pinot Noir, Merlot, Cabernet Sauvignon, Shiraz (Syrah) and Grenache.
- White: Muller Thurgau, Chasselas, Chardonnay, Riesling, Semillon, Sauvignon Blanc, Chenin Blanc and Palomino.

JUICE GRAPES In the USA, in particular, grapes for juice constitute an important and significant outlet. One cultivar, Concord, accounts for over 90% of the production. This cultivar, plus others such as Niagara and Iona, is mainly derived from *Vitis labrusca*.

DRIED GRAPES (RAISINS) These grapes are *V. vinifera* cultivars and include Sultana (synonyms: Sultanina, Thompson's Seedless) and Currants (synonyms: Zante, Zante Currant, Black Corinth), both of which are seedless, and seeded cultivars such as Muscat Gordo Blanco (Muscats, Muscatels). Dried grapes need warm conditions (over 1950°C degree days) for economic production.

DESSERT GRAPES (TABLE GRAPES) Dessert grapes are produced most economically in warm to hot temperate climates. Good appearance, large size and thin skin, together with good texture and flavour are required. Seedless grapes are an advantage, but often their size is too small; treatment with the plant hormone gibberellic acid can make small seedless grapes, such as Sultana, acceptably large.

Further Reading

Adsule, P.G., Sawant, I.S. and Shikhamany, S.D. (eds) (2008) International Symposium on Grape Production and Processing. *Acta Horticulturae* 785.

Bouquet, A. and Boursiquot, J.M. (eds) (2000) VII International Symposium on Grapevine Genetics and Breeding. *Acta Horticulturae* 528.

Bravdo, B.A. (ed.) (2000) V International Symposium on Grapevine Physiology. *Acta Horticulturae* 526.

Hajdu, E. and Borbas, E. (eds) (2003) VIII International Conference on Grape Genetics and Breeding. *Acta Horticulturae* 603.

Nuzzo, V., Giorio, P. and Giulivo, C. (eds) (2007) International Workshop on Advances in Grapevine and Wine Research. *Acta Horticulturae* 754.

Peterlunger, E., Di Gaspero, G., and Cipriani, G. (eds) (2009) IX International Conference on Grape Genetics and Breeding. *Acta Horticulturae* 827.

Reynolds, A.G. and Bowen, P. (eds) (2004) XXVI International Horticultural Congress: Viticulture – Living with Limitations. *Acta Horticulturae* 640.

Sequeira, J.C. and de Sequeira, O.E. (eds) (2004) I International Symposium on Grapevine Growing, Commerce and Research. *Acta Horticulturae* 652.

Williams, L.E. (ed.) (2005) VII International Symposium on Grapevine Physiology and Biotechnology. *Acta Horticulturae* 689.

16 Berry Fruit

GRAHAM THIELE

Blueberries and Other *Vaccinium* Species

The genus *Vaccinium* contains over 100 species of deciduous and evergreen shrubs and small trees. Many of these have considerable economic importance, particularly in North America, where blueberries and cranberries are produced on acid land, much of which would otherwise be rather barren.

The northern highbush blueberry, *Vaccinium corymbosum* (2–3 m), is the most common species grown commercially in North America (Table 16.1). The earliest record of the highbush or tall blueberry is in the eastern USA about 1765. Other cultivated blueberry species include the southeastern highbush blueberries, *Vaccinium atrococcum* and *Vaccinium australe*; the rabbiteye blueberry, *Vaccinium ashei*; the lowbush blueberry, *Vaccinium angustifolium*; the sourtop or Canadian blueberry, *Vaccinium myrtilloides*; and the dry land blueberries, *Vaccinium hirsutum* (the huckleberry) and *Vaccinium vacillans*.

Other species of note include the bilberry or whortleberry, *Vaccinium myrtillus*; the cowberry or mountain cranberry, *Vaccinium vitis-idaea*, a low-growing (15–30 cm) evergreen with creeping stems, often found growing wild in Europe and Scandinavia; and the American cranberry, *Vaccinium macrocarpon*, also a creeping evergreen shrub with ascending shoots, cultivated extensively near both coasts of the USA in so-called 'cranberry bogs'.

It is difficult to determine accurately world production figures because of the extent of public harvesting from natural (wild) areas. For example, parts of Eastern Europe and Scandinavia have abundant natural resources of cowberries or lingonberries (*V. vitis-idaea*), small cranberries (*V. oxycoccus*), bilberries (*V. myrtillus*) and bog whortleberry (*Vaccinium uliginosum*).

North America, including Canada, the USA and Mexico, dominates world blueberry production. Each decade production increases by at least 25% and is now over 100,000 ha. South American production, particularly Chile and Argentina, is increasing. Australasia, the Netherlands, France, Poland, Spain, Germany, Sweden and Japan are other blueberry producers.

Key points

Cultivars and plant breeding

One of the most difficult problems of the blueberry industry is that many of the developed cultivars have not performed consistently once they have been released. This is despite good productivity performance under small-scale plant research. It is suggested that the planting of large blocks commercially is creating pollination problems. A similar picture is developing with rabbiteye plantings.

For many years there has been interest in combining the two elite gene pools of highbush (tetraploid) and rabbiteye (hexaploid). The F_1 is a pentaploid with relatively low pollen fertility, but backcrossing has been successful. The gene pool within the *Vaccinium* genus is extensive and has offered a wide range of materials for the plant breeder. As a result, many new hybrids are originating in North America, but all require commercial testing under the conditions prevailing in different countries. The breeding objectives include: the incorporation of cold tolerance, with or without low chilling requirements; early, mid-season and late maturity; high fruit set; high yields and high fruit quality; and resistance to pests and

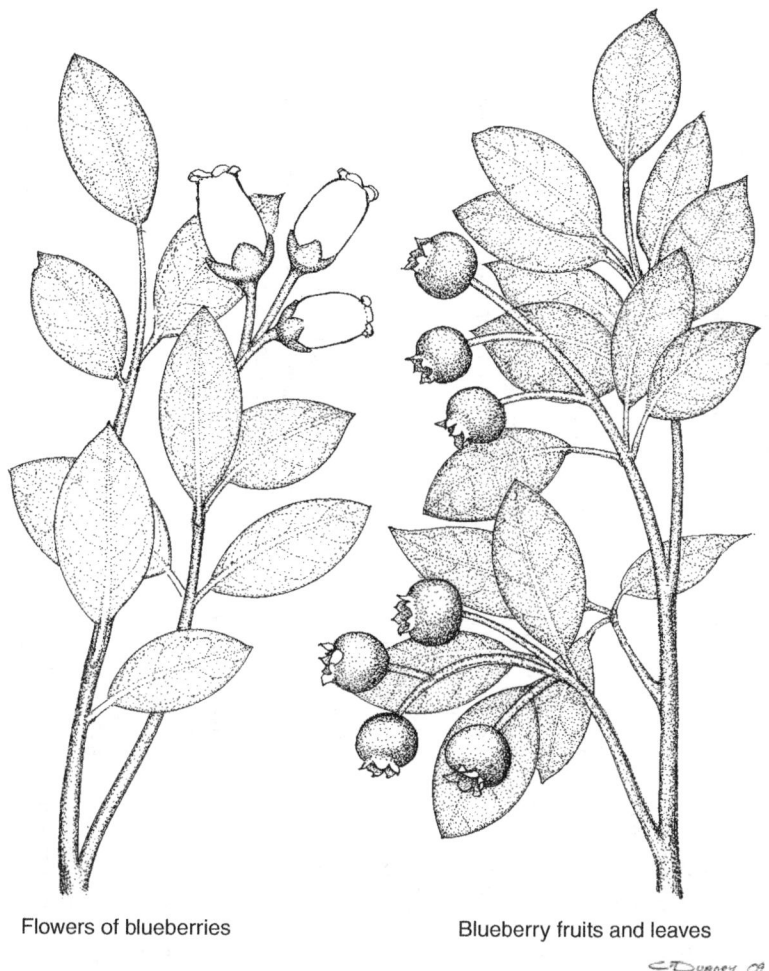

Flowers of blueberries   Blueberry fruits and leaves

C. Durney. 09.

Fig. 16.1. Blueberries.

diseases. High fruit quality requires good flavour, firmness and a small, dry picking scar.

Highbush blueberries are classified into northern, intermediate and southern types (referring to the northern hemisphere). About 75% of the world highbush blueberry plantings are in North America, with yields varying between 8 and 10 t/ha.

The planting of southern-type blueberries is increasing rapidly in Florida and Mexico. O'Neal and Misty are favoured cultivars, along with Bluecrisp and Star. O'Neal is suitable for mechanical harvesting and also responds well to mechanical topping.

The intermediate blueberry has a wide range of chilling requirements, from 400 to 800 h below 7°C. Suitable conditions occur in Arkansas, North Carolina and parts of the Pacific Northwest. Cultivars include Legacy, Ozarkblue and Reveille.

The northern blueberry tolerates temperatures down to –20°C and performs best where there are 800–1000 h of annual chilling below 7°C. Cultivars include Duke (probably the most widely planted worldwide), Elliot, Bluecrop, Reka and Brigitta. Plantings of these are found in the USA (Michigan/ New Jersey), Canada, Australia and New Zealand, France, the Netherlands Germany, Spain and other European countries, and in Chile.

Rabbiteye (*V. ashei* Reade) is grown in southeastern USA. They require 400–700 h chilling below 7°C. Cultivar examples are Ocklockonee, Alapaha and Tifblue.

Table 16.1. Blueberries and other *Vaccinium* species.

Botanical, anatomical and physiological aspects	Climatic, geographic, soil and water requirements	Cultural aspects
Common name Highbush blueberry	*Temperature needs* Will grow in most temperate regions, although winter chilling may not be sufficient in warm temperate regions. The lowbush is more tolerant of cold conditions than the highbush, and the rabbiteye more tolerant of warm conditions	*Propagation* Normally from cuttings taken in winter. Can be produced under mist propagation from softwood cuttings. One-year-old rooted cuttings will be less expensive but 2-year-old planting material is preferred
Botanical name Vaccinium corymbosum		*Rootstocks* Not used
Botanical name of related and useful species Vaccinium angustifolium (lowbush) Vaccinium ashei (rabbiteye) Vaccinium myrtillus (bilberry) Vaccinium vitis-idaea (cowberry) Vaccinium macrocarpon (cranberry)	*Frost tolerance* Late spring frosts reduce yield. Crop is seldom completely destroyed by frost, and frost control not usual	*Spacing* 2.5–3.0 m between rows; 0.9–1.2 m between plants
Type of plant and size Woody perennial. Very variable in size. Hybridizes freely with lowbush blueberry. Normally 1–4 m in height and 2 m wide	*Water needs* Irrigation mostly necessary. Very much a surface-rooting plant	*Training and pruning* Flowers are borne at the tips of previous season's growth. Remove spindly growth and prune to balance growth and cropping. Remove old wood whenever possible
Sexuality Hermaphrodite	*Tolerance of wet soils* Do not succeed in wet, poorly aerated soil. The fine, fibrous roots prefer moist soil, high in peat or other organic matter	*Thinning* Not required
Pollination Although partially self-fertile, crop yields have been increased markedly by cross-pollination, particularly where cultivars have not been planted in blocks	*Drought tolerance* Will not tolerate either waterlogged or drought conditions	*Tillage* Blueberries are very shallow rooting and cultivation should be minimized. Sod culture with chemical weed control along the rows can be used where soil moisture is adequate. Mulches of sawdust or bark are beneficial
Flower buds Produce fruit from buds formed on the previous season's growth	*Humidity tolerance* Full sunlight preferred. Fungus diseases may be troublesome in humid conditions	*Time to first harvest* Some fruit on 3-year-old plants, but establishment of a strong bush is first requirement
Growth of fruit Double-sigmoid growth curve. A berry may increase 25% in volume after it has turned blue. Fruit size is related to number of seeds	*Wind tolerance* Shelter essential in windy areas	*Time to full production* 6–7 years
		Expected yield Year 3: 0.25 t/ha, varies between highbush and rabbiteye (slightly higher yield) Year 4: 1.5 t/ha Year 5: 3.0 t/ha Year 6: 7.0 t/ha Lowbush yields: 2 t/ha
		Normal productive life 20–30 years

G. Thiele

Time of bud burst
Early spring

Time of flowering
Wide variation from spring to early summer, depending on cultivar and district

Time of fruit maturity
Early summer to early autumn

Edaphic features
Light blue bloom on the berries may be lost in poorly sheltered areas. Flat land preferred. To minimize frost risk, avoid areas with poor air drainage

Soil needs
Acid soils (pH 4.0–5.0) are essential, with peat preferred. Sandy soils within the recommended pH range are suitable, provided high soil moisture levels are maintained. Addition of bulky organic matter such as sawdust is useful. Blueberry roots cannot penetrate compacted soils

Nutrient requirements
Responds to nitrogen, but do not apply excessive amounts (see text). Standard concentrations (%) for foliar analysis are:
N: 1.80–2.10
P: 0.12–0.40
K: 0.35–0.65
Ca: 0.40–0.80
Mg: 0.12–0.25

Method of harvest
Dessert fruit with a desirable 'bloom' is still picked by hand but machine harvesting is common in USA (40%) for processing. Selective harvesting important for dessert fruit. Berries continue to enlarge for 3–6 days after turning blue. Hand-held shaker units can be used

Storage
Can be held in punnets for 10–14 days at 1–2°C. Longer storage life than other berry fruit

Main pathogens
Pests and diseases are not generally serious problems. Botrytis stem blight, stem canker, mummy berry, fruit rot, rust and leafroller may need control measures. Birds are the most serious problem and netting will be needed, particularly in small plantations. Phytophthora root infections occur in compact and sodden soils

Soil and planting systems

Although low-fertility swamp peat with a low pH is the natural habitat for most *Vaccinium* species, countries with limited areas of peat have had success with blueberry production on other substrates. Europe, for instance Germany and the Netherlands, has substantial blueberry production on mineral soils.

The pH can be lowered on mineral soils with organic mulches such as sawdust or with chicken and other animal manures. The use of ammonium sulphate will also help create or sustain acid soil conditions. Applying sulphur to the soil prior to planting is effective. To reduce the pH from 6.5 to 5.5 about 100–250 kg of sulphur will probably be needed per hectare. For soils with a higher clay content, the upper end of the scale should be used; for sandy soils, the lower end.

Fertilizer applications should be based on soil and leaf analysis. For mature bushes, 10–20 kg P/ha and 20–30 kg K/ha should be sufficient as maintenance dressings. Nitrogen to the extent of 30–40 kg/ha should be applied each year, depending on the soil type and the age of the bushes. On mineral soils, an additional amount of 20 kg N/ha should be applied in summer, as either a single or split application. The greatest demand for nitrogen is when the crop is maturing. Flower number, fruit set, fruit size and yield are all reduced by nitrogen deficiency. It is preferable to apply nitrogen in forms other than as nitrate. Sometimes peat soils are deficient in magnesium and the base dressing should have magnesium incorporated at about 20 kg/ha magnesium sulphate. Iron deficiency is likely to be induced by high pH.

The importance of ericoid (heath) mycorrhizal fungi to blueberry growth and yield has been clearly demonstrated. These fungi are present naturally in most (but not all) peat soils but not in most mineral soils. It is suggested that mycorrhizal fungi greatly increase the uptake of nitrogen and phosphorus and can break down the organic soil nitrogen to an available form for *Vaccinium* spp. crops. Research suggests that mycorrhizal infection of the roots is particularly beneficial on higher pH soils and that it can increase yields by as much as 20%. Further research is needed to determine the exact role of ericoid mycorrhizas on nutrient uptake and metabolism and also on their interaction with soil-borne root disease.

Blueberries grown in peat soils are often planted on raised rows 25–50 cm high to prevent waterlogging after rain or irrigation. This is not required in lighter, free-draining soils. Planting systems for highbush and rabbiteye cultivars vary markedly, but closer spacing is now generally favoured. Spacings recommended are 0.9–1.2 m in the row and 2.5–3.0 m between rows, depending on the soil type, vigour of the cultivar and whether or not the block is machine harvested. Closer spacing in the row results in a hedge after several years, and plants can be thinned if necessary as the block matures. It also increases early returns and maximizes benefits of expensive bird-netting enclosures. Plant density generally varies between 1600 and 4500 bushes/ha, with an average of about 2500/ha.

The importance of quickly establishing bush size for a good early yield means that good cultural practices are needed in the early stages. To control weeds and improve moisture retention, some growers establish a plastic mulch when planting. If mulches are not used, chemicals are commonly applied to control weeds within the rows.

Pests and diseases

Stem blight (*Botryosphaeria dothidea*) has become a serious disease of both highbush and rabbiteye blueberries in parts of North America. New strains of stem blight seem to appear as soon as 'resistant' cultivars are developed.

Stem canker (*Botryosphaeria corticis*) severely infects early-ripening highbush cultivars, such as Weymouth, and rabbiteye cultivars such as Tifblue and Climax. A number of resistant cultivars have been released, including Reveille and Bladen.

Blueberry stunt disease has been identified as a mycoplasma. So far no inherent resistance has been found and control has concentrated on the sharp-nosed leafhopper vector. Other diseases of importance include anthracnose or mummy berry (*Monilinia vaccinii-corymbosi*), twig blight and fruit rot (*Phomopsis vaccinii*) and botrytis (*Botrytis cinerea*). Rust causes leaf damage in warmer areas. Root rot (*Phytophthora cactorum*) occurs in poorly drained soil conditions.

For production of high-quality berries for fresh consumption, particularly in small plantations, bird control, in the form of netting, is essential.

G. Thiele

Currants and Gooseberries

The blackcurrant cultivars grown commercially today are derived from the wild parent *Ribes nigrum*, growing from Europe to the Himalayas (Table 16.2). The *Ribes* genus has about 150 species of deciduous and evergreen shrubs, originating mainly in the cool and temperate regions of the northern hemisphere. Redcurrants and whitecurrants are also native to Europe and are normally considered to be *Ribes sativum*, although sometimes mistakenly said to be *Ribes rubrum*. Varieties of *R. rubrum* are grown for their fruit in Scandinavia and known there as the northern redcurrant.

The cultivated gooseberry, *Ribes grossularia*, a deciduous spiny shrub and native of the mountainous regions of Europe and northern Africa, is also a member of this genus.

Plant breeders have successfully crossed selected varieties of *R. nigrum* with *R. grossularia*, producing a wide range of characteristics in the progeny. One example is the jostaberry, a gooseberry-sized blackcurrant fruit on a spiny bush with gooseberry-like leaves. The worcesterberry, a natural hybrid between the blackcurrant and gooseberry, grows wild in North America.

Poland is the largest *Ribes* producer in the world and produces 70% of the EU total. The production of black/redcurrants is probably above 70%.

Russia, the UK, New Zealand, Denmark and Germany are other main producers.

Key points: blackcurrants

Cultivars and breeding

There are ongoing breeding programmes in the leading blackcurrant-producing countries. Winter chilling in *Ribes* spp. is important. Breeding requirements aim at 1320–2380 chill units (one chill unit = 1 h less than 7°C). There is also emphasis on ascorbic acid (vitamin C) levels. Levels exceeding 300mg/100ml of juice are acceptable. Disease resistance in new cultivars is also important.

Each country seems to have its new favourite cultivars. New blackcurrant cultivars in Poland include Tiben, Tisel, Ruben, Ores and Tines; in Russia, Zúsha and Zagadka; in Scotland, Ben Tran, Ben Hope and Ben Gairn; in the Czech Republic, Otelo and Vebus; in England, Black Dawn and Ben Tirran, as well as Ben Hope; in New Zealand, Murchison and Blackadder are two recent releases. Murchison is gall-mite resistant and Blackadder has a low chilling requirement and very good flavour.

One of the main cultivars used in Poland and other major producing countries has been Ojebyu. Titania has been favoured in Germany and the

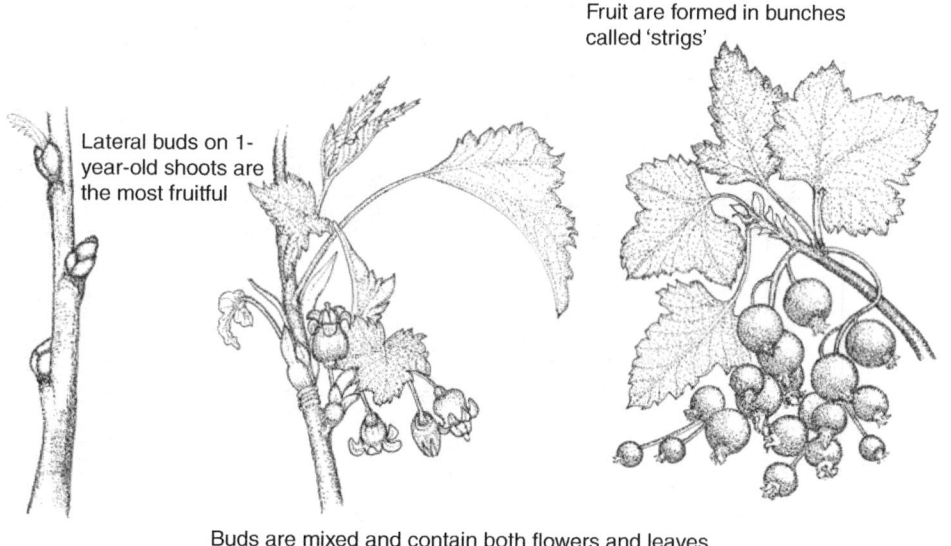

Fruit are formed in bunches called 'strigs'

Lateral buds on 1-year-old shoots are the most fruitful

Buds are mixed and contain both flowers and leaves

Fig. 16.2. Blackcurrants.

Table 16.2. Blackcurrants.

Botanical, anatomical and physiological aspects	Climatic, geographic, soil and water requirements	Cultural aspects
Common name Blackcurrant	*Temperature needs* Temperate climate. Short days required for flower-bud initiation. High chilling requirement for breaking dormancy	*Propagation* Normally by hardwood cuttings. Modern method is to plant *in situ* in winter with cuttings 20–45 cm long
Botanical name Ribes nigrum		*Rootstocks* Not required
Botanical name of related and useful species Ribes grossularia (gooseberry) Ribes sativum (redcurrants and whitecurrants)	*Frost tolerance* Tolerate –20°C dormant, –2 to –3°C grape bud stage, –0.5 to –1°C flower and young fruit. Specific frost control seldom used	*Spacing* 2.5 m between rows, 15–30 cm between plants
Type of plant and size Bush up to 2.0 m in height and width	*Water needs* Irrigation normally essential	*Training and pruning* Thin dense, thick growth keeping young 1 year-old shoots
Sexuality Hermaphrodite	*Tolerance of wet soils* Will tolerate heavier soils than most fruit but adversely affected by poor drainage	*Fruit thinning* Not necessary
Pollination Self-fertile, but cross-pollination in poor weather improves set. Bee hives recommended	*Drought tolerance* Water essential for cutting establishment *in situ* and for adequate fruit development. Growth and following season's fruiting seriously affected by drought	*Tillage* If cultivated, only 5–10 cm depth because of shallow fibrous roots. Grass/clover sward between rows becoming more popular where irrigation is adequate: use weed control chemicals along rows in 0.5–1 m band. Planting cuttings through black polythene aids weed control and moisture retention
Flower buds Mixed buds produced laterally on previous season's shoots. Some fruit spurs on older wood. Each bud produces one bunch of fruit. Most fruits are borne on young wood of blackcurrants but on older wood of redcurrants and whitecurrants	*Humidity tolerance* Disease problems (*Botrytis*) greatest during high humidity. Leaf spot troublesome in high-rainfall conditions	*Time to first harvest* 15–18 months after establishment from cuttings
		Time to full production 4–5 years
Growth of fruit Double-sigmoid growth curve		*Expected yield* 2nd year: 0.5–2 t/ha 3rd year: 2–5 t/ha 4th year: 4–8 t/ha
		Normal productive life Depending on pruning technique, productive life can be 15–20 years

G. Thiele

Time of bud burst
Early spring

Time of flowering
Spring

Time of fruit maturity
Midsummer, earlier or later according to climate

Wind tolerance
Growth and cropping affected by wind. More tolerant than *Rubus* spp.

Edaphic features
Undulating ground can cause mechanical harvesting difficulties. Slight slope can aid air drainage and reduce frost risk

Soil needs
Tolerant of a wide range of soils, given adequate irrigation and drainage. Deep, fertile soil preferable, pH 6.0–6.5, not below 5.5

Nutrient requirements
Varies according to soil nutrient status. Suggested foliage analyses (%), taken after harvest:
N: 2–3.5
P: 0.3–0.8
K: 2–4
Ca: 2–3.5
Mg: 0.3–0.5
Typical recommendations are based on 5kg N/t of fruit anticipated per hectare, 5kg K/t/ha and 3kg P/t/ha. High nitrogen promotes growth. Nitrogen applications sometimes split over the growing season

Method of harvest
Mechanical harvesting used. Gate sales and 'pick-your-own' (PYO) in small areas

Storage
Fresh sales insignificant. Normally frozen

Main pests and diseases
Aphids, mites, root weevil, currant clearwing, botrytis, leaf spot, mildews

Czech Republic for mechanical harvesting, although it can be too vigorous.

Fruit set

Although naturally self-fertile, some varieties set better with cross-pollination. Normally, four to five hives of bees per hectare are recommended to ensure a good set, but the requirement seems to vary from season to season. Try to eliminate or minimize alternative nectar sources during the pollination period. There is no set planting pattern recommended, but where fruit set is a problem it is advisable to alternate cultivars at four- to six-row intervals.

Planting systems

The planting system for blackcurrants has changed dramatically with the advent of mechanical harvesting. From a square spacing of about 2 × 2 m, designed for hand picking and cultivation in both directions, the plants are now spaced 40 cm apart in the row and about 3–4 m between rows, depending on the machinery used.

Cuttings, 20–30 cm in length, are planted *in situ* with about half to two-thirds of the cutting beneath the soil surface. Plant establishment is normally in excess of 90%, particularly when a black plastic mulch is laid before the cuttings are inserted. This mulch maintains soil moisture and controls weed growth. Plastic is laid by machine in well-prepared ground with no perennial weeds. Some blackcurrant plantations have been established directly into pasture by strip spraying with herbicides and laying plastic over the sprayed area. Cuttings are pushed directly through the plastic. Plastic may not be necessary where irrigation is laid at the time of establishment and where weeds are not a problem.

The key factor in the establishment decision is the production of adequate fruiting wood in the initial years. Close planting encourages upright growth, resulting in a narrow plant base and ensuring that the loss from berries dropping through the catching plates with mechanical harvesting is minimized. This system allows a low labour input with little or no pruning. After about 5 or 6 years of cropping the bushes are renewed by mowing back to within about 10 cm of ground level. This is the so-called short-term cropping system. To minimize silver leaf disease, *Stereum purpureum*, heavy cut-

ting such as this is best carried out during the growing season, soon after completion of harvest. (See also notes in Chapter 6.)

Pests and diseases

Reversion, a virus disease causing sterility in flowers, can seriously reduce yield and is considered the most important disease in Europe. It is spread by the blackcurrant gall mite, *Cecidophyopsis ribis*, and a routine spray programme to control the mite is required, along with roguing of infected bushes. The gall mite is also responsible for the swelling of buds, known as 'big bud'. Infested buds fail to produce leaves and flowers.

The larvae of the currant clearwing, *Synanthedon tipuliformis*, tunnels along the pithy centre of young shoots. Although it may reduce yield by damage to buds at the point of entry and by weakening of branches causing breaking, the main significance is the effect on growth and rooting of infested cutting material.

Other damaging pests include the currant aphid, *Hyperomyzus lactucae*, the two-spotted spider mite, *Tetranychus urticae*, mealybug, *Pseudococcus obscurus*, and root weevil, *Otiorhynchus sulcatus*.

Diseases include leaf spot (or anthracnose), *Drepanopeziza ribis*, white pine blister rust, *Cronartium ribicola*, and powdery mildew, *Sphaerotheca mors-uvae*.

Juice quality, time of picking

Blackcurrants have one of the highest levels of ascorbic acid (vitamin C) of any fruit. Because of the nutritional value of blackcurrant juice and other products processed from this fruit, it is not surprising that considerable analytical work has been done to determine the factors influencing vitamin C content. Harvesting time, variety, climate, soil, fertilizer and other cultural practices have a marked influence on the biochemical content of the fruit. Tests on ascorbic acid levels have shown variations between 100 and 400 mg/100g, related to seasons and varieties (300 mg/100g is considered acceptable).

On the other hand, processing companies seem to place more emphasis on juice quality indicators such as colour and sugar content. Unfortunately, sugar content, as measured by °Brix level, is difficult to assess accurately with field refractometers. Colour charts are also difficult to apply accurately

G. Thiele

to the determination of optimum harvesting time. Since fruit size increases rapidly in the last 7–10 days prior to optimum harvest, careful monitoring of fruit growth may be the most useful approach to ensuring correct harvesting time.

Key points: redcurrants and whitecurrants

Although redcurrants and whitecurrants are less well known than blackcurrants, they are popular fruits in parts of Europe. They are used for blending in jellies and incorporated with other fruit juices as a beverage and health food.

The growth habit of *R. sativum* is more sprawling than that of blackcurrants. The most significant difference is that it fruits on spurs on 2-year-old and older wood, whereas blackcurrants fruit predominantly on 1-year-old wood. This means that older wood must be retained on the redcurrant bush to obtain good yields. Closer spacing of bushes in the row is also becoming the norm with redcurrants and whitecurrants, although not as close as the blackcurrant. Within-row spacing is normally 0.5–1 m, with rows about 2.5 m apart, depending on the training method. To allow for effective straddle harvesting, a light supporting structure has been adopted in some countries to allow training on an espalier or hedgerow system.

The two redcurrant cultivars most widely grown are Red Dutch and Random; White Dutch and Olin, a more recent cultivar, are favoured whitecurrants.

Key points: gooseberries

The gooseberry is less popular in commercial fruit growing than the currants, perhaps because of its spiny nature. It is one of the earliest fruits to ripen in the spring (Table 16.3). Gooseberry pie was a well-known and popular dessert in the 20th century, but fresh strawberries and raspberries and cream are more popular spring desserts nowadays. Other berry fruits are preferred for jams, jellies and juices. World demand for gooseberries is less than for other berry fruits.

Attempts to modernize the commercial culture of gooseberries have concentrated on closer spacing in the row (30–40 cm) and development of bushes from a leg 15–30 cm in height (see below) to keep the branches off the ground and to facilitate mechanical harvesting. A modified fan or espalier-trained system has been developed in some

European countries with wire and post support. Closer spacing has allowed more accurate mechanical harvesting, up to 85% collection, on cultivars such as White Smith and Lady Delaware. White Smith is the main cultivar grown in Poland.

To produce a 'legged' bush it is necessary to use hardwood cuttings 30–40 cm in length, with all buds except the top three to four removed. Another method being tried utilizes the intercompatibility of *Ribes* species, by grafting gooseberry varieties such as Farmers Glory on to stocks of the most common flowering currant, *Ribes sanguineum*.

Cultural methods and disease and pest control of gooseberries are similar to those for blackcurrants but 'American gooseberry mildew' (*Sphaerotheca mors-uvae*) is a severe problem in some countries.

The gooseberry has a longer shelf life than other berry fruit and transports well.

Raspberries and Brambles

The genus *Rubus* includes raspberries, blackberries and dewberries. Raspberries, subgenus *Idaeobatus*, are distinguished from blackberries, subgenus *Eubatus*, in that the mature raspberry fruit separates from the 'torus' or plug.

The European raspberry is *Rubus idaeus*, one of over 400 species of the genus found in most European countries and the North American continent (Table 16.4). Two related species *Rubus occidentalis*, with black or yellow fruit, and *Rubus strigosus*, with light red and sometimes white and yellow fruit, have provided much of the genetic material for the development of many of the American cultivars of raspberry.

The raspberry of greatest importance in commercial production is a deciduous perennial with biennial stems, the upright primocanes being produced in the first year of growth and the fruit-bearing laterals produced from them in the second year, when the canes are known as 'floricanes' (Fig. 16.4). However, there are autumn-bearing raspberries that fruit on current-year primocanes. Primocane-fruiting varieties are becoming increasingly important in extending the fruiting season. The potential for greater mechanization in autumn-bearing raspberry production is an important advantage and is also driving the ongoing improvement of these cultivars by fruit breeders.

The major raspberry-producing countries are Russia, Poland and the countries on the Balkan peninsula.

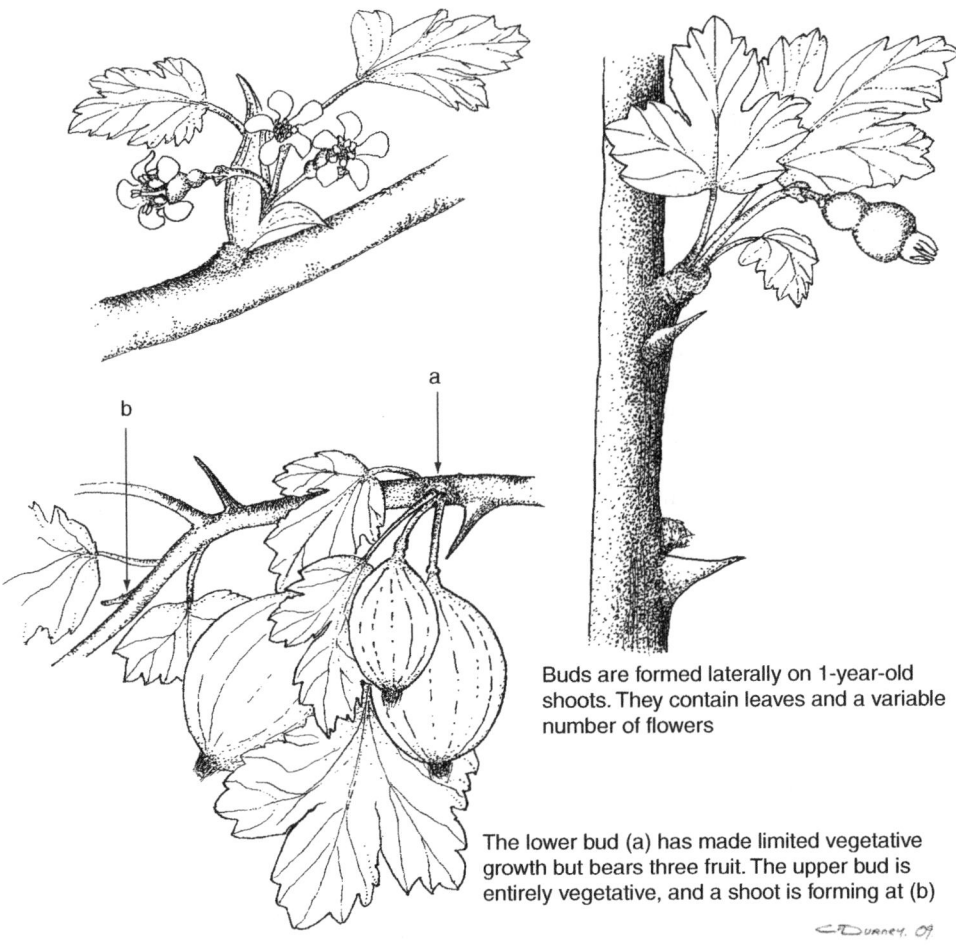

Buds are formed laterally on 1-year-old
shoots. They contain leaves and a variable
number of flowers

The lower bud (a) has made limited vegetative
growth but bears three fruit. The upper bud is
entirely vegetative, and a shoot is forming at (b)

C. Durney. 09

Fig. 16.3. Gooseberries.

Cultivated fruits commonly known as 'brambles'
usually differ from raspberries in having creeping,
rather than upright, stems (Table 16.5). Like raspberries, they are deciduous perennials, usually with
biennial stems. As implied in the common name,
brambles have more thorns than most raspberry
cultivars, but plants vary from being thorn-free to
extremely spiny.

Two original thornless blackberry varieties were
Smoothstem and Thornfree. These cultivars are
genetically thornless, due to a recessive gene, and can
be propagated without fear of losing thornlessness.
Occasionally thornlessness is the result of a chimaera,
and plants propagated from roots could have thorns,
for example Thornless Evergreen blackberry,

Thornless Loganberry and Youngberry. Tissue culture
propagation can be used but reversion may occur.

The brambles may also vary in their deciduous
nature. The evergreen blackberry is the main type
of bramble in cultivation in Washington and
Oregon in the USA. Evergreen types often have
canes which persist for more than 2 years, new
laterals being produced each year.

Hybridization is very common within the various
Rubus species and hence the exact origin of many
cultivated forms is unclear. It is likely that many of the
blackberries in cultivation have originated from the
wild blackberries of England and Scotland, *Rubus
laciniatus*, cut-leaved or parsley-leaved blackberry,
and *Rubus schlechtendalii*, a large-fruited form

G. Thiele

Table 16.3. Gooseberries.

Botanical, anatomical and physiological aspects	Climatic, geographic, soil and water requirements	Cultural aspects
Common name Gooseberry	*Temperature needs* Temperate, similar needs to blackcurrants	*Propagation* Hardwood cuttings. Root formation may be slower than with blackcurrants, particularly if long cuttings are used for semi-standard production. Best rooting with only two or three buds left above ground level
Botanical name Ribes grossularia	*Frost tolerance* Early flowering causes susceptibility to late spring frosts. However, frost protection seldom used and serious losses are not common	*Rootstocks* Not required
Botanical name of related and useful species Ribes nigrum (blackcurrants) Ribes sativum (redcurrants and whitecurrants)	*Water needs* Moderate – see under Drought tolerance	*Spacing* 2.5–3 m between rows, 0.5–2 m between plants
Type of plant and size Woody shrub to 1 m high, branches upright and often reflexing with spikes	*Tolerance of wet soils* More tolerant of excess water than raspberries, similar to blackcurrants	*Training and pruning* Prune to eliminate old and weak wood and to encourage new growth. Prune to upward-pointing buds in weeping varieties to avoid contact of fruiting wood with the ground
Sexuality Hermaphrodite	*Drought tolerance* Moderately good. Early fruit maturity avoids main period of summer drought. Flower-bud development and growth affected by prolonged dry periods	*Fruit thinning* Not required
Pollination Self-fertile but setting improved with cross-pollination under poor weather conditions during flowering		*Tillage* Surface rooting. Avoid deep cultivation. Grass/clover sward common, where adequate irrigation available, with chemical weed control in rows
Flower buds Mixed buds produced laterally on previous season's wood and on spurs on older wood	*Humidity tolerance* Not severely affected by high humidity. Subject to fruit cracking with excessive rainfall at maturity	*Time to first harvest* 2nd year from cuttings *Time to full production* 3rd to 4th year *Expected yields* 2nd year: 2.5 t/ha 3rd year: 5.0 t/ha 4th year: 7.0 t/ha
Growth of fruit Double-sigmoid	*Wind tolerance* Tolerant, apart from young growth, which is subject to breakage. Damage to fruit occurs from wind rub against thorns	*Normal productive life* 15–20 years
Time of bud burst Early spring	*Edaphic features* No specific requirements	*Method of harvest* Still predominantly hand picked. With upright, hedgerow training, mechanical shakers can be used but thorns may puncture the fruit during shaking. Usually harvested in 'green' condition and not allowed to ripen to produce a yellow or red skin coloration
Time of flowering 4–6 weeks after bud burst		

Continued

Table 16.3. Continued.

Botanical, anatomical and physiological aspects	Climatic, geographic, soil and water requirements	Cultural aspects
Time of fruit maturity 8 weeks after flowering, i.e. early summer. May be harvested immature for culinary use	*Soil needs* Tolerant of a wide range of soil types. pH range 6.0–6.5 is best *Nutrient requirements* Similar to blackcurrants. Gooseberries have a high potassium requirement	*Storage* Can be frozen but may crack with rapid reduction in temperatures. Can store at room temperature for about 2 weeks. Mainly sold fresh *Main pests and diseases* *Botrytis* spp., anthracnose, leaf spot (American gooseberry mildew). Silver leaf and verticillium wilt can cause loss of plants. Mites, scales and aphids sometimes troublesome

Table 16.4. Raspberry.

Botanical, anatomical and physiological aspects	Climatic, geographic, soil and water requirements	Cultural aspects
Common name Raspberry	*Temperature needs* Temperate plants with requirement for winter chilling. Do not do well in subtropical and tropical climates	*Propagation* Normally propagated by lifting root suckers in a cane nursery. Possible to build stock more rapidly from root cuttings. Rapid increase of new cultivars possible with tissue culture
Botanical name *Rubus idaeus*	*Frost tolerance* Withstands −20°C in winter if wood is hardened, but subject to frost damage at flowering and small fruit stage below −2°C. Total loss is not common and frost protection not usual	*Rootstocks* Not applicable
Botanical name of related and useful species More than 400 species of *Rubus* have been described, some alpine and arctic, but mainly found in temperate regions. *Rubus idaeus* is the European raspberry, but many cultivated types have been developed from *Rubus occidentalis* (black raspberry), *Rubus strigosus* (North American raspberry) and *Rubus neglectus* (purple raspberry)	*Water needs* Needs regular water supply in growing season *Tolerance of wet soils* Roots are sensitive to poor soil aeration. Avoid poorly drained areas	*Spacing* Hedgerow, 2.5–3.0m between rows, depending on equipment; 20 canes per metre of row after pruning *Training and pruning* Remove floricanes after fruiting and thin out excess primocanes. Normally trained in a hedgerow maintained to a width of 40cm. Alternative methods developed for mechanical harvesting and for autumn bearing cultivars. Support is necessary – posts and wire (see Chapter 6) *Fruit thinning* Not required
Type of plant and size Biennial canes: primocanes in the first year, floricanes in the second year. Growing interest in cultivars that fruit in the autumn on primocanes. Height 1.5–3m. Width of plant depends on extent to which primocanes are allowed to develop from adventitious root buds	*Drought tolerance* Dry soil restricts development of primocanes and affects sizing of fruit *Humidity tolerance* Subject to *Botrytis* infection during fruit ripening. Fungus diseases on the canes are troublesome in high-humidity regions	*Tillage* Shallow rooting. Avoid cultivation deeper than 10cm. Grass/clover sward common inter-row with chemical weed control at base of canes *Time to first harvest* Second season *Time to full production* 3rd or 4th season
Sexuality Hermaphrodite	*Wind tolerance* Crops reduced by hot, dry winds during fruit ripening. Fruiting laterals are subject to breakage. Primocane growth affected by wind	*Expected yield* 2nd season: 2 t /ha 3rd season: 6–12t /ha 4th season: 7.5–15t/ ha
Pollination Bees help in producing high drupelet set	*Edaphic features* Not important unless mechanical harvesting is affected. Slopes <10° recommended for mechanical harvesting	*Normal productive life* 10–15 years

Continued

Table 16.4. Continued.

Botanical, anatomical and physiological aspects	Climatic, geographic, soil and water requirements	Cultural aspects
Flower buds Flower initiation takes place in the late summer–early autumn preceding the cropping year. Buds on the primocanes form fruiting shoots on laterals the next spring	*Soil needs* deal soil is a deep, sandy loam, well supplied with humus, and with good water-holding capacity, yet well drained. Avoid compact, impervious subsoils. Slightly acid soil preferable, pH 5.8–6.3	*Method of harvest* Hand picking slow and expensive. Mechanical harvesting increasing mainly for processing fruit *Storage* Fresh berries will store for several days at 0–1°C. Commonly frozen in block or IQF (individually quick frozen) form for long storage. Raspberries for jam can be preserved in solution with sulphur dioxide
Growth of fruit Aggregate fruit consisting of a collection of drupelets. Three growth stages (double sigmoid)	*Nutrient requirements* Standard optimum concentrations for foliar analyses (%): N: 2.75 P: 0.30 K: 1.50 Ca: 0.6–2.5 Mg: 0.4	*Main pests and diseases* Regular budmoth control essential. *Botrytis cinerea* attacks ripening berries under humid conditions. Cane spot, cane wilt, silverleaf, rubus bushy dwarf virus, raspberry mosaic virus, leafroller, aphids, mites and root nematodes may cause problems
Time of flowering 3–4 weeks after bud burst	Fertilizer applications should be based on estimated crop removal, i.e. use 5kg N per tonne of crop, 3kg K p/t, 3kg P p/t	
Time of fruit maturity 8 weeks after flowering		

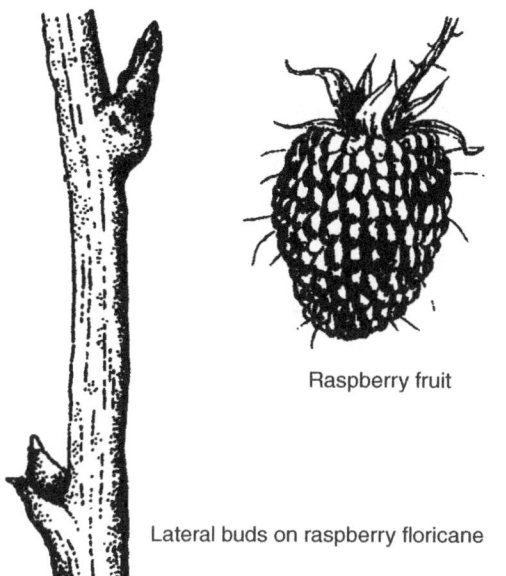

Raspberry fruit

Lateral buds on raspberry floricane

Fig. 16.4. Raspberries.

commonly found growing wild in roadside hedges. The American blackberry, *Rubus ursinus vitifolius*, is likely to be one of the parents, along with the raspberry, of the large, red-fruited loganberry.

More popular has been the boysenberry, with its large, sweet, black fruit. Other hybrid selections in cultivation include the youngberry and marionberry, both arising from controlled breeding programmes. For instance, the youngberry is derived from Phenomenal and Austin Mayes parentage. The olallieberry is now more popular among the hybrids in California and the marionberry in Oregon.

The European dewberry, *Rubus caesius*, and the American dewberry, *Rubus canadensis*, with black, juicy, sour fruit are also known commercially.

Key points

The fruit of *Rubus* species consists of a number of one-seeded drupes covering a cone-shaped receptacle, or plug. Brambles are harvested with the receptacle still in place whereas raspberries come free from this plug. Cultivated *Rubus* species are self-fertile but can be small and misshapen if all the drupes do not set. More even setting in raspberries and brambles is induced by introducing two bee hives per hectare at the start of flowering. Brambles flower later than raspberries and are less subject to frosts, but hot, drying winds can reduce set and yield in all *Rubus* species.

Propagation

As suggested in Chapter 11, raspberries are normally propagated by lifting suckers from established plants. These suckers arise from adventitious buds on the roots. Since viruses can be a serious problem, clean mother stock is required and it would be wise to know the history of the block from which the suckers are taken. It is usually preferable to buy from nurseries which are known to have clean cane nursery beds and where cropping does not occur. An alternative method is to use root cuttings.

Some brambles, such as boysenberries and youngberries, root readily where the tip growth touches the soil. This method of propagation is preferred. Another way to obtain rooted plants is to cause a wound at a stem node, pin this section of the stem down and lightly cover it with soil. Two-node cuttings taken in early autumn will also produce satisfactory plants for field planting the following spring. Whatever the method, care should be taken to select plants for propagation during the fruiting season to ensure that good strains free from virus and other diseases are used. This is particularly important with boysenberries and youngberries, where considerable variation occurs in the field.

Tissue culture is increasing as the preferred method particularly where rubus bushy dwarf virus (RBDV) is a problem.

Cultivars

Much of the ongoing research work on raspberries around the world is directed towards plant breeding to improve the yield and quality of fruit, select for disease and pest resistance, and to satisfy the requirements for summer or autumn fruiting and for climatic variables such as severe winters.

Talisman is a popular, standard, floricane-fruiting cultivar, along with Chilliwack, Malahot and Qualicum.

Glen Ample is popular for tunnel production in Europe, and Polka and Polona are suitable for short-summer conditions in Scandinavia and Central Europe.

Cultivars developed for mechanical harvesting include Coho, Cascade Delight and Octavia.

Most new cultivars have been bred for resistance to raspberry bushy dwarf virus (RBDV), root rot (*Phytophthora* spp.) and the aphid vector of raspberry mosaic virus (RMV), *Amphorophora agathonica*.

Table 16.5. Brambles.

Botanical, anatomical and physiological aspects	Climatic, geographic, soil and water requirements	Cultural aspects
Common names Boysenberry, loganberry, youngberry, blackberry, marionberry, ollalieberry, auroraberry, dewberry, tayberry	*Temperature needs* Require winter chilling. Normally grow best in a temperate climate, although some will crop well in subtropical regions, e.g. boysenberries	*Propagation* Normally by stem cuttings or tip layering. Root suckers and root cuttings can be used for rapid increase. Also propagated by tissue culture
Botanical name The subgenus *Eubatus* is a very heterogenous series of species and hybrids with polyploidy ranging from diploids to dodecaploids. Chromosome numbers are in multiples of seven. Boysenberries and youngberries have 49 chromosomes. Most *Rubus* spp. are hybrids. Parentage is not clearly known in the case of boysenberries and loganberries. Later cultivars are the result of controlled breeding programmes but often involve crosses of existing hybrids	*Frost tolerance* Will stand −20°C when dormant. Later flowering than raspberries and not usually damaged by frost at the flowering stage. Young shoots are frost sensitive	*Rootstocks* Not applicable *Spacing* 2–3 m between rows; 1–2 m between plants *Training and pruning* Trained on 2 m-high post-and-wire fences in a variety of ways. Espalier or fan and 'rope' methods. Remove fruiting canes at ground level after harvest. Remove unwanted and weak primocanes before training on the fence. Best to leave them until midwinter, when the canes are less brittle (see Chapter 5)
	Water needs Irrigation required	*Fruit thinning* Not required
	Tolerance of wet soils Will not tolerate waterlogged soils	*Tillage* As for raspberries. Grass/clover sward can cause intertwining problems with primocanes. Non-cultivation, with overall use of herbicides is preferred
	Drought tolerance Will not tolerate drought or excessive periods of low humidity. Prefer soils with good water-holding capacity kept near field capacity	*Time to first harvest* Second season
Botanical names of related and useful species *Rubus idaeus*	*Humidity tolerance* Berries very susceptible to *Botrytis* at maturity. Downy mildew (dryberry) will almost eliminate the crop if not controlled in high-humidity conditions	*Time to full production* Third or fourth season *Expected yield* 2nd year: 1–2 t/ha 3rd year: 5–8 t/ha 4th year: 8–25 t/ha

G. Thiele

Type of plant and size
Biennial canes: primocanes in the first year, floricanes in the second. Vigour varies between hybrids but may produce canes up to 4 m in length. May be erect or semi-erect (some blackberries) or trailing (boysenberries and youngberries). Usually thorny canes, but thornless hybrids have been bred and are in cultivation (e.g. Smoothstem and Thornfree blackberries, thornless boysenberries and youngberries)

Sexuality, pollination, flower buds, growth of fruit
Similar to raspberries

Time of bud burst
Spring

Time of flowering
Normally 4–6 weeks after bud burst. Flowering period may be extended according to the position of the flower in the truss and on the cane. Youngberry is 7–10 days earlier than boysenberry, which is 3–4 days earlier than marionberry

Time of fruit maturity
Midsummer onwards

Wind tolerance
Very susceptible to wind and cannot be grown successfully without adequate shelter

Edaphic features
Not important, unless machinery usage affected. Slopes exposed to drying winds should be avoided

Soil needs
Ideal soil is a deep silt loam with a high water-holding capacity. Will stand heavier soils than raspberries but impervious subsoils should be avoided. Slightly acid soils preferable, pH 5.8–6.3

Nutrient requirements
Similar to raspberries

Normal productive life
15–20 years

Method of harvest
Dessert fruit are hand harvested. Mechanical harvesting increasing for processing fruit

Storage
As for raspberries

Main pests and diseases
Dryberry, *Botrytis*, cane spot, cane blight, raspberry budmoth, leafroller, two-spotted mite, foliar nematode

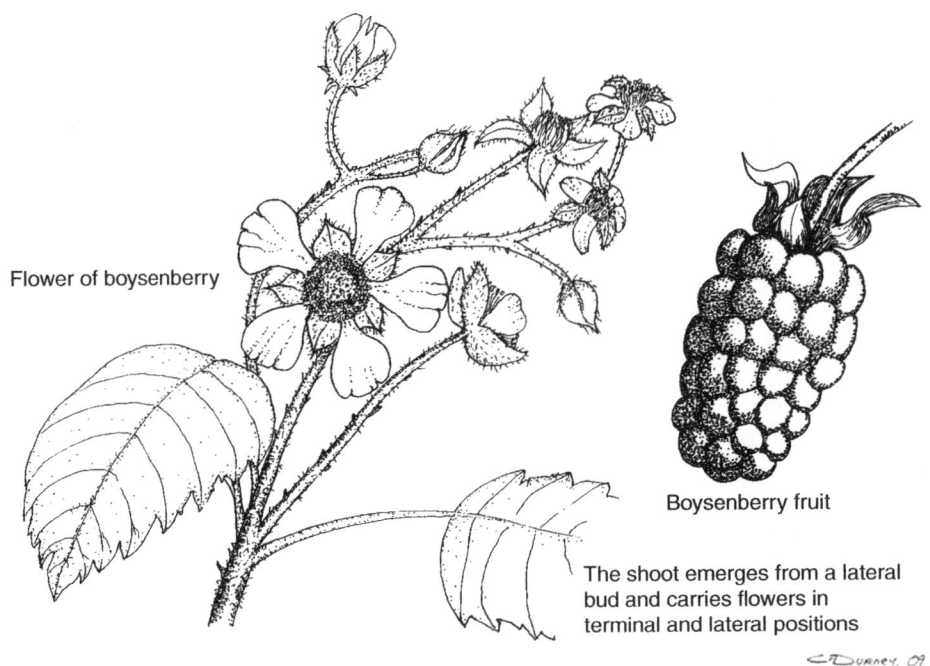

Flower of boysenberry

Boysenberry fruit

The shoot emerges from a lateral
bud and carries flowers in
terminal and lateral positions

C Durney. 09

Fig. 16.5. Boysenberries.

Black (as distinct from red) raspberries have received some attention commercially. Cultivars include Mauger, Jewell and the late-ripening Mac Black.

Nova is a major primo-fruiting cultivar in California due to its early spring crop. Caroline, Autumn Bliss and Autumn Britten are primocane-fruiting cultivars in North America and Europe suitable for extending the season into the autumn.

Autumn cropping

Primocane-fruiting raspberries are those which fruit mainly on the end of the current season's growth but are cultivated in a similar way to summer-fruiting raspberries. The key difference is that all growth can be pruned close to ground level during the winter, instead of having to remove the canes which have fruited during the summer and leaving the primocanes for fruiting the following summer. Autumn-fruiting raspberries require a similar spray programme for insect and fungus control as summer-fruiting plants, although the overwintering of cane-borne disease can be less of a concern.

Growth control

In commercial raspberry and bramble plantations, primocane growth begins in early spring and excess growth can become a problem by harvest time. This is particularly so where machine harvesting is being carried out, as the excess growth can stop catch plates fitting tightly at the plant base. Chemicals such as Reglone (diquat) are still used and are directed towards the base of the plants to burn off the new growth early in the growing season. Subsequently the re-emerging primocanes are allowed to grow to provide fruiting wood for the following season. Another chemical used for de-suckering in brambles is carfentrazone ethyl, which can be applied up until the first week in June in the northern hemisphere (first week in December in the southern hemisphere).

Pests and diseases

All *Rubus* species in commercial cultivation require a regular spray programme, although the range of pests varies from region to region. Insect pests of importance include the strawberry rootworm, *Paria fragariae*, the strawberry root weevil, *Brachyrhinus ovatus*, and the fruit-feeding

G. Thiele

Consperse stink bug, *Euschistus conspersus*. The raspberry cane borer, *Oberea bimaculata*, the raspberry crown borer, *Bembecia marginata*, the raspberry sawfly, *Monophadnoides geniculatus*, and the strawberry leafroller, *Ancylis comptana fragariae*, are other important pests of the *Rubus* species fruits in many countries. Mites can build up on leaves during hot, dry periods, turning leaves bronze and reducing their ability to photosynthesize. The dryberry mite, *Phyllocoptes gracilis*, afflicts fruit of some *Rubus* spp. cultivars. Both foliar- and root-feeding nematodes are known to be pests of brambles.

Important diseases of brambles include *Botrytis* spp. fruit rots, a range of viruses such as raspberry mosaic virus and rubus bushy dwarf virus, and root rots caused by *Phytophthora* spp. fungi. Verticillium wilt, caused by a fungal infection of roots and stems, and crown gall, caused by *Agrobacterium tumefaciens*, can reduce yields of bramble plantations.

A key concern of plant breeders is developing resistance to these various diseases and pests. The ones receiving considerable attention at present are:

- *Botrytis* spp. fruit rot.
- Phytophthora root rot.
- Root lesion nematodes such as *Pratylenchus penetrans*.
- Raspberry mosaic virus (developing resistance to aphid vectors).
- Raspberry bushy dwarf virus, which is pollen and seed borne.
- Somatic mutations causing crumbling fruit.

Strawberries

The common garden strawberry has an interesting history. The original parentage appears to involve *Fragaria virginiana* from eastern North America, with its small aromatic fruit, and *Fragaria chiloensis* from the west coast of North and South America, with its large fruit.

It is written that a French sailor carried five plants of *F. chiloensis* from Chile to France on a 6-month voyage and eventually interplanted them among *F. virginiana*. From seed obtained from cross-pollination between these species the 'Fraisier-ananos' or pineapple strawberry was selected and subsequently named *Fragaria × ananassa*.

It is from *F. virginiana* and *F. chiloensis* and the first-generation hybrids between these species that most of the modern large-fruited strawberry varieties have been derived. *F. virginiana glauca*, a subspecies from the high plains of western North America, has been used as a parent to produce the modern day-neutral (DN) cultivars.

The strawberry is unique among cultivated temperate fruits. Being an herbaceous perennial, it requires different culture from the cane, bush and tree fruits (Table 16.6). It is the quickest plant to produce fruit from the time of planting and is the earliest fruit on the market in the spring.

World production of strawberries has increased dramatically in the latter part of the 20th century and is now about 3.8 Mt. Five countries produce about 70% of world production. These are, in descending order of importance, the USA, Russia, Spain, Japan, Italy and Poland. Expansion has occurred primarily in areas with mild winters and about 50% of world production now comes from these regions. Spain has increased production by over threefold in the last decade or so, to about 306,000 t/year (FAO, 1998). The development of day-neutral varieties has been mainly responsible for the trend towards warm-climate production.

Key points

Propagation

Vegetative multiplication by runners, or 'stolons', is the common method of propagation. The stolon is a creeping branch produced from a leaf axil on the crown. Runner plants are produced at each successive node on the stolon, forming root initials where the node touches the ground. The practice of taking runners from fruiting beds is quite unsound, since the most common causes of plant decline in fruiting areas are the viruses yellow edge and crinkle, and these can be transmitted to new fruiting areas by taking runners from infected plants.

Strawberry viruses are spread by aphids. Most countries now operate a certification scheme whereby nurseries obtain 'mother' plants, indexed for freedom from viruses, for use in propagation beds.

Micropropagation, or 'tissue culture', is receiving increasing attention in the major producing countries, although research suggests that plants raised by tissue culture may not produce as high a

Flower inflorescence and the leaves
of strawberries. These originate from a small crown –
or main stem – situated in the centre of the plant

Fig. 16.6. Strawberries.

yield in the first season as runner plants. The negative effect on genotype stability of a high number of subcultures has been confirmed. A number of countries, for example Italy, have rules limiting the number of subcultures. Thereafter, micropropagation must be followed by open-field propagation. Micropropagation is especially important for bulking up new varieties for rapid distribution, for freeing stock from virus diseases and for storage and transportation throughout the world.

Breeding

Larger berry size, firm fruit, high yield, good shelf life and disease resistance are the traditional strawberry breeding objectives. A bright, but not too dark, fruit colour is being demanded by consumers.

Interplant competition has also become an important focus of plant breeding work, concentrating on both vegetative growth and reproductive parameters.

All the major strawberry-producing countries have breeding programmes, making it difficult to summarize the key cultivars in worldwide production. Much emphasis has been placed on developing early-ripening, short-day cultivars, suitable for mild climates with low chilling requirements.

One of the University of California short-day cultivars which has gained universal approval is Comarosa. Many European countries still use Elsanta. This cultivar is considered suitable for organic production in spite of being susceptible to *Phytophthora* spp. Chandler still has widespread use in the USA, Italy, the UK and other countries.

In Spain, popular short-day cultivars are Medina and Merina, with Aquedilla a recent development. China has its own cultivars in Toyonaka, Tudla and Dorselect, but it also favours Newstar (selected from the USA Allstar) and New Earliglow (selected from the USA Earliglow). Cultivars favoured in Poland include Comarosa (mid-season), Vikat (late) and Pandora (very late).

The UK has developed a useful ever-bearing strawberry in Flamenco, which produces a high proportion of its crop in autumn, although another ever-bearer, Everest, is still favoured.

Day-neutral cultivars emerging from the University of California include Aromas and Diamante.

Harvesting, handling and postharvest physiology

Although there have been numerous attempts to develop strawberry production for mechanical harvesting, most strawberries are harvested by hand. The cost of hand harvesting has been reduced by the development of large-fruited varieties, but a size limit of about 60 g has been reached in terms of public acceptance. 'U-pick' or 'Pick-your-own' operations are common. Families enjoy the 'entertainment' of travelling into the country to pick strawberries.

Rapid postharvest cooling is the best way to maintain fruit quality. Although reduction of oxygen and ethylene levels will reduce fruit respiration rate, loss of flavour and an acid taste can result from controlled atmosphere storage. None the less, a short period of CA storage, not exceeding 3 days, can help maintain quality during transport.

G. Thiele

Table 16.6. Strawberries.

Botanical, anatomical and physiological aspects	Climatic, geographic, soil and water requirements	Cultural aspects
Common name Strawberry	*Temperature needs* Grown under a wide range of temperature conditions throughout the world but needs a period of colder conditions to break dormancy. High temperatures (30°C) may cause fruit damage. Selection of varieties to suit temperatures and day lengths is important	*Propagation* By separation of adventitiously rooted runners from parent plant. Meristem propagation is receiving increased attention for rapid increase of new cultivars
Botanical name *Fragaria × ananassa*		*Rootstocks* Not required
Botanical name of related species *Fragaria vesca* (alpine and perpetual strawberries) *Fragaria virginiana* *Fragaria moschata*		*Spacing* See under Key points. Depends on intended bed life and other factors
Type of plant and size Herbaceous perennial 15–20 cm high, 20–40 cm wide. Main stem is short and called the crown. Produces runners up to about 1.5 m in length in propagation bed	*Frost tolerance* Will stand –15°C to –20°C while dormant. Flowers and young fruit damaged by –1 to –3°C. Often grown under plastic tunnels to counteract low spring temperatures. Nevertheless, most are grown without frost protection	*Training and pruning* Necessary to restrict runner growth in fruiting areas by hand or with herbicides such as diquat. Removal of old leaves during winter is advisable if plants are to be retained for a further season
		Thinning Not required
Sexuality Hermaphrodite	*Water needs* Irrigation usually required. Water stress affects size and quality of berries	*Tillage* Normally grown on ridges 10–15 cm high at 0.8–1 m centres. Ridges can be covered with black plastic laid by machine. Straw mulch used between rows. Must be free of perennial weeds
Pollination Self-fertile, but bees advisable	*Tolerance of wet soils* Intolerant of poor drainage. Particularly subject to red stele (*Phytophthora fragariae*) in water-logged conditions	*Time to first harvest* Depending on climate, 3–6 months after planting
Flower buds Formed within axillary buds on crown of plant. May require short day (SD) to stimulate flowering – these strawberries flower once only, in spring. Many new cultivars are day-neutral or 'everlasting' and will produce more than one crop per year	*Drought tolerance* Poor, need good water-holding-capacity soils or good irrigation to maintain growth and ensure adequate fruit size	*Time to full production* In cooler climates, higher production occurs in the second year *Expected yield* Commercial yields of 50 t/ha recorded under good conditions, 30 t/ha common *Normal productive life* Three years. Depends on region: warmer areas 1 year, cooler areas up to 4 years
Growth of fruit Sigmoid. Size at maturity depends on whether primary, secondary or tertiary position on the flower truss. Fruits in the primary position are larger	*Humidity tolerance* Highly susceptible at ripening stage to *Botrytis* under humid conditions	*Method of harvest* By hand with calyx for dessert, without calyx for processing. In some areas in the USA they are machine harvested for processing

Continued

Table 16.6. Continued.

Botanical, anatomical and physiological aspects	Climatic, geographic, soil and water requirements	Cultural aspects
Time of bud burst First sign of growth is about 2–3 weeks prior to flowering	*Wind tolerance* Shelter essential. Fruit distortion may result from strong winds during flowering, affecting pollination. Marked reduction in plant growth and yield in poorly sheltered areas	*Storage* A fruit with high respiration rate. Rapid deterioration if field heat is not removed quickly. Fruit free from botrytis will hold 7–10 days at 1–2°C
Time of flowering In warm areas, flowering may be in early spring, especially under polythene tunnels. In cooler areas, flowering occurs mid-spring	*Edaphic features* Gradual slope facing the sun useful for early production	*Main pathogens* Root diseases such as *Phytophthora* spp. and *Verticillium* spp. the subject of ongoing breeding for resistance. Fruit and leaf diseases such as *Botrytis* spp. and *Colletotrichum* spp. troublesome
Time of fruit maturity 6 weeks after flowering, but depends on district and temperature	*Soil needs* Avoid heavy, poorly drained soils and light, stony soils. Friable silt loam with good water-holding capacity is preferable	
	Nutrient requirements Standard optimum concentrations (%) for foliar analyses: N: 2.6–3.5 P: 0.25–0.35 K: 1.0–2.0 Ca: 0.7–1.5 Mg: 0.25–0.40; do not use chloride forms (e.g. KCl). Application rates kg/ha for annual applications: N: 50–100 P: 20–50 K: 80–100 For fertigation, rates can be halved	

G. Thiele

Cultural systems

There are three main methods of growing strawberries: the matted bed system, the hill system and bag culture.

MATTED BED This system is normally used where plants are grown for more than one season (2–4 years) and in regions with severe winters, where crown injury can occur. Plants are set out at 25,000–50,000/ha, either in rows or in beds, and runners are allowed to develop and root to form fruiting crowns for the second and subsequent seasons. Although higher initial plant numbers will increase yields in the first and second seasons, runner control becomes a greater problem with higher initial plant densities.

HILL METHOD This system is commonly used for annual production, using more plants per hectare than the 'bed' system (up to 120,000 plants/ha). Rows or beds are usually raised and covered with black polyethylene, through which the plants are inserted. Depending on the climate and target market, planting may be during the summer (cool-stored plants) or winter (fresh-dug plants). The 'hill' method may be used in the open field or under tunnels. Tunnel protection methods are widely used in Europe and Japan but seldom in America.

BAG CULTURE This method is common in northern Europe, where alternatives to *in situ* cultivation have become necessary since soil sterilants, such as methyl bromide and chloropicrin, have been banned. The system is capital intensive and normally used under plastic tunnel or greenhouse cultivation. While hydroponics and the nutrient film technique have been used, the most accepted method is 'bag' culture using peat as the growing medium. The bags usually contain 8–12 l of a mixture of 80% peat and 20% perlite, with nutrients supplied through fertigation. Plant densities vary between 12 and 16 plants per square metre, with the plants either standing alone (the usual method) or suspended. It is possible to have more than one crop per year on a 3- to 5-month cycle, using cold-stored or fresh-dug plants according to the time of the year.

Further Reading

General

About every 4 years, international symposia for specific crops and related horticultural research are conducted under the auspices of the ISHS (International Society of Horticultural Science). The proceedings of these symposia are published as *Acta Horticulturae*. Some of the more recent berry fruit symposia proceedings are listed for further reading.

Bañados, P. and Dale, A. (eds) (2008) IX International *Rubus* and *Ribes* Symposium. *Acta Horticulturae* 777.

Brennan, R.M., Gordon, S.L. and Williamson, B. (eds) (2002) VIII International *Rubus* and *Ribes* Symposium. *Acta Horticulturae* 585.

Hepp, R.F. (ed.) (2002) VII International Symposium on *Vaccinium* Culture. *Acta Horticulturae* 574.

Hicklenton, P. and Maas, J. (eds) (2003) XXVI International Horticultural Congress: Berry Crop Breeding, Production and Utilization for a New Century. *Acta Horticulturae* 626.

Hietaranta, T., Linna, M.M., Palonen, P. and Parikka, P. (eds) (2002) IV International Strawberry Symposium. *Acta Horticulturae* 567.

Hummer, K.E. (ed.) (2009) IX International *Vaccinium* Symposium. *Acta Horticulturae* 810.

Krüger, E., Carlen, C. and Mezzetti, B. (eds) (2009) Workshop on berry production in changing climate conditions and cultivation systems. COST-Action 863: Euroberry research: from genomics to sustainable production, quality and health. *Acta Horticulturae* 838.

Lopes da Fonseca, L. and Romero Muñoz, F. (eds) (2006) VIII International Symposium on *Vaccinium* Culture. *Acta Horticulturae* 715.

López-Medina, J. (ed.) (2009) VI International Strawberry Symposium. *Acta Horticulturae* 842.

McGregor, G.R., Hall, H.K. and Langford, G.I. (eds) (1999) VII International Symposium on *Rubus* and *Ribes*. *Acta Horticulturae* 505.

Simpson, D.W. (ed.) (2004) Euro Berry Symposium – COST-Action 836 final workshop. *Acta Horticulturae* 649.

Waite, G. (ed.) (2006) V International Strawberry Symposium. *Acta Horticulturae* 708.

17 Citrus

MICHAEL MORLEY-BUNKER

Citrus are believed to be native to the tropics, subtropics and drier monsoon areas of South-east Asia. Their use and cultivation in China is recorded in 2200 BC and the first Chinese book on citrus appeared in AD 1178.

The further spread of cultivated citrus plants probably owes much to Arab traders, who are thought to have brought citrus, beginning with the citron, *Citrus medica*, from Asia to the Middle East and Palestine. The Hebrews use the citron in their rituals for the Feast of the Tabernacle. Roman contact with Palestine led to continued spread of the citron into the Mediterranean basin. Other citrus plants, such as the sour orange and the lemon, followed the path of the citron. The sweet orange may have reached the Middle East during the time of the Crusades (1099–1291); however, many writers prefer to date its arrival in Europe at around 1480.

European voyagers and colonization spread citrus to South America, North America and the Caribbean. Columbus is reported to have carried the sweet orange, the lemon and the citron to Haiti on his second voyage (1493). The Dutch settlers established citrus at the Cape of Good Hope after 1652. The presence of wild citrus trees in the Zambezi river basin may be evidence that there was an earlier introduction, south of the Sahara, brought about by Arab merchants, who traded along the east coast of Africa as far as Sofala prior to European establishment at the Cape of Good Hope.

Citrus species hybridize freely: both interspecific and intergeneric hybrids have been described. As a result, the taxonomy of citrus has considerable problems. In practical terms, this implies a continuation of the addition of 'new' fruits to the existing range, either by conscious breeding, hybridization and selection or by chance hybridization or mutation.

One important citrus crop, the grapefruit, *Citrus × paradisi*, is known to be of recent origin. It was described in the West Indies around 1750. It is presumed to have arisen as either a chance bud mutation of *Citrus maxima* (the shaddock or pummelo) or as a chance hybrid between *C. maxima* and *Citrus sinensis* (the sweet orange).

Citrus General

Citrus taxonomy

There are many distinctive cultivar selections of citrus arising from hybridization, mutation and polyploidy within citrus species. The origin and parentage of many citrus selections is uncertain. As a consequence, the grouping and classifying of citrus cultivars has proved difficult. The genus *Citrus* has been divided into two subgenera, *Papeda* and *Eucitrus*. The subgenus *Eucitrus* contains all the edible fruit-producing species. Some of the citrus species, hybrids and citrus relatives are listed in Table 17.1.

Polyembryony

Many citrus species exhibit the condition of multiple embryos within a single seed. Only one of these embryos will have resulted from the fusion of the sexual gametes. The other embryos will have formed in the nucellus tissue and are called nucellar embryos. The nucellus tissue is entirely maternal in origin and therefore these embryos, and the plants that may be produced from them, have an identical genetic composition to the mother plant. This leads to the possibility of producing clonal populations from seed, providing, of course, the sexual embryo and the resulting seedling can be identified and

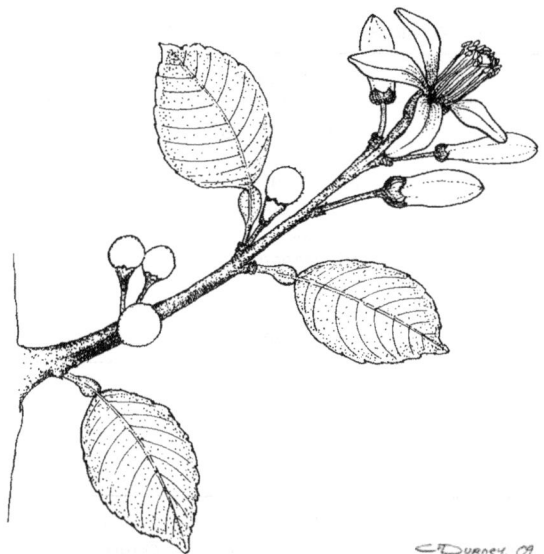

Fig. 17.1. Flowers on young shoot of citrus. Flowers are formed terminally and/or laterally on shoots normally formed in spring. These shoots are produced from buds which are laterally placed on the previous year's wood or as a continuation of growth from spurs.

Choice of plant material

The successful establishment of a citrus orchard, as indeed with any orchard, should utilize plant material which is healthy and is chosen to suit the growing conditions to be experienced. Market demands should also be taken into account. The temperature requirements for citrus that might be grown in cool climate conditions can be viewed from two positions: winter cold tolerance and tolerance of cool summer conditions (Table 17.2).

Rootstocks

Various citrus seedlings may be used for rootstocks, as may other citrus relatives that are not ordinarily used as cropping plants, for example trifoliate orange, *Poncirus trifoliata*. Trifoliate orange does not yield edible commercial fruit but the plant offers good cold tolerance when used as a rootstock. This property reflects its deciduous nature. Trifoliate orange offers other attributes, including a tolerance of relatively heavy soil types, a generally dwarfing effect on tree growth and yield of good-quality fruit. By way of contrast, rough lemon rootstock has poor cold tolerance, requires light soil types and produces a vigorous, large tree with very good yield levels. However, the fruit tends to be of poor quality. The consideration of which rootstock to use for any selected citrus cultivar should take into account a range of known attributes or properties for any rootstock under consideration. There is a very wide range of rootstocks cited in the world literature. Some of the commonly used rootstocks and their recognized properties are shown in Table 17.3.

rogued out. The growth of the nucellar shoots in the seedbed tends to be uniform, and if abnormal and non-vigorous plants are rogued there is a near certainty of 100% clonal population.

The production of clonal populations from seedlings is usually restricted to the production of rootstocks for working over to the required scion. In this way, the root performance of certain clones can be exploited. Some plants that do not commonly produce polyembryonic seeds are the Meyer lemon, the Clementine mandarin and the pummelo.

Table 17.1. Botanical and common names of some *Citrus* species and relatives – adapted from *Hortus III*.

Botanical name	Common name	Cultivar example(s)
Citrus aurantifolia	Lime	Mexican (syn. West Indian)
Citrus aurantium	Sour orange	
Citrus limon	Lemon	Eureka, Lisbon
Citrus maxima	Pummelo or shaddock	
Citrus medica	Citron	Etrog
Citrus × paradisi	Grapefruit	Marsh
Citrus reticulata	Mandarin	Clementine
Citrus unshiu	Satsuma mandarin	Okitsu
Citrus sinensis	Sweet orange	Valencia
Citrus × tangelo	Tangelo	Seminole
Poncirus trifoliata	Trifoliate orange	

Table 17.2. Tolerance to cool summer and cold winter conditions of some *Citrus* crops and rootstocks.

	Winter cold	Cool summers
Most tolerant	Trifoliate orange	Trifoliate orange
	Mandarin	Lemon
	Tangelo	Mandarin
	Sweet orange	Meyer lemon
	Meyer lemon	Tangelo
	Lemon	NZ grapefruit
		Sweet orange
Most sensitive	Lime	Lime
	Grapefruit	Grapefruit

Table 17.3. Some rootstock characteristics of commonly used *Citrus*.

	Trifoliate orange	Rough lemon	Sweet orange	Sour orange	Cleopatra mandarin	Troyer citrange
Winter hardiness	Good	Poor	Fair	Fair	Good	Fair–good
Vigour and tree size	Fair vigour, smaller than standard tree	Very vigorous, large tree	Fair vigour, standard size tree	Fair vigour, standard size tree	Early slow growth, fair vigour, standard size tree	Standard to large, vigorous growth
Fruit yield and quality	Precocious, good yields, high fruit quality, fruit slightly small	High yield, poor fruit quality, large-sized fruit	Good yield, high fruit quality and size	Good yield, fruit slightly small	Good yield, fruit slightly small	High yields, good fruit quality, large fruit size
Tolerance of heavy soils	Good, except for highly calcareous soils	Poor, best on sandy soil	Fair. Prefers well-drained loam soils	Good	Good	Good
Tolerance of root diseases	Good	Poor	Poor. Susceptible to root rot	Good	Poor	Good
Virus tolerance	Tolerates tristeza, sensitive to exocortis	Tolerates tristeza and exocortis	Tolerates tristeza and exocortis. Psorosis sensitive	Tolerates exocortis and psorosis, highly sensitive to tristeza	Tolerates tristeza, sensitive to exocortis	Tolerates tristeza, sensitive to exocortis

Sweet Oranges

The sweet orange, *C. sinensis*, is probably native of central China and north-eastern India and did not reach the Mediterranean basin until some time after the lemon, perhaps as late as the 15th century.

Seeds were taken by Columbus to Haiti (1493), and the plant then spread through the West Indies, South America, and Central and North America.

The sweet orange is used as a dessert fruit. The juice may be extracted and marketed in various

M. Morley-Bunker

forms, for example canned, frozen, concentrated and dried.

The major world producers are Brazil, the USA and Mexico.

Key points

Fruiting and harvesting

The cultivars fall into harvest season groups early-, mid- and late season. Using various cultivars the harvest season can range over a 6-month period.

Fruit set is not usually a problem. Some cultivars, such as Washington navel, set fruit parthenocarpically. Flower production may occur through much of the year on new growth. However, the spring growth flush is the most usual flowering period.

Fruit ripening does not occur off the tree. However, the skin colour can change from light green to orange.

When harvesting, care should be taken not to bruise or scratch the rind, and gloves may be worn. The fruit may be clipped or plucked from the tree. A stem button should be left on the fruit – if this is removed the fruit is more subject to rotting. Plucking or 'snap picking' is acceptable if the fruit can be removed without damaging the tree, removing the button or tearing the fruit skin. If fruit is clipped from the tree it is also feasible to clip the fruit-bearing shoot back to the closest and newest lateral shoot. This is often all the pruning that need be carried out on mature plants.

After harvesting, the fruits are washed and may also be waxed. The fruits are size sorted and substandard fruit downgraded or rejected. Short-term storage can be obtained in cool (below 15°C), ventilated conditions. Several months of storage can be obtained at 3–8°C with humidity at 85–90% RH.

General culture

The best-quality fruits are produced in areas with hot, dry summers, where irrigation is available to ensure fruit growth and development. Water deficits can be severe in these conditions, and irrigation application must make allowance for the extra demand from competing vegetation between the trees or else the vegetation should be removed. Regular tillage may not be a satisfactory method for controlling weed and grass growth, since continual exposure of moist soil depletes both soil moisture levels and organic matter. Green manuring, cover crops and permanent swards may be used if moisture levels can be maintained.

Pest and disease control requires careful selection of plant material before planting. Plants should be virus and mycoplasma free. Young trees should come from nurseries free of known soil-borne pathogens. Disease vectors must be controlled in the orchard after tree planting.

Cultivars

There are many available cultivars among which are:

- Round. (i) Hamlin – heavy crops, early season (autumn–midwinter), poor quality, mainly used for juice. (ii) Valencia – moderate to high cropping, late (late winter to spring), excellent quality.
- Navel. Washington navel – mid-season, lower yields than Valencia, not good in areas with high temperatures, especially at night. Suits Mediterranean climates. Many improved strains available.
- Pigmented. Several cultivars, e.g. Tardocco, Sanguinello, Moro. They need Mediterranean climates.

Lemons

Lemons, limes, citrons and Rangpur lime are generally placed in a group referred to as acid citrus fruit.

The lemon is used particularly for juices and food preparations and is not sold in large volumes as a fresh fruit. It originated in South-east Asia and in particular the eastern Himalayan region of India and also Burma (Myanmar). The lemon was known and traded by the Arabs in the 10th century. It is closely related to the citron. The lemon was being grown in Europe by the time of the European explorations of South America. It was soon taken to the West Indies (Columbus in 1493) and South America.

Although lemons are tolerant of cool summers, they are very sensitive to extreme and sudden temperature variations. Hardening of the plant tissue is not easily induced and this is probably the cause of the lack of tolerance to extreme conditions (Table 17.5). The best climates for lemon growing

Table 17.4. Sweet oranges.

Botanical, anatomical and physiological aspects	Climatic, geographic, soil and water requirements	Cultural aspects
Common name Sweet orange	*Temperature needs* Growth can occur between 13 and 40°C. Summer months should have average temperatures in excess of 15.5°C. Most grow in latitudes 23.5–40.0°	*Propagation* Budding and grafting of selected scions on to seedling stocks is usual. Seedling stocks may be true to type if nucellar seedlings are used
Botanical name Citrus sinensis		*Rootstocks* Varied, see Table 17.3
Botanical names of related and useful species See Table 17.1	*Frost tolerance* In an inactive growth condition may tolerate −5°C (some hardening in winter months). In active growth conditions, 0°C causes severe damage	*Spacing* Traditionally at 3–5 m squares, but recent experience suggests closer in-row spacing of as little as 1.3 m between plants could be used
Type of plant and size Small tree, evergreen up to 10 m, maintained in orchard between 3 and 7m	*Water needs* Well-distributed rainfall of approximately 850 mm required	*Training and pruning* The tree naturally forms a round shape. In the early years prune to form framework. Use little or no pruning of bearing trees but remove dead wood and water shoots. At picking, may cut fruited growth back to new lateral. More severe pruning may later be required to maintain tree at allotted size. Mechanical hedging sometimes used
Sexuality Usually hermaphrodite	*Tolerance of wet soils* Poor – see also humidity tolerance. Sensitive to waterlogging	*Thinning* Not normally practised
Pollination Usually by insects. Cross-and self-pollination possible. Some cultivars do not produce viable pollen and parthenocarpic seedless fruit may develop, e.g. navels, in the absence of cross-pollination. Some, such as Valencias, have low seed numbers	*Drought tolerance* Varies with rootstock but generally fair	*Tillage* Normally clean, bare surface (herbicide or cultivate). If soil moisture is adequate the use of green manures and cover crops is possible. The tree base should be kept clear
	Humidity tolerance Fair – grown in areas from humid Brazil and Florida to Mediterranean California and Australia. Unsuited to high-humidity tropics, as skin quality low and disease incidence high	*Time to first harvest* Small yields in third season
Flower buds Bud initiation occurs in midsummer; flower initiation occurs just prior to the late-spring growth flush	*Wind tolerance* Fair, shelter belts advisable for wind-prone areas	*Time to full production* Varies with climate and rootstock, often 8–12 seasons after planting
		Expected yield 3–4 years: 2.5–5 t/ha 8–12 years: 20–40 t/ha

M. Morley-Bunker

Growth of fruit
Sigmoid growth curve

Time of bud burst
Evergreen, so no bud burst as such. There are usually at least two growth flushes per year

Time of flowering
May continue over a prolonged period; main flowering flush spring–early summer

Time of fruit maturity
Midwinter to late summer, depending on cultivar

Edaphic features
Slopes and gullies for air drainage utilized and equator-facing aspect used in cool districts. Otherwise, flat sites preferred

Soil needs
Preference for light, well-drained soils. Choice of rootstock may extend soil range: pH range 5–8 preferred

Nutrient requirements
Leaf analysis increasingly used to determine mineral needs. Nitrogen requirement is high. Application split into two-thirds and one-third in winter and late summer, respectively. Rate interpreted for tree age, vigour and soil. General recommendation for each mature tree is:
N: 0.40 kg
P: 0.45 kg
K: 0.25 kg

Normal productive life
20–30 years

Method of harvest
Hand pick after sampling for sugar content. Handle with care (gloves). Snap-pick off the tree or clip fruit from tree close to stem button. Separation of button from fruit promotes rot

Storage
Ventilated, cool temperature conditions may be used to keep fruit 3–10 weeks. Ethylene may be used to de-green fruit

Main pathogens
Large pest and disease range, e.g. black aphid, scales, leafrollers, mealybugs, thrips, brown rot, root and collar rot. Some important virus diseases occur, e.g. tristeza and exocortis virus. When they do, rootstock choice is important (see Table 17.3). Specific disease problems may occur in particular regions

Table 17.5. Lemons.

Botanical, anatomical and physiological aspects	Climatic, geographic, soil and water requirements	Cultural aspects
Common name Lemon	*Temperature needs* See Table 17.4. Lemons are more tolerant of cool summer temperatures than sweet oranges. Not tolerant of tropical climates	*Propagation* Selected scions are budded or grafted on to seedling stocks
Botanical name Citrus limon		*Rootstocks* Trifoliata orange is not acceptable for Eureka, Villa Franca and Genoa, but Lisbon and Meyer may be satisfactory on this rootstock. Sweet and sour orange are suitable, also rough lemon (citronelle)
Botanical names of related and useful species See Table 17.1	*Frost tolerance* Lemons, except Meyer lemons, are one of the least frost-tolerant citrus	*Spacing* Traditionally wide spacings up to 7.5 m squares were used; now often reduced to 1.3–3.0 m between plants
Type of plant and size Small tree, evergreen, up to 10 m, but usually 3–7 m in height. Spiny branches, and vigorous early growth relative to other citrus species	*Water needs* A well-distributed rainfall of 800–950 mm needed	*Training and pruning* See Table 17.4. Tree tends to make long shoots that hang and tangle with fruit weight. However, it is probable that regular pruning to keep a clear framework would be too costly. Hedging and topping to keep trees to required size is sometimes practised. Undercutting is practised to remove low 'skirt' branches, allowing easy mowing, spraying and ventilation
	Tolerance of wet soils Poor: sensitive to waterlogging. Fruit quality becomes coarse and juice quality reduced with high water levels	
Sexuality Hermaphrodite		*Thinning* Not common
Pollination Usually by insects; cross- and self-pollination are effective. Parthenocarpic fruit may develop even if pollination does not lead to ovule fertilization	*Drought tolerance* Varies with rootstock but generally fair. Drought causes leaf and fruit drop and small fruit	*Tillage* See Table 17.4. Clean soil surface may be of assistance in reducing frost hazard
		Time to first harvest Small yields begin in third season
Flower buds See sweet orange	*Humidity tolerance* Not good. Disease incidence increases; fruit quality is reduced. Dislikes humid conditions, especially in warm areas	*Time to full production* Varies with climate and rootstock. Full yield: 8–12 seasons after planting

M. Morley-Bunker

Growth of fruit
No bud burst as such. Three main growth flushes per year: in spring, summer and autumn. Flush periods are longer and timing is less synchronized than with other citrus fruit

Time of flowering
May continue over most of the year. Main season: spring

Time of fruit maturity
May be harvested through year as fruit matures. Main crop season: winter–spring

Wind tolerance
Fair, wood may be brittle and damage likely in high winds

Edaphic features
Slopes and gullies for air drainage plus equator-facing aspect used in cool districts. Sheltered location from cool and prevailing winds desirable in some districts

Soil needs
Rootstock choice will affect soil range. A reasonably drained, aerated soil, 1 m in depth is preferred

Nutrient requirements
See Table 17.4. Lemons have a higher K requirement than other citrus. Fruit quality can be reduced with high nitrogen applications

Expected yield
3–4 years: 5–10 t/ha
8–12 years: 50–70 t/ha

Normal productive life
20 years

Method of harvest
Pick most lemons when full yellow colour has developed. For export, can pick before full colour to allow for transport time. May test for acid content but fruit on same tree may have very variable levels. Handle with gloves. Clip from trees (see Table 17.4)

Storage
See Table 17.4

Main pathogens
See Table 17.4. Lemons very susceptible to many specific citrus diseases, e.g. verrucosis

seem to be dry with hot summers. World production of lemons and limes is about 20% that of oranges. Mexico, India and Argentina are the largest producers (Table 1.2). Other countries that are listed as major producers of lemons and limes include Brazil, Spain, China, the USA and Turkey.

The lime (*Citrus aurantifolia*) originated in more tropical environments in South-east Asia than the lemon. The tree tends to be more restricted in its environmental range since it is the least cold-tolerant citrus. Limes are best grown in tropical and warm subtropical regions. Limes are divided into two groups: the acid-less limes and the more important acid limes.

Key points

Plant material

There are many named lemon cultivars, although some of the named lemons are not true lemons. For example, the Meyer lemon is a hybrid, probably between *C. sinensis* and *C. limon*. The Meyer lemon is notable because it is more suited to cooler districts than the standard lemon cultivars such as Lisbon and Eureka.

There are three groups of cultivars:

- The Femminello group, mainly grown in Italy, having consistent crops over most of the season.
- The Verna group, grown in Spain, has a main season in late winter to midsummer and a smaller one winter–spring.
- The Sicilian group, also popular in California and Australia, has two main cultivars – Lisbon and the less frost-tolerant Eureka. The harvest is mainly winter to spring.

Most lists of recommended rootstocks for lemons (and other acid fruit) do not include trifoliate orange (*P. trifoliata*). This rootstock is not normally recommended because of the potential for stock–scion interaction if there is any latent virus in either or both trifoliate orange and lemon. Fruit quality is also said to be poor. Lemons are normally budded on to sweet orange rootstocks.

Harvesting and curing

Maturity in acid citrus fruit is sometimes difficult to detect visually, and sampling fruit 6 cm in diameter or greater for juice and acid content may be undertaken. The fruit will also 'cure' when kept for a short period in a well-ventilated place with a cool, even temperature. During curing, skin thickness is reduced and the percentage juice content increased.

By way of contrast, if the fruits are left on the tree long after maturity is reached, they become less juicy and the pith under the skin increases in thickness.

Pruning fruit-bearing laterals

The lemon tree bears fruit on long lateral shoot growth which is of recent origin. There is no subsequent flower development on this shoot, but there may be flowering on new lateral shoots that will arise behind the maturing fruit. Fruit harvesting can, therefore, be combined with some shortening of the lateral shoot. Secateurs (pruners) are used to clip back the fruited lateral to the newest and nearest lateral growth, thereby shortening the existing shoot. The fruit is also harvested at the same time by clipping the fruit stalk and leaving the 'button' part of the stalk attached to the fruit.

The weight of the fruit on the lateral also tends to result in branch bending and for branches to hang downwards. Branches may overlay each other around the 'skirt' of the tree. Undercutting is a procedure used to shorten outer branches and may be combined with the removal of internal branches, with the objective of opening and lifting the 'skirt' from the inside. This allows access to the trunk and under-tree canopy area, which may assist management practices such as irrigation and fertilization.

Grapefruit and Pummelos

Grapefruit (*Citrus × paradisi*) is probably a hybrid between the pummelo (*C. maxima*) and the sweet orange (*C. sinensis*) and is believed to have originated in the West Indies (Table 17.6). The grapefruit is typically larger than an orange in size, and the relatively acid nature of the fruit makes it less popular as a fresh fruit compared to oranges. This is also true for the even larger-sized pummelo. Nevertheless grapefruit and pummelo are used fresh and in fruit mixes. The juice of grapefruit and pummelo can be blended with other juices or used on their own. Both fruits are more heat demanding than most other citrus, except perhaps

M. Morley-Bunker

Table 17.6. Grapefruit.

Botanical, anatomical and physiological aspects	Climatic, geographic, soil and water requirements	Cultural aspects
Common name Grapefruit	*Temperature needs* Tropical and warm subtropical	*Propagation* See Table 17.4
Botanical name Citrus × paradisi	*Frost tolerance* Low	*Rootstocks* See Table 17.3
Botanical name of related and useful species See Table 17.1	*Water needs* Must be adequate	*Spacing* Grapefruit are vigorous trees and are usually planted 4–7m on the square
Type of plant and size Large and vigorous tree with spreading habit	*Tolerance of wet soils* Poor – low crops under waterlogged heavy soils	*Training and pruning* May have to limit height by trimming
Sexuality Usually hermaphrodite	*Drought tolerance* Fair	*Thinning* Not common
Pollination Self- and cross-pollinated. Fruit vary from seedy to almost seedless	*Humidity tolerance* Fair to good	*Tillage* Grass usually maintained between rows
Flower buds Borne in clusters	*Wind tolerance* Fair – the trees usually have a heavy framework	*Time to first harvest* Three to four seasons
Growth of fruit Sigmoid	*Edaphic features* Flat land is easier to manage	*Expected yield* 3–4 years: 5–10 t/ha 8–12 years: 30–60 t/ha
Time of bud burst See Table 17.4	*Soil needs* See Table 17.4	*Normal production life* See Table 17.4
Time of flowering See Table 17.4	*Nutrient requirements* See Table 17.4	*Method of harvest* See Table 17.4
Time of maturity Late summer to midwinter		*Storage* See Table 17.4
		Main pathogens See Table 17.4

the lime. Grapefruit and pummelo are mostly found in tropical and warm subtropical, humid regions. World interest in grapefruit and pummelos is presently of the order of 5 Mt. Important grapefruit-producing countries include China, the USA and South Africa.

Pummelo is more common in tropical rather than subtropical climates. The pummelo is larger than grapefruit and has less acidity and bitterness and is easier to eat fresh. The main production and consuming region is South-east Asia, with Thailand and China particularly involved. Other producing countries include Taiwan, Japan, Indonesia and Malaysia.

Key points

Plant material and cultivars

- Grapefruit (*Citrus × paradisi*). There are two groups of grapefruit, the white- and the red-fleshed cultivars. Duncan and Marsh are the main white-fleshed cultivars. Duncan is of better shape and quality than Marsh but has the disadvantage of having a large number of seeds compared with the seedless Marsh. Marsh holds on the tree better than Duncan. Red-fleshed grapefruit have been developed predominantly in the USA during the 20th century. Some recent and popular cultivars are Ruby Red, Ray Ruby,

Flame, Rio Red and Star Ruby. They have proved popular with the public.

- Pummelos (*C. maxima* syn. *C. grandis*). The availability of white- and pink-fleshed pummelo selections mirrors the situation with grapefruit. Most of the pummelo selections are regional in importance. There is an American selection named Chandler, which is a hybrid of two pummelos, one named 'Siamese Sweet', which is indicative of where many other selections may be found.
- Pummelo–grapefruit hybrids. There are hybrids between the two fruit crops. Work at Riverside, California (USA) produced cultivars named Melogold and Oro Blanco. Siamese Sweet pummelo was the seed parent of Oroblanco and Melogold pummelo–grapefruit hybrids.
- Rootstocks for grapefruit and pummelo. Pummelos and grapefruit may be grafted on to rootstock selections of Cleopatra mandarin, rough lemon, sweet orange, sour orange, Carrizo and Troyer citranges and citrumelo.

Mandarins and Related Citrus

The common mandarin probably originated in north-eastern India and spread, as a fruit crop, both westward and into the east. The subsequent development of the fruit crop has occurred in several geographical areas. The result is that the mandarin group of citrus fruit is now quite diverse (Table 17.7). Many people accept that there are two major divisions within the mandarin group – the common mandarins and the satsuma mandarins. Mandarins are capable of hybridizing with other mandarins and also with other *Citrus* species. There are naturally occurring interspecific hybrids with mandarin-like character and flavour, such as Temple, and the man-made hybrids such as the tangelos Seminole and Robinson.

Most botanical descriptions of the mandarin give the crop the name *Citrus reticulata*. However, the satsuma mandarin group has been named *Citrus unshiu* by some authorities. The naming recognizes that a distinctive version of the mandarin has developed in the Orient, particularly China and Japan.

The mandarin fruit resembles the orange, although the shape may be more flattened and is typically much smaller in size. The rind colour may also have a more reddish tinge. Mandarin oranges are usually eaten plain or in fruit salads. The reddish orange mandarin selections of Mediterranean origin have been named and marketed as tangerines. The name tangerine comes from Tangier, Morocco, indicating an historical shipping route into northern Europe. There are no botanical grounds though to differentiate tangerines as different from mandarins.

Mandarins are widely cultivated in cool and warm subtropical regions and even warm Mediterranean areas. The wide range of cultivars – mandarins and hybrids – allows a choice to suit various climatic conditions. Some mandarin cultivars are particularly cold tolerant and able to develop cold-hardiness.

The appealing character of some of the mandarins is their loose skin, allowing easy peeling. The fruits are often small with small segments, depending on the cultivar. The fruit may be seedless.

The first synthetic hybridization between the pummelo and the mandarin was made by Webber and Swingle in 1897. The name tangelo was coined to describe the hybrids between these two parents. Tangelo plants have highly coloured, aromatic, distinctive and richly flavoured fruit. Repeated hybridization with these two parents has produced a range of character combinations. It is now usual to apply the name tangelo to the more mandarin-like hybrids and to find other names for hybrids whose character is more pummelo-like.

China is the largest producer of mandarin fruits. Other countries with significant production include Spain, Brazil, Japan, Turkey and Italy. Total world production is currently of the order of 26 Mt (Table 1.1) and has doubled in the last 40 years (Table 1.3).

Key points
Plant material

- The common mandarins. These are generally given the botanical name *C. reticulata*, although some included in this group are more correctly hybrids. Compared with satsuma, common mandarin fruits are smaller with less-easily removed skin. The 'clementine' is the most widely planted of all mandarins; it has good flavour, few seeds and high yields, although small size can be a problem. It performs well in Mediterranean areas. Selections of clementine give variation in fruit and date of maturity. Of lesser importance is the cultivar 'Darcy', which is suited to hot, humid conditions.
- Satsuma. Most development of the satsuma has occurred in Japan, where it is the most widely

M. Morley-Bunker

Table 17.7. Mandarins (tangerines) and related citrus.

Botanical, anatomical and physiological aspects	Climatic, geographic, soil and water requirements	Cultural aspects
Common names Mandarin, tangerine, satsuma, tangelo *Botanical name* *Citrus unshiu* (satsuma), *Citrus reticulata* (mandarin) *Botanical name of related and useful species* See sweet orange: note also the natural hybrids *C. reticulata* × *C. sinensis* Temple; *C. reticulata* × *C. maxima* Tangelo *Type of plant and size* Small tree, evergreen, height usually 2–8 m. Sometimes have spiny branches. Some variability in habit and appearance, e.g. mandarins are more upright than satsumas *Sexuality* Usually hermaphrodite *Pollination* Both self- and cross-pollination may occur. Seedless fruit may be produced after weak and aborted ovule growth, and after fertilization failure from incompatible pollen. Male sterility does occur in some cultivars and, for these, cross-pollination should be encouraged. Some are naturally seedless *Flower buds* See Table 17.4 *Growth of fruit* Sigmoid growth curve. Mandarins are the smallest and have a less-easily removed skin than satsumas *Time of bud burst* Two growth flushes occur per year *Time of flowering* Mostly in spring *Time of fruit maturity* Mostly winter; time may be extended by planting an appropriate range of types	*Temperature needs* Probably intermediate between lemon and sweet orange in their response to cool summer temperatures. Fruit quality and sweetness are improved by warm summers *Frost tolerance* More frost tolerant than lemon and sweet orange. The satsuma mandarin group is more cold tolerant than mandarin/tangelo fruits *Water needs* See Table 17.4 *Tolerance of wet soils* Poor *Drought tolerance* See Table 17.4 *Humidity tolerance* See Table 17.4 *Wind tolerance* See Table 17.4 *Edaphic features* See Table 17.4 *Soil needs* See Table 17.4 *Nutrient requirements* See Table 17.4	*Propagation* See Table 17.4 *Rootstocks* See Table 17.1 *Spacing* Variable, depending on cultivar, rootstock and climate. Compared with other citrus, many mandarins produce small trees. Planting on the square at about 5 m is giving way to rectangular and closer in-row spacings *Training and pruning* See Table 17.4 *Thinning* May be necessary to avoid small fruit size and/or avoid biennial bearing *Tillage* See Table 17.4 *Time to first harvest* Small yield beginning in third season *Time to full production* Yields near maximum at end of 12 years *Expected yield* 3–4 years: 2.5–5 t/ha 8–12 years: 7.5–15 t/ha *Normal productive life* 20 years *Method of harvest* See Table 17.4 *Storage* See Table 17.4. Usually expect poorer storage ability than sweet orange *Main pathogens* See Table 17.4

planted citrus. The plant is the most cold resistant of all citrus and does well in the cooler subtropical regions. There are many selections within the satsuma and these have been made to give better quality and extended harvesting season, in particular earliness.

- Mandarin hybrids. Temple and Murcott are naturally occurring hybrids of tangerine and sweet orange (this cross is given the generic name tangor). They ripen after midwinter, and selections within the Temple and Murcott hybrid plants are being made, particularly for improved husbandry characteristics. They are grown in Florida and South Africa.
- Tangelos. These are man-made hybrids of C. *reticulata* and C. *maxima*, and several cultivars, such as Orlando and Minneola, are common and popular in North America.
- Robinson. This is a cross between tangelo and clementine. The fruit is most like the common mandarin; the quality is excellent and most fruit ripen in early winter.

Many of the mandarin hybrids have specific requirements for cross-pollination and growers should enquire locally for suitable pollenizers.

Further Reading

Albrigo, L.G. and Galán Saúco, V. (eds) (2004) XXVI International Horticultural Congress: Citrus and Other Subtropical and Tropical Fruit Crops: Issues, Advances and Opportunities. *Acta Horticulturae* 632.

Oh, D.-G. (ed.) (2008) XXVII International Horticultural Congress – IHC2006: International Symposium on Citrus and Other Tropical and Subtropical Fruit Crops. *Acta Horticulturae* 773.

M. Morley-Bunker

18 Kiwifruit

MICHAEL MORLEY-BUNKER AND PETER LYFORD

Actinidia deliciosa, the kiwifruit, previously called Chinese gooseberry, originated in the Yangtse Valley in China (Table 18.1). The genus *Actinidia* is of east Asian origin. About 36 species have been recorded, several of which have been utilized for food, for example *Actinidia arguta*, which has smaller fruit without the hair covering of *A. deliciosa*. Chinese records refer to the use of *Actinidia arguta* as a fruit as early as AD 770. *Actinidia chinensis* is widespread in China and is the main type collected in the wild and processed rather than eaten fresh.

Specimens of *A. deliciosa* were taken from China to the UK by Robert Fortune in 1847. Seed obtained from China was introduced into New Zealand in 1906. These seedlings, grown by Alexander Allison of Wanganui, began fruiting in 1910, and the progeny from that introduction are the foundation of the present-day New Zealand cultivars. In about 1924, using this material, Hayward Wright of Auckland grew about 40 seedlings and selected one, which he called 'Wright's Giant'. This cultivar has been renamed 'Hayward' and has become the world's first main commercial cultivar.

Key Points

Kiwifruit cultivars

The cultivars presently used originated as seedlings, the naming of which was standardized in New Zealand in 1958. Hayward has become the main variety of world commerce and is grown mainly in Italy, New Zealand, Chile, China, the USA, France, Greece, Japan and Korea. The recent variety Zespri Gold, owned by Zespri International, is now 20% of New Zealand production and is planted on a commercial scale in Italy, Chile, Japan and Korea.

Advances in plant breeding since 1980 have seen a number of new varieties being tested by researchers and growers in a number of countries. Small volumes of *A. arguta* strains are produced in several countries. This is a small, green-skinned kiwifruit that can be eaten whole and has a short shelf life, similar to grapes and some berry fruit.

A. chinensis types are being selected and bred by plant breeders in most kiwifruit-growing countries. New Zealand is leading in the commercial development of *chinensis* strains, with the launch in 1999 of a new golden-yellow-flesh variety on to world markets. This variety, named Zespri Gold, has plant variety rights protection. The fruit is sweet tasting, being picked at 12°Brix and is 18–20°Brix when ripe. The vine is more subtropical than Hayward, with bud break and flowering occurring 4–5 weeks earlier. The vine continues to grow well into autumn, unlike most *deliciosa* types, which stop growth in late summer. Since 2005, about 20 new *chinensis* and *deliciosa* varieties have been tested by commercial growers in the main kiwifruit-growing countries. There are also experimental trials being done with rootstocks to improve flowering and yields.

Propagation and establishment

Kiwifruit are propagated by cuttings and by grafting. Cuttings produce smaller plants with a more fibrous rooting system and can be taken from softwood growth, semi-hardwood growth and from plant roots. Softwood cuttings are taken from midsummer to early autumn and should exclude the very soft shoot tips which occur above the leaf with a diameter of 4 cm. The top two leaves are left on the 10–15 cm-long cutting and are cut in half to reduce moisture loss. Cuttings are cut just below a

Table 18.1. Kiwifruit.

Botanical, anatomical and physiological aspects	Climatic, geographic, soil and water requirements	Cultural aspects
Common name Kiwifruit, Chinese gooseberry, Yang-Tao (China) *Botanical name* Actinidia deliciosa *Botanical name of related and useful species* Actinidia arguta Actinidia kolomikta Actinidia chinensis *Type of plant and size* Trailing deciduous vine. Up to 9 m length or height with support. Large cordate leaves of up to 20 cm *Sexuality* Dioecious (separate male and female plants) *Pollination* Insect pollinated. Beehive introduction to orchards recommended at up to 8 hives/ha beginning at 15% female vine flowering. Male vines are interspersed through the planting at male:female ratios of between 1:3 and 1:8 *Flower buds* Flowers found on first one to six nodes of current season's extension growth from lateral buds on previous season's canes. Conditions in previous season influence plant's ability to initiate flowers, e.g. heavy shading of leaf and axil reduces flowering	*Temperature needs* Requires frost-free growing period of 8–9 months. Winter chilling (600–1100 h below 7°C) necessary for good floral development and bud burst *Frost tolerance* Young spring growth damaged at –2°C. Woody growth damaged at –8°C *Water needs* High leaf transpiration rates have been reported. Estimated irrigation requirement 5 l/m^2 *Tolerance of wet soils* Will not tolerate wet feet. Tolerates high rainfall if drainage is good *Drought tolerance* Low tolerance. Reduction of yield is likely *Humidity tolerance* Good *Wind tolerance* Poor. Essential to have good shelter in windy sites; young shoots snap and bruise in wind; fruit may wind scar *Edaphic features* Will benefit from warm slopes with good air drainage and wind shelter in cooler districts	*Propagation* Seedlings may be raised for grafting. Seed is sown in mid- to late spring. Leafy cuttings and root cuttings may be used. T-budding, cleft, whip-and-tongue and rind grafts may be used for working over stocks (see Key points) *Rootstocks* Bruno seedlings have commonly been used as rootstocks. Other cultivars and seedlings have been developed *Spacing* Variable with cultivar, rootstock, trellis or pergola. Traditional spacings have been at 5 m between rows and 5.5–8 m between plants. There is a general trend for closer spacings, down to 3 m, to improve early yields *Training* Trained on to T-bar or pergola trellis. Permanent arms supported at 1.8–2 m height. Temporary fruiting arms are renewed every 2–4 years but can be left longer on some pergola systems; laterals on temporary arms are summer pruned to control leaf vigour and shading (for details see Chapter 6) *Thinning* May be required to assist in obtaining good fruit size and reducing biennial tendency and abnormal fruitlets that produce deformed fruit. Studies indicate thinning immediately after petal fall has best effect. *Tillage* Use herbicide strip in row; grass between rows. Sometimes swing-arm mowers are used instead of herbicides *Time to first harvest* Hayward 2–4 years after first planting, *chinensis* types 1–2 years *Time to full production* Hayward 7–9 years, earlier if close planted; *chinensis* Gold types 4–5 years

M. Morley-Bunker and P. Lyford

Growth of fruit
Overall, kiwifruit has a sigmoidal growth curve. Some additional small growth spurts occur and a triple-sigmoid growth curve has been reported

Time of bud burst
Spring, slightly before apples and grapes. Bud swell begins approximately 10 days before bud burst

Time of flowering
Late spring

Time of fruit maturity
Autumn to midwinter. Fruit will ripen satisfactorily after picking if soluble solids exceed 6.2% at harvest. Fruits store best if picked at 7–9% soluble solids and also ripen to give best flavour. Fruit dry matter content may also be used to determine maturity.

Soil needs
Deep, well-drained soils. A good water-holding capacity desirable for summer growth of shoots and fruit

Nutrient requirements
Heavy fertilization rates currently used. Aim to promote vine vigour and quick canopy cover. Nutrient analysis of leaf samples now common. Generally mature vines given 170–220 kg N/ha, 30–60 kg P/ha, 200–300 kg K/ha. Spread fertilizer dressings evenly around the plant. Apply in spring but add a third of N in early summer. pH of 6 satisfactory

Expected average yield:
A. deliciosa types, e.g. Hayward
3 years: 6 t/ha
5 years: 20 t/ha
8 years: 30 t/ha

A. chinensis types, e.g. Zespri Gold
3 years: 20 t/ha
5 years: 30 t/ha
8 years: 45 t/ha

Normal productive life
Similar to grapes, over 50 years.

Method of harvest
Hand pick. Sample for soluble solids (above 6.2%) to establish degree of maturity. Some marketing agents pay sliding-scale bonuses based on fruit dry matter content. Dry matter content can vary between 12 and 20%, although most fall in a range of 14–17%.

Storage
The Hayward variety of A. deliciosa has perhaps the longest storage and shelf life of any major commercial fruit. 0°C and 90% relative humidity will keep fruit approx. 9 months. Very sensitive to ethylene gas. Do not store with fruit that release ethylene, e.g. apples. For quality, best-storing fruit have 7–9% soluble solids. A. chinensis Gold varieties are stored at 1.5°C and have a storage life of 6 months.

Main pests and diseases
Leafroller caterpillar, greedy scale, white peach scale (Asia), thrips, passionvine hopper, rootknot eelworm, botrytis, Pseudomonas viridiflava, Pseudomonas syringae (Asia), Sclerotinia, phytophthora root rot

bud and treated with rooting hormone before placement in a mist propagation unit with bottom heat at 20–25°C. Rooting occurs about 3–4 weeks later and plants should be moved into a protected environment to continue growth; they can then be set out in the nursery the following spring.

Semi-hardwood cuttings can also be taken from midsummer to early autumn. Cuttings should be 15–25 cm long and all the leaves should be removed except the top leaf, which is cut in half. The cuttings are treated with rooting hormone and then placed under mist. Bottom heat is not required.

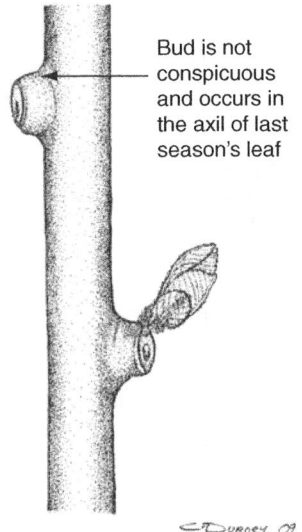

Bud is not conspicuous and occurs in the axil of last season's leaf

Fig. 18.1. Bud burst.

Root cuttings are taken from roots ranging in size from 10 to 30 mm in diameter, which are cut into pieces 5–10 cm long; they are then placed horizontally in trays of a free-draining medium, just beneath the surface. The best time for regenerating shoots from root pieces is winter/spring. Once shoots are formed they are separated from the old root section, which is then discarded. The shoots are placed in a mist environment with a bottom heat of 21–25°C to produce new roots.

Seed extraction is commonly achieved by removing the fruit pulp with the seed and then liquefying it in a blender. Providing the speed is not set too high and the length of time spent in the blender is not too long, the seed should be undamaged. The pulp can be washed through a sieve and the seed allowed to dry. The extracted seed should be kept in a sealed polythene bag at 4–5°C for 2–3 weeks (stratification) before sowing, followed by fluctuating temperatures between 10 and –2°C during germination. Seed can be soaked in gibberellic acid (up to 5000 ppm for 20 h). This will promote germination and may substitute for the stratification and fluctuating temperature treatment.

The young seedlings are transplanted into seedling trays or tubes as soon as they are large enough. They are then placed in a cool (unheated) greenhouse or a shade house before placing outside. Transplanting into the open ground is usually left until spring. The plants are usually grown on for 1 year before either being moved into the orchard or being grafted. Most orchardists grow their 1-year-old nursery seedlings in the orchard for a further

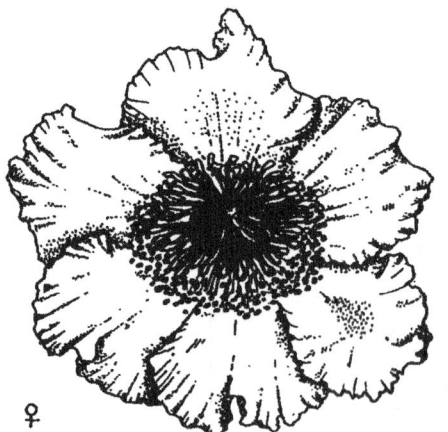
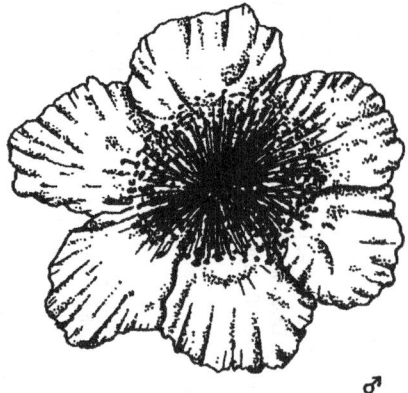

♀   ♂

Fig. 18.2. Flowers of kiwifruit, female on left, male on right. Although the female flower has styles (bulbous) above and filaments/anthers below, self-fertilization is not possible.

M. Morley-Bunker and P. Lyford

year, attempting to get the seedling as high up the trellis as possible, so that they can graft over with the scion piece set high up on the seedling stem.

Grafting, usually performed in late winter to spring, is the usual means of working over the seedling. Cleft grafts and whip-and-tongue grafts are both acceptable. The scion pieces should be selected from good-bearing vines of proven ability and good character. They should be held in a cool place (e.g. in damp moss or sawdust) or in a sealed polythene bag in a refrigerator at about 4–5°C.

The new *chinensis* gold varieties are often established by grafting mature Hayward vines using a cleft graft on stumps.

Pollination and fruit set

There are two main reasons for the considerable concern being felt about pollination. First, the kiwifruit cultivars presently grown are dioecious and pollen must be transferred from one male plant to another female plant. Second, fruit size is correlated with seed number, which in turn depends on the transfer of adequate pollen to the female flower. A good-sized Hayward fruit contains about 1150 seeds, which requires the deposit of about 12,000–15,000 viable pollen grains on the stigmatic surface.

Wind can play a role in kiwifruit pollination in warm climate areas like south Italy. Most research confirms that insects carry out the most effective pollination of kiwifruit. Honey bees are the main insect used commercially. Honey bees, from a well set-up hive with high levels of larval brood and sugar feeding, will forage kiwifruit flowers for their pollen. As with most fruit crops, competing sources of nectar and pollen from citrus or clover flowers can reduce pollination levels in the kiwifruit.

Male vines should be well distributed throughout the planting (see Chapter 6, Fig. 6.18). Most modern orchards now have a 1:5 or 1:6 ratio of male to female vines.

The female flower is receptive to pollen for 7–9 days from flower opening. An individual male flower only produces viable pollen for 2–3 days after opening. In warm weather, 80–90% of male pollen is released on the day of flower opening. However, most male varieties grown commercially produce large numbers of flowers, opening successively over 2–3 weeks. The common early-flowering Matua male clone is being replaced with improved male types. The main males used today

are Chieftain and M56. Growers who rely on bees for pollination need to test several male types in a new orchard planting to find which variety provides the best overlap in flowering with Hayward.

A. chinensis varieties have their own male types that flower at the same time as the female vines.

New Zealand, Chile and France are the main countries that rely mainly on honey bees for pollination. In many other countries, such as Japan, China and Korea, hand pollination is common practice.

There are a number of techniques to apply pollen by hand and machine to achieve pollination. There has been wide acceptance in New Zealand since 1995 of applying extra pollen in conjunction with using bees, to improve fruit size, fruit shape and dry matter. New Zealand pollen producers now export pollen to many overseas countries, mainly Italy, Japan and Korea. A wet spray application using pollen in suspension has proved valuable for use in seasons when the weather over pollination time is very wet.

Factors affecting flowering and crop yield

Crop yield is the culmination of many events and processes, especially those factors that influence bud break and flower number. Kiwifruit can only be grown where there are no early-spring or early-autumn frosts. Flower induction occurs in the mid- to late-summer period, and good induction is favoured by high sunshine levels and good vine health and nutrition. Flower development inside buds occurs in the late dormant and early bud break period.

Lack of winter chilling leads to delayed bud break and fewer flowers per shoot. Although fruits have been grown in areas with only 500 Richardson chill units (= no. of hours below 7°C), most observers believe kiwifruit need 700–800 Richardson chill units to have an adequate bud break and flower number. Sharp fluctuations in temperature during the bud break period can also lower flower numbers. The use of bud break-enhancing sprays, such as hydrogen cyanamide, is common in many kiwifruit-growing areas around the world, to ensure adequate bud break after a mild winter and help male to female flowering overlap.

Climatic factors such as excess wind, dry conditions and excessive wet weather over flowering can severely reduce crop yields. The correct management of winter and summer pruning (Chapter 6) is the grower's main tool to ensure good light levels into the vine to maximize the current crop volume as well as prepare the vine for the next season's flowering.

In New Zealand many growers are using trunk girdling to improve yield and fruit quality. A girdle is applied at 4 weeks from fruit set, which improves fruit size. A summer girdle in mid-February is used to raise dry matter and get an earlier harvest.

Harvesting and storage

Kiwifruit are harvested by hand. The fruit may be snapped off at the base of the fruit, leaving the stalk on the vine. Although the fruit is quite hard, it should still be handled carefully. Wounded fruit may undergo over-rapid ripening.

The fruits are normally picked into bags and then emptied into bulk bins, similar to apple bins, holding about 300kg. Picking can be organized as a once-over pick, but some growers do pick over twice, leaving the smaller fruit until last to encourage any further size increase. The fruits are normally graded on the basis of weight and inspected carefully. Any fruits with blemish or poor shape are downgraded according to company or industry grade standards.

Fruits are packed in a variety of pack types. The most common in use are the 10 kg pack and single-layer trays of between 3 and 4 kg per tray.

Kiwifruit have three major needs for satisfactory storage:

- A temperature range of −0.5 to 0°C for *deliciosa* varieties and +0.5 to +1.5°C for *chinensis* varieties.
- A relative humidity of 95%.
- No ethylene.

They will also keep well in a controlled atmosphere at 5% CO_2 and 2% O_2.

Pest and disease control

High-quality and export fruit must be clear of pests and diseases and be of good size and sound appearance. Spray schedules have been designed with control and chemical residue levels in mind. There is currently an emphasis on minimal spray usage while maintaining pest and pathogen control. Sprayers must be properly calibrated and chemicals applied at the correct rates and times. The follow-ing notes refer to the most common and trouble-some pests and diseases.

There are various leafroller species. Several generations will appear in a year. The second generation usually occurs during blossom but, because of bee activity, spraying must be delayed until the bees have left the vicinity. An immediate post-bloom spray is vital for control. The caterpillar usually works its way into a space between two surfaces, for example two leaves or leaf and fruit, which it then webs together. The caterpillar may scar fruit. It is both the presence of the pest and the scarring of the fruit rather than growth reduction that makes control important.

Greedy scale is also a problem that is more related to the presence of the pest or the remains of the pest than its effect on plant growth. There are two or three generations of scale in a year. The first generation is usually found on the foliage in spring. Later generations are released over the remaining season and will crawl on to the fruit. The scale is best controlled either when overwintering or with sprays timed to coincide with the crawler stage in the life cycle, which is the most susceptible stage to chemical applications. White peach scale is the main scale problem in Asian countries.

Diseases occur both in the orchard and in the fruit store. In the crop cycle the first concern is bacterial blossom rot, *Pseudomonas viridiflava*. The bud or the flower parts will turn brown and fail to develop. Fruits from infected flowers fail to develop properly and will remain small. Bactericidal sprays have not proved particularly effective. These sprays have been administered in the early season before and after bud burst. There is now good evidence to suggest that infection may take place in the preceding summer and autumn. Spray schedules may need reassessment in the light of this information. The incidence of this disease has been irregular, which makes determining the methods and economics of control rather unclear.

Later on in the crop cycle, sclerotinia (*Sclerotinia sclerotiorum*) may infect young fruitlets or cause dieback of lateral shoots which have been girdled by an infection. The infections can spread from rotting flower clusters or from orchard floor litter. The disease is favoured by warm, wet conditions and may infect the plants from October through to February. The sprays used for botrytis control can also be used for sclerotinia.

Botrytis (*Botrytis cinerea*) can gain entry to the fruit from sclerotinia infections at blossom time or from harvest damage to the fruit. Botrytis causes storage loss. It may lie dormant in the fruit until changes in the stored fruit allow it to become aggressive. Botrytis damage may also show up at harvest in the form of scar tissue following an infection lesion. Botrytis infection is encouraged by wet conditions. Since thick, dense growth on the trellis may also encourage infections, pruning to give an open vine canopy aids botrytis control. Chemical control includes sprays at blossom and further protective sprays up to harvest, providing residues will not exceed set limits.

Further Reading

Ferguson, A.R., Hewett, E.W., Gunson, F.A. and Hale, C.N. (eds) (2007) VI International Symposium on Kiwifruit. *Acta Horticulturae* 753.

Huang, H. (ed.) (2003) V International Symposium on Kiwifruit. *Acta Horticulturae* 610.

Sfakiotakis, E. (ed.) (1997) Proceedings of the 3rd International Symposium on Kiwifruit. *Acta Horticulturae* 444.

Warrington, I.J. and Western, G.L. (eds) (1990) *Kiwifruit Science and Management*. Ray Richards, Auckland, New Zealand.

19 Warm Temperate and Subtropical Fruit

MICHAEL MORLEY-BUNKER

Tamarillos

The tamarillo is one of several 'tree tomato' species (Table 19.1). Traditionally the tree tomatoes were placed in the genus *Cyphomandra*. The name tamarillo was coined in New Zealand for plants of *Cyphomandra betacea* – elsewhere the tamarillo is likely to be simply known as tree tomato. However, examination of the phylogenetic relationships in the *Solanaceae* family has led to a family re-organization, and tamarillo is now named *Solanum betaceum*. Tree tomatoes come from the Andean region in South America, where it is extensively cultivated but not in large plantations. It has been grown as a minor and novelty crop in other world regions. New Zealand, where it has been cultivated since the 1890s, is one of the few locations where it has an established commercial market status. It has been particularly useful as a domestically available subtropical fruit with a harvest season that does not coincide with many other fresh fruits.

The rather tart, sharp flavour typical of most red cultivars can be reduced by stewing. Some people eat the fresh fruit with a spoon to remove the contents and leave the unpalatable skin. The fruit can be used in pickles, preserves and chutneys, and the pulp can be used as a dessert flavouring or incorporated into jellies. Only cultivars without red colouring can be canned, as the redness signifies the presence of acidity, which will corrode tin cans. The yellow cultivars are generally milder in flavour and sharpness. Amber-coloured fruits are intermediate in flavour intensity between the red and yellow types.

Key points

Selection of propagation material

The tamarillo may be propagated by seed or cuttings. Micro-propagation is also feasible. Grafting seedlings over with selected scion material is feasible but not practised commercially. Seedlings, which are normally virus free, come fairly true to type, but some rogueing will be required to maintain reasonable uniformity.

Plant material is selected from high-yielding plants with good fruit characters, especially fruit size, absence of sclerids (hard, stony inclusions) and a good flavour, without undue astringency. Virus-infected plants should be avoided as a propagation source for cuttings. Viruses identified in tamarillo include tamarillo mosaic virus, cucumber mosaic virus, alfalfa mosaic virus, potato aucuba mosaic virus and Arabis mosaic virus. Virus infection can exhibit, depending on strain and environmental conditions, mild or severe symptoms. Viruses can also exist as a latent (non-visible symptom) infection.

Some of the named plant selections in New Zealand are Laird's Large, Oratia Red, Red Beau, Andy's Sweet Red, Secombes Red, Goldmine (yellow). Other growing regions may have their own named selections – for example the selection Rothamer originated in San Rafael, California.

Training and pruning

Seedlings will produce a single, long, straight stem. The plant may be encouraged to form a branched head by pinching out the main stem at about 1–1.2 m high. Cuttings may produce several shoots, and selection of one of these does not always result in a straight stem. Furthermore, the junction between the older wood of the cutting and the new growth of the new stem may be weak and liable to break under stress.

The head may be supported with wires. The growth habit and large, soft leaves of the tamarillo make it particularly susceptible to wind damage.

Table 19.1. Tamarillos.

Botanical, anatomical and physiological aspects	Climatic, geographic, soil and water requirements	Cultural aspects
Common name Tamarillo, tree tomato	*Temperature needs* Can be grown only in areas with no, or very slight, winter frosts. Will grow in districts with 1100°C days, but better with 1500 and above (see degree days, climatic zones, Chapter 2)	*Propagation* For seedlings, start under glass after midwinter, plant out spring. Use cuttings from virus-free plants with 1–2-year-old wood, 45 cm long, 2 cm thick; set out autumn or spring in nursery. Grafting techniques are successful
Botanical name Solanum betaceum Cyphomandra betacea	*Frost tolerance* Severe plant damage by frost of –2°C	*Rootstocks* Selected seedlings and cuttings. Most plants grown on their own roots
Botanical name of related and useful species Tamarillo is in the *Solanaceae* family (tomato, potato, etc). About 40 species were formerly placed in a genus *Cyphomandra*, including a related fruit plant, the casana	*Water needs* Requires ample water	*Spacing* 1.5 m between plants; 4.5 m between rows
Type of plant and size Brittle softwood shrub or small tree, growing to 3–4 m in height and 1–3 m spread. Has large leaves, brittle branches, shallow roots	*Tolerance of wet soils* Very intolerant to excess soil moisture	*Training and pruning* Plants may need staking and trellis may be used to support the head. Plants encouraged to branch at 1.2 m by heading. Bearing tree pruned to remove dead, diseased and crowded old wood. Prune to produce new growth and flowers distributed throughout plant head
	Drought tolerance Shallow root system, has poor ability to withstand drought	*Thinning* Not normally performed
Sexuality Hermaphrodite	*Humidity tolerance* Fair; disease incidence increased	*Tillage* Usual to grow in weed-free strip. Green manure or cover-crops of grass and clover can be grown between rows
Pollination Self-pollinating but also insect-assisted cross-pollination. Tamarillo is self-compatible	*Wind tolerance* Very poor, weakly rooted and anchored; large leaves and brittle branches may break off	*Time to first harvest* 18 months after setting out
Flower buds Borne on current season's growth	*Edaphic features* Use warm, sun-facing slopes with good wind shelter and cold air drainage, especially in marginal areas	*Time to full production* 3 years after setting out
Growth of fruit Single sigmoid curve	*Soil needs* Light, well-drained soil required	*Expected yields* 18 months: 6.5 t/ha 3–4 years: 15–17 t/ha
Time of bud burst No bud burst as such. Phases of increased growth may be induced by pruning as well as by seasonal conditions		*Normal productive life* Often limited by virus build up – symptoms show as undesirable flecks on fruit surface. Normal life expectancy is 7–10 years

Continued

Table 19.1. Continued.

Botanical, anatomical and physiological aspects	Climatic, geographic, soil and water requirements	Cultural aspects
Time of flowering Flowering may be influenced by pruning and season. Main period November/April	*Nutrient requirements* General recommendations: split applications winter, spring, late summer of total NPK/annum: N = 110–170 kg/ha P = 35–55 kg/ha K = 50–100 kg/ha	*Method of harvest* Pick by colour assessment: fruit must be uniformly at correct colour, e.g. red right up to the calyx. When fully mature, calyx goes light green or even a yellowy colour
Time of maturity Commercially mature 21–24 weeks after flowering		*Storage* Fruit will keep up to 12 weeks at 3.5–4.5°C with 7-day shelf life afterwards
		Main pathogens Looper caterpillar, white fly, nematode, green vegetable bug, bacterial blast, powdery mildew, leaf spot (*Colletotrichum acutatum*), root rot, various viruses including cucumber mosaic, tamarillo mosaic, arabis mosaic

M. Morley-Bunker

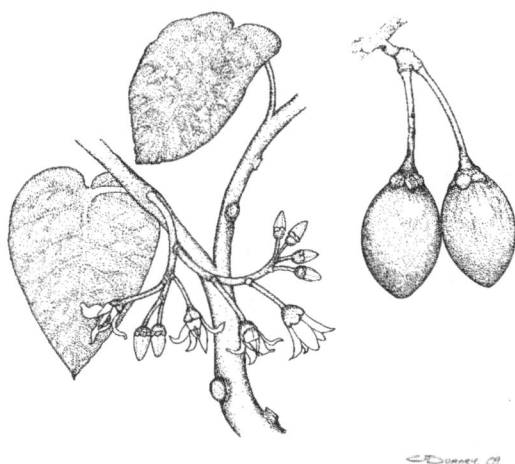

Fig. 19.1. Flower and fruit of tamarillo: they are formed in the axils of shoots which grow in the current season.

The plant is easily blown over because of its fibrous, shallow rooting. The wood breaks even in quite light winds, as the stems are brittle and the large leaves tear. For these reasons shelter, staking and trellis work may all be needed.

The flowers and fruit are produced on new extension growth. Usually 3 or 4 leaves are produced and then a terminal compound inflorescence with up to as many as 50 flowers. Lateral growth behind the inflorescence can produce further extension growth. If left unpruned, the plant head bears fruit further and further from the tree centre. Branches can be pruned to manage and select the lateral growth for further flowering. Dead, diseased and crowded wood is easy to cut and remove. While mechanical pruning and hedging is feasible, hand pruning allows the grower to be selective and shape the head to produce a better distribution of young and structural growth and an appropriate canopy density. With careful attention to pruning and disease control, the plant can be kept in production for 8 years. Pruning may be performed between early spring and midsummer, after harvesting. Hard pruning and late pruning delays flowering and therefore fruit maturity.

Maturity and postharvest treatment

Tamarillos reach horticultural (commercial) maturity approximately 21–24 weeks after anthesis, depending on cultivar and production area. Spring and summer flowering leads to ripening in a period stretching from autumn through to spring. Regular, successional picking is required. Fruit is picked on the basis of size and coloration typical for the selection. The fruit can be pulled or cut from the tree, with the fruit stalk left on the fruit. The fruits are then graded for size and evenness of shape, colour and lack of blemish. Ethylene treatments can advance fruit ripening (mostly softening) after harvest but ethylene treatment does not improve eating quality. Fruits can be held for up to 10 weeks in temperatures of 3–4°C and with a humidity of between 90 and 95%. There is the potential for fruit to succumb to postharvest rots, and treatments including a hot water dip at 50°C for 8 min can minimize infection.

Passionfruit

The *Passiflora* L. (*Passifloraceae*) genus consists of about 500 species widely distributed throughout tropical and subtropical regions, commonly known as the passion flower (Table 19.2). They are mostly climbing plants (sometimes called lianas), usually with showy flowers and conspicuous fruit.

In many regions of the world various passionfruit species have, following introduction, escaped from cultivation and become troublesome endemic weeds. For example, *Passiflora mollissima* (Kunth) L. Bailey (also known as *Passiflora tarminiana* and *Passiflora tripartita* var. *mollissima*) is known as banana poka in Hawaii and regarded as a serious weed, while in New Zealand the same plant species is listed as a 'national surveillance' pest species.

Passiflora edulis forma *edulis*, commonly called the purple passionfruit, originated in the tropical highlands of South America. It has been introduced to many subtropical and tropical regions and is often grown commercially on a small scale, especially in developing countries. A yellow form, *Passiflora edulis* forma *flavicarpa*, is also popular, and this tends to be larger and more globular, while the purple passionfruit is ovoid. Hybrids between the two forms are grown commercially in Australia. New hybridization research between the many *Passiflora* species offers opportunities for new and novel plants – both fruit crop and ornamental plants. Production occurs in South America, including Brazil, Ecuador and Colombia. Africa, Kenya, Nigeria, Uganda and South Africa also have commercial plantings. Other world areas with passionfruit production include India, Indonesia, Sri Lanka, Vietnam, Taiwan, Hawaii, Australia, New Zealand and some island states in the Pacific, such as Fiji and Niue.

Table 19.2. Passionfruit.

Botanical, anatomical and physiological aspects	Climatic, geographic, soil and water requirements	Cultural aspects
Common name Passionfruit, granadilla	*Temperature needs* Natural environment is in tropical highlands; therefore it needs cool winters and frost-free warm summers. In areas with hot summers plants have shorter bearing life	*Propagation* Seedlings sown inside in spring, planted out when 20 cm high in midsummer. Cuttings – use tip cuttings of virus-free, matured laterals after growth flush. Use mist with bottom heat in well-drained compost
Botanical name *Passiflora edulis* – forma *edulis* – purple passionfruit; forma *flavicarpa* – yellow passionfruit	*Frost tolerance* Very frost sensitive. Plant damage severe at −2°C	*Rootstocks* Various rootstocks used to increase tolerance to root rot and cankers. In warm/hot regions, yellow passionfruit is used
Botanical name of related and useful species *Passiflora quadrangularis*, giant granadilla	*Water needs* Well-distributed rainfall, 750–1250 mm	*Spacing* 3–6 m between plants; 2–5 m between trellis rows, depending on machinery, management and soil fertility
Passiflora mollissima, banana granadilla *Passiflora ligularis*, sweet granadilla *Passiflora caerulea*, blue passion flower	*Tolerance to wet soils* Poor tolerance, with increasing susceptibility to disease	*Training and pruning* Trained on to two-wire fence trellis or variations thereof (Fig. 19.4). One or more leaders may be taken up to wires running along the row. A curtain of fruiting laterals is produced from the wire-supported leaders. When the curtain becomes too long, dense and tangled, laterals are shortened to 20 cm long and new laterals encouraged
Passiflora incanata, maypops	*Drought tolerance* Not high due to shallow fibrous roots. Poor flower set and fruit drop may result; some defoliation, also occurs	*Thinning* Not normally practised
Type of plant and size Vigorous, semi-woody perennial climber, up to 15 m long. Stems have tendrils; leaves are ovate or three-lobed, approx. 10 × 18 cm. Flowers are showy (5 cm across), solitary. Fruit is a many-seeded berry. Seed is surrounded by juicy, pulpy aril	*Humidity tolerance* Fair, causes increased susceptibility to leaf and fruit diseases	*Tillage* Plant rows are kept weed free by shallow cultivation or herbicide. Grass strips or green manures may be used between rows unless drought conditions are common
Sexuality Hermaphrodite	*Wind tolerance* Poor. Shelter essential in windy climates to prevent vine damage through breakage and tangling. Also fruit bruising and marking may occur	*Time to first harvest* 15–18 months after planting out *Time to full production* 24 months after planting

M. Morley-Bunker

Pollination
Fruit set is not usually a problem. Self- and cross-incompatibility has been reported for some species. Bees and wasps can act as pollination agents. Hand pollination is sometimes used

Flower buds
Flowers are produced on current season's growth. Flowers are axillary. No information is available regarding initiation

Growth of fruit
Fruit has respiration climacteric and will release ethylene

Time of bud burst
In non-tropical areas a spring growth flush can be expected and one or two later flushes may occur

Time of flowering
Prolonged flowering periods for most of the year

Time of maturity
Approx. 8–12 weeks from fruit set. Main season autumn, also spring, but there may be fruit at other times

Edaphic features
Warm, sun-facing aspect, sheltered from strong and cold winds, with good air drainage needed in cooler districts

Soil needs
Wide range suitable, except heavy clay. Good drainage essential, sandy loams very suitable

Nutrient requirements
Apply according to soil fertility. Split application in spring and summer recommended. Annual total per hectare approximately 50–200 kg N, 49 kg P, 29 kg K

Expected yields
15–18 months: 1.5–2.5 t/ha
24 months: 3–7 t/ha. Nevertheless higher yields up to 25 t/ha are possible in some areas

Normal productive life
3 years in tropical climates, up to 8 years in subtropical climates

Method of harvest
Common to allow fruit to fall and collect from ground every 1–3 days. Fully coloured fruit may be picked from plant. Remove perianth (flower remains) if present

Storage
No recommendation available. Will store reasonably well at room temperature

Main pathogens
Grease spot, brown spot, *Sclerotinia*, septoria blotch, bitter rot, *Cladosporium*, *Fusarium* and other root rots, woodiness virus

The majority of the world crop is processed into pulp, juice and juice concentrates. Processed fruit concentrate is used for producing generic drinks and for blending with other fruit flavours. Concentrate juice and pulp may be used for flavouring confectionery and dairy products like ice cream. The fruit is rich in carotene, vitamins C and A, and potassium. The fruit rind can be processed into animal feed. There is some interest in medicinal properties that have been identified in *Passiflora* species.

Fig. 19.2. Young shoot of passionfruit: note the position of tendrils in the leaf axils.

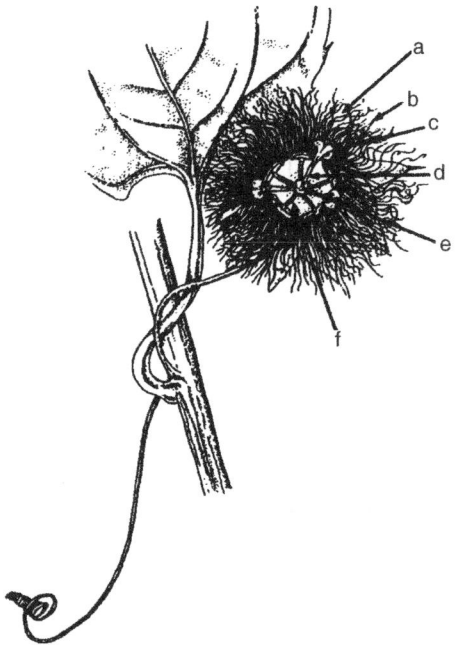

Fig. 19.3. Passionfruit flower. The flower of passionfruit is showy and formed laterally on growing shoots during the summer period. Sepals (a) are white; petals (b) are modified to fine filaments and purple at the base. Stigma (c), style (d) and ovary (e) are green and tripartite, while there are five anthers and filaments (f).

Key points

Disease problems

The passionfruit crop is susceptible to a wide range of plant diseases. In many areas disease problems limit or prohibit the production of passionfruit. Consequently disease recognition and management, using all the available options such as the selection of appropriate plant material, use of appropriate cultural techniques, the adoption of hygiene practice and use of therapeutic sprays should be considered.

The roots may be attacked by various soil fungi, such as *Sclerotinia* and *Phytophthora* species, as well as nematodes, but dramatically the soil-borne disease fusarium wilt of passionfruit *Fusarium oxysporum* f. sp. *passiflorae* will at first manifest itself through a slight wilt of the branch tips, followed by sudden death of the plant within 4–14 days. In New Zealand a condition that causes collar rots and root death is called passionfruit crown canker and probably involves a fungal complex including one or more *Fusarium* species. All soil-borne diseases pose problems regarding control. As root diseases build up in the soil, growers may need to move production to a 'clean' area. Tolerant and resistant rootstocks have been sought, and the yellow passionfruit has been used as a rootstock for the purple passionfruit in South Africa. Unfortunately *P. edulis* forma *flavicarpa* is unsuitable for the cooler subtropical climates. In Australia hybrid passionfruit varieties have a varying degree of resistance to Fusarium wilt. Some reports suggest *Passiflora caerulea* and *Passiflora incarnata* may be suitable as rootstocks or for hybridizing with yellow passionfruit to produce rootstocks; work on this approach has been reported in Australia. Rootstock compatibility would need to be determined; for example, a New Zealand report states *P. caerulea* and purple passionfruit are not compatible.

M. Morley-Bunker

The leaves and fruit may be infected by bacterial grease spot (*Phytomonas passiflorae*), bacterial blast (*Pseudomonas syringae*) brown spot (*Alternaria* spp.) and Septoria blotch (*Septoria passiflorae*). A consequence of these diseases is reduced effective leaf cover and shoot death, which reduces yield, whilst infected fruits become unsaleable. Fruit with thick, hard, distorted woody rinds, symptoms that are often also accompanied by scabs and cracks, are typical of fruit affected by woodiness virus. Passionfruit woodiness virus (PWV) may be a viral complex rather than a single virus. Isolates from infected plants have been examined and relationships have been shown with cowpea aphid-borne mosaic virus (CABMV) and cucumber mosaic cucumovirus (CMV). The vector of PWV is thought to be aphids.

A regular spray schedule is required for many of the foliage- and fruit-infecting diseases. However, some of the effective chemicals, for example copper-containing compounds, reduce vine vigour and crop production.

Trellises: training and pruning

The plant produces flowers on the current extension growth, or 'laterals'. The laterals bear tendrils, and consequently tend to tangle together. The usual approach when growing passionfruit is to let the fruiting laterals hang downwards in a 'curtain' from a trellis.

A wide range of trellis types have been employed to carry a fruit-bearing curtain. The simplest approach is a single wire held at a height of 2 m above the ground. The plant is trained up to the wire, with laterals hanging from the wire (Fig. 19.4). Variations of trellis type include a two-wire assembly, with either two wires held apart by a short T-bar or a second wire lower down, at a height of 1 m. Leaders are trained on these additional wires.

Eventually, the fruit laterals will hang too low and become too thick and tangled. The lateral growths should be pruned back to renew the fruit curtain. Diseased, weak and unfruitful laterals may also be pruned for renewal purposes. The lateral should be cut back to leave about 20 cm of lateral growth (about two buds). Old leaders growing on the wires can be renewed by cutting back to a vigorous new shoot chosen as a leader replacement. Some growers try to minimize lateral tangling by removing lateral tendrils. However, this is a con-

tinuous and time-consuming activity. Lateral pruning is usually done just before the spring growth flush commences. Some of the hanging winter crop fruit may have to be sacrificed in order to prune at this time.

Recent Australian work has looked at new crop production systems, which included different trellis options. One trellising option was the use of A frames with a peak at 3.7 m, and a second option was an overhead canopy akin to the pergola structures used with kiwifruit. While both systems performed well, disease pressure may be difficult to resist with less ventilation for the vines.

Avocados

The avocado, *Persea americana*, is related to cinnamon, sweet bay (*Laurus nobilis*) and sassafras. The avocado has undergone evolution into three distinct ecotypes or races, named Mexican, Guatamalan and West Indian (Table 19.3). The geographic naming signifies the ecological requirements rather than origin. The West Indian race is thought to have developed in the tropical lowlands of Central America. Many authorities identify the ecotypes as botanical varieties, which are named: *P. americana* var. *americana* (West Indian variety – abbreviated as WI), var. *guatemalensis* (Guatemalan variety – G) and var. *drymifolia* (Mexican variety – M).

Consumption of the avocado is thought to have occurred as far back as 8000–7000 BC based on archaeological sites in Coxcatlán (in the region of Tehuacán, Puebla State, Mexico). Human selection and crop domestication may have begun between 4000 and 2800 BC. The early Spanish explorers recorded cultivation in several different regions – Peru as well as Mexico. Evidence suggests domestication occurred in three separate locations, leading to the development of the three ecotypes recognized today. The avocado has spread throughout the subtropics and tropics beyond Central and South America as both a commercial and a subsistence crop. The main producing countries now include Mexico, Indonesia, the USA, Columbia, Chile and Brazil. Countries known for exporting avocados include Israel and South Africa. World production is approximately 3.3 Mt (Table 1.1).

The fruit exhibits a somewhat unusual chemical composition: it is rich in oil and vitamin B, and is low in sugar. It remains an important item of diet in Central America.

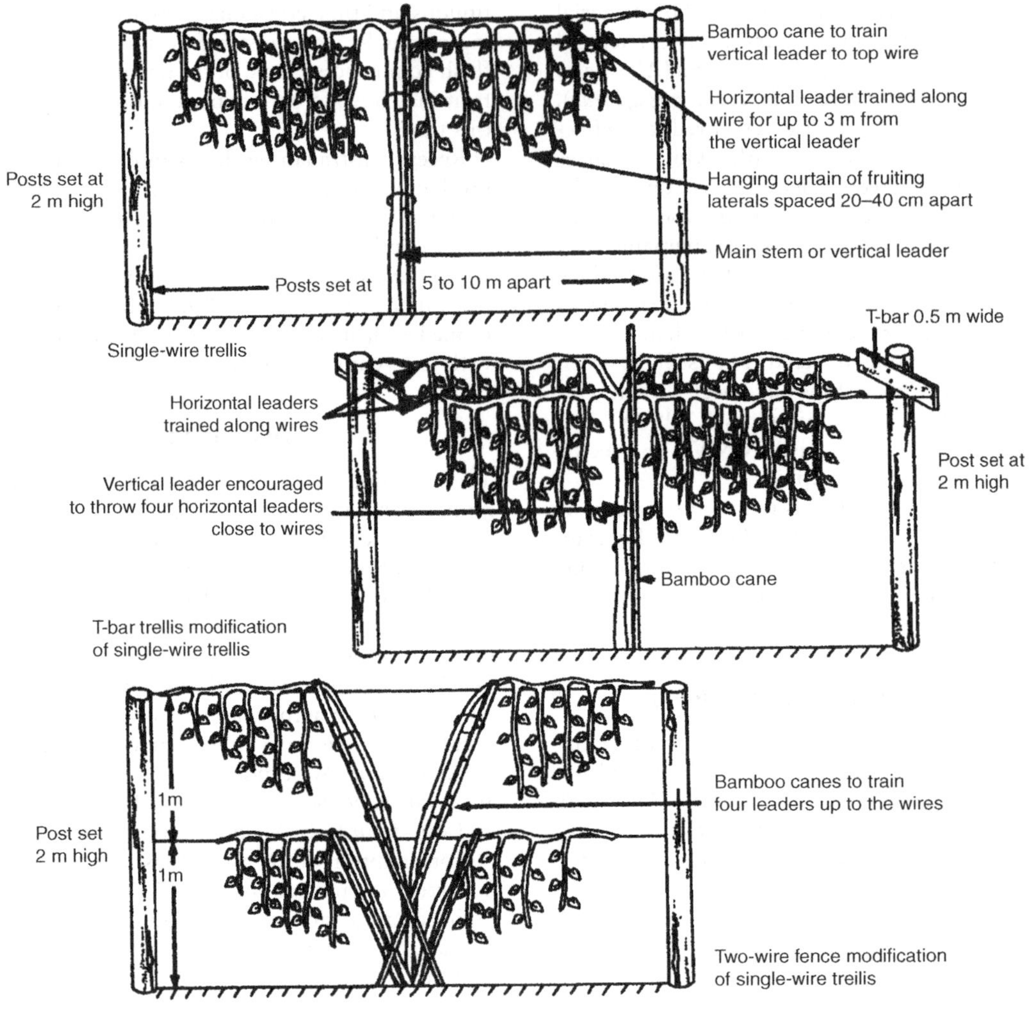
Bamboo cane to train
vertical leader to top wire
Horizontal leader trained along
wire for up to 3 m from
the vertical leader
Posts set at
2 m high
Hanging curtain of fruiting
laterals spaced 20–40 cm apart
Main stem or vertical leader
Posts set at
5 to 10 m apart
T-bar 0.5 m wide
Single-wire trellis
Horizontal leaders
trained along wires
Post set at
2 m high
Vertical leader encouraged
to throw four horizontal leaders
close to wires
Bamboo cane
T-bar trellis modification
of single-wire trellis
1m
1m
Bamboo canes to train
four leaders up to the wires
Post set
2 m high
Two-wire fence modification
of single-wire treilis

Fig. 19.4. Three passionfruit trellises.

Key points

Crop cultivars and climate

The range of avocado plant material is normally described on, what is presumed to be, climatic adaptation and origin. Hence the three ecotypes or races of avocado are labelled Mexican, Guatamalan and West Indian.

The Mexican race originates from the Mexican highlands and is most tolerant of cold. Unfortunately, the fruit from trees with a pure Mexican race background are the least desirable, typically being small and soft skinned with large seeds. There are cross-bred cultivars that are of value. Fuerte and Zutano are Mexican × Guatamalan hybrids. They have medium-sized fruit with the smooth skin associated with Mexican race. Fuerte also shows some hardiness to frost. Other notable cross-bred cultivars include Hayes, Hopkins and Hass. These three cultivars are largely Guatamalan in character, for example the cultivar Hass is a G × (G×M) backcross.

The Guatamalan race originated in highland areas but is not as tolerant of cold as the Mexican race. Fruits most typical of the Guatamalan heritage are large, thick skinned with a rough pebbled texture. The skin colour may change from green to black as the fruit matures. The seeds are small and tightly held in the fruit.

Table 19.3. Avocados.

Botanical, anatomical and physiological aspects	Climatic, geographic, soil and water requirements	Cultural aspects
Common name Avocado, alligator pear *Botanical name* *Persea americana* Three races within the species are Mexican, Guatamalan and West Indian *Botanical name of related and useful species* Family *Lauraceae*, which includes: *Laurus nobilis* – sweet bay, *Cinnamomum zeylanicum* – cinnamon *Type of plant and size* Large evergreen tree 8–20 m high, shallow rooted and young shoots succulent. Branches of fair structural strength. Leaf size 10–30 × 5–15 cm. There are many inflorescences (panicles), each with hundreds of flowers, crowded at the ends of the shoots. Individual flowers are small. Fruit is a large, single-seeded, fleshy berry; skin varies in thickness, texture and colour; flesh is yellowish, with a butter-like consistency when ripe *Sexuality* Hermaphrodite, but have separate male and female opening periods (see Key points) *Pollination* Bees principle agents. Most cultivars are self-fruitful but cross-pollination is beneficial	*Temperature needs* Depends on genetic background: the Mexican race is most cold tolerant; the West Indian race is most suited to hot tropics and least cold tolerant; Guatamalan intermediate. Young growth, flower initials and dormant buds may be harmed by temperatures just above 0°C. Therefore, frost-free cool/warm winters and warm/hot summers required. Average minimum summer temperature not below 14°C *Frost tolerance* Very susceptible to frost. Fruit on tree during winter may be harmed by frosts *Water needs* Ample moisture preferred, providing there are no soil drainage problems *Tolerance to wet soils* Poor. Excessive water may increase likelihood of root death *Drought tolerance* Fair. Fruit, flower and particularly leaf fall increased by high water deficit *Humidity tolerance* Fair. Blossom may be affected by fungal and bacterial infections *Wind tolerance* Poor. In marginal climates, warm, sheltered micro-climates are required. Wind breaks branches and encourages flower and fruit fall, as avocado growth is brittle	*Propagation* Usual to graft on seedling rootstocks. Use semi-ripe scion material with two to three buds with whip-and-tongue or cleft graft. Care must be taken to stop drying out. Topworking of mature trees is feasible *Rootstocks* May be chosen for tolerance to cold weather and root rots. Selected seed is raised in a sandy, free-draining medium, in protected seedbeds; scions are grafted on when stocks are 30–45 cm tall. *Phytophthora*-tolerant stocks should be used if available (see Key points) *Spacing* Avocados are large trees. Double spacing with later tree removal is feasible. Mature trees should be at 8–12 m spacing *Training and pruning* Pruning is minimal. Removal of dead, diseased, overhanging and excessively crowded branches may be beneficial *Thinning* Not normally practised *Tillage* Herbicide strips with grass between rows is common. Mulching young trees and cover crops in the rows may be practised *Time to first harvest* 3–6 years *Time to full production* 12–18 years *Expected yield (Hass)* 3–6 years: 1–6 t/ha 12–18 years: 5–20 t/ha *Normal productive life* 25–35 years

Continued

Table 19.3. Continued.

Botanical, anatomical and physiological aspects	Climatic, geographic, soil and water requirements	Cultural aspects
Flower buds Avocado exhibits biennial bearing. Floral initiation occurs just a short period before full bloom. Uniform cool temperatures (20°C or less) without higher peaks promote flower formation. Short day length may accelerate flower development but reduces the total flower number	*Edaphic features* Warm, sheltered, sun-facing aspect, with good air and soil drainage needed in marginal climates *Soil needs* Deep, well-drained soils essential. Root death in poorly drained and poorly aerated soils may be severe. Light- or medium-textured soils of at least 1 m depth recommended	*Method of harvest* Clip from tree with stem (0.5–1 cm) attached to fruit. Most growers rely on experience of specific cultivars to gauge harvesting time *Storage* Fruit may be held at room temperature until ripe. Fruit may be damaged by low temperatures, even though this can slow respiration and ripening. Once ripe, cool store at 4.5°C; however, fruit skin may go brown even at this temperature
Growth of fruit The chemical composition of the fruit is different from most fruit and the development features are also rather different. Cell division occurs up to maturity. Ripening occurs off the tree, with a rapid respiration rate just before flesh softening	*Nutrient requirements* Apply according to foliage and/or soil analysis. Regular rather than annual applications should be given. Excessive nitrogen will stimulate vegetative rather than reproductive growth	*Main pathogens* Root rots are a major cause of ill health and death. These include *Phytophthora*, *Armillaria* and *Verticillium*. Sunblotch virus has fruit spot and streak symptoms. Pests include leafroller caterpillar, greedy scale, thrips, mealybugs. A lygus bug may cause bud drop
Time of bud burst Not applicable, growth flushes may occur in spring and late summer		
Time of flowering Varies with cultivar. Cool climate delays flowering. Most avocados flower in late spring		
Time of maturity Maturity time varies with cultivar and covers almost the whole year; variable maturity occurs on the same tree		

M. Morley-Bunker

The West Indian race is native to the lowlands of Central America, despite the name. Plants with this genetic background are suited to the low-altitude, hot and humid conditions that might be expected of the West Indies. Cultivars from this race have little or no cold tolerance. Fruits from West Indian cultivars are large and the skin is smooth, leathery and not as thick as Guatamalan fruits. Pollock, Booth and Simmonds are examples of West Indian cultivars that are grown in Florida.

The avocado tree is sensitive to frost and wind damage. Growers should try to choose warm, sheltered sites that have a low frost risk. This is especially necessary for juvenile and newly planted trees. Trees with matured growth, especially cultivars with Guatamalan and Mexican characteristics, can withstand and recover from slight frosting. The summer average minimum temperature should be in excess of 14°C for growth and fruiting. A critical period is at flowering: wet and cold spring conditions tend to reduce fruit set. Cool conditions can also delay ripening but do not affect fruit quality. However, delayed ripening can adversely affect next-season flowering, and irregular flowering encourages biennial-bearing patterns in trees.

Prior to the commercial dominance of the Hass cultivar, the typical avocado fruit was represented as pear shaped with a dull green, smooth, leathery skin (Table 19.4). The Hass cultivar altered that perception of avocado fruit appearance. Hass has a skin that is black (when the fruit has ripened) and has a pebbled texture. Producers and marketers of avocado have to inform and encourage consumers to appreciate the range of avocado cultivars, allowing sale of different cultivars as the year passes. Some production areas have a wide enough range of cultivars to supply fruit for practically 12 months of the year. Cultivar selection is required for each production area.

Crop propagation and rootstocks

The world avocado industry has been based on seedling rootstocks. Avocado scions have traditionally been grafted on to seedlings that were grown from seed taken from selected mother trees. Clonal rootstocks have been used in a wide range of commercial fruit crops as part of the strategy to standardize and improve tree performance. Avocado production would be improved if these objectives could be realized and also if susceptibility to root pathogenic fungi such as *Phytophthora* spp. could be overcome. Unfortunately clonal propagation is technically difficult with avocado. Techniques, including the use of nurse stocks, have been developed to produce clonal rootstock material. However, setting up the facilities for clonal avocado propagation, the rate of success and the time involved in producing rooted clonal pieces still makes the alternative use of seedling rootstocks attractive.

Rootstocks are regularly being evaluated for avocado production. Interest focuses mostly on rootstock responses to soil pathogens and especially *Phytophthora cinnamomi* root rot. Plant material commonly cited as offering prospects of root-disease tolerance include selections of Duke and Barr Duke, Thomas, Canyon, Martin Grande (a hybrid between *P. americana* and *Persea schiedeana*) and Merensky. 'Thomas', 'D9', 'Duke 7', 'Merensky 1' and 'Merensky 2' rootstocks are reputed to have *Phytophthora* tolerance. Other rootstock properties that are of interest are control of plant vigour, cropping efficiency, plant size (and in particular opportunities for using dwarfing rootstocks), tolerance of soil salinity and cool temperatures. The rootstock Colin is reputed to be more dwarfing than most selections. Duke 7 is reputed to be more cold tolerant than most others.

Fruit set and pollination

Considerable interest has been aroused by the unusual opening and functioning of avocado flowers (Table 19.5). Cultivars are grouped as either A or B flowering patterns. In both groups stigma receptivity precedes the release of pollen. The female/male phase is separated by a single night, when the flower closes. In type A the time interval between the female/male phases is approximately 24 h, beginning on Day 1 in the morning; in type B it is 12 h, beginning on Day 1 in the afternoon.

Cross-pollination is dependent on the coincident pattern of flowering for the two groups: a flower belonging to a cultivar in group B is releasing its pollen on Day 2, while a flower belonging to a group A cultivar is beginning its Day 1 pattern by being functionally female. Cool temperatures (between 16 and 21°C) at flowering appear to diminish the morning and afternoon pattern. As the timing becomes more irregular the potential for flowers to open over both morning and afternoon increases. Cooler temperatures reduce pollen and ovule activity but self fruit set can and does occur in these conditions.

Table 19.4. Summary of characteristics of main varieties of avocados.

Variety	Main season	Shape	Size	Colour when ripe	Skin texture	Skin thickness	Flavour	Eating quality	Tree growth	Flower type (Table 19.5)	Blossom period	Comments
Hass	Summer	Ovate	Medium	Black	Pebbled	Medium–thick	Nutty–rich	Excellent	Rounded	A	Mid	Most important and main export variety
Zutano	Winter	Pear	Large	Light green	Smooth	Very thin	Mild	Fair	Tall, upright	B	Early	Cold tolerant – first of the season
Fuerte	Late winter	Pear	Medium	Green	Leathery	Medium	Mild–rich	Excellent	Spreading	B	Early	Very sensitive to temperature at fruit set. An errant cropper
Hayes	Spring	Ovate	Medium to large	Black	Pebbled	Thick	Nutty–rich	Good	Rounded	A	Late	Fills in between Fuerte and Hass
Hopkins	Autumn	Pear	Large	Green	Smooth–leathery	Very thick	Rich	Good	Broadly rounded	A	Late	Very biennial. Declining in popularity
Reed	Late summer	Round	Large	Green	Slightly rough	Thick	Mild–rich	Good	Angular	A	Late	Regular cropper. Useful interplant

M. Morley-Bunker

Table 19.5. Dichogamous flowering in avocado.

	Day 1				Day 2		
	9 am	noon	3 pm	midnight	9 am	noon	3pm
Flowers of group A cultivar	Female receptive			Flower closed			Pollen release
Flowers of group B cultivar			Female receptive	Flower closed	Pollen release		

Avocado diseases

The biggest problem in avocado growing is the high susceptibility of the root system to phytophthora root rot. This organism is not native to Central America and therefore there has been no evolutionary selection for resistance. When the root system of the avocado is affected, the feeder roots are killed and this severely impacts on tree growth. The observable orchard symptoms include leaf discoloration (turning pale green and yellow), wilting, leaf fall and new growth failing, producing a characteristic 'stagshead' appearance. The tree quickly declines in health and the disease is sometimes called 'quick decline'. The disease has affected avocado crops worldwide and various measures have been suggested to control the disease. It is probable that the best approach is a combination of measures. Soils that are of light to medium texture, with good aeration and drainage, will help reduce the risk of 'wet feet' and root rot infections. Soils and nursery stock selected for orchards should have no history of *P. cinnamomi* incidence and only high-health planting material should be planted out. Rootstock choice can affect susceptibility.

The use of chemicals to protect or treat root systems is problematical for all fruit crops, largely because of the soil volume that needs treatment if the tree roots are to be thoroughly exposed to the treatment compounds. South African and Australian use of trunk injection with the compound Aliette (an alkyl phosphate) has been effective and has now been adopted widely. The timing of phosphonic compound injections should coincide with the beginning of root flushes.

There are other disease problems, such as verticillium wilt and anthracnose. A viroid disease, sunblotch, may cause tree and fruit distortions.

Sunblotch can be transmitted through infected seed, pruning and grafting. Virus-free material should be used for propagation purposes. Various fungi can affect the fruit in store, and stem-end rot, which may be caused by a complex of fungi, may be of concern.

Harvesting, ripening and storage

Fruit ripening to a stage suitable for consumption does not occur on the tree and fruit must be picked and ripened off the tree. Only mature fruit develop good eating quality. Sample testing for the level of maturity is necessary to gauge whether fruit are ready. This may entail size and colour assessment, or forced ripening and testing fruit quality. Testing for either oil content or dry matter content (which can be related to oil content) is used in various countries. In New Zealand fruit must reach at least 20.8% dry matter for harvesting. Harvested fruit are size-sorted, and picking the larger fruit early allows the later fruit to size-up more. Fruits are often described by tray count size – for example 20s, 23s and 25s. Most retail fruits have weights between 100 and 250 g.

Avocado fruit ripening may take from less than 7 days at 27°C to 1 month at 5°C. Regimes using set temperature levels and ethylene exposure have been developed to manage the timing of the supply of ripe fruit. However, lower-range temperatures may not be conducive to good flavour and internal quality.

Avocado storage and shelf life can be managed by using the appropriate environment temperature in the holding areas. The optimal temperature varies for cultivar and accompanying treatments and conditions. Chilling injury can occur with temperatures below 10°C, so investigation to determine the appropriate level is required. Postharvest disease control is important. The use of atmosphere regulation, chemical treatments such as 1-MCP (1-methylcyclopropene) to delay fruit ripening and the application of waxes and other coating materials can also be used to extend the postharvest life of avocado.

Persimmons

The persimmon, *Diospyros kaki*, belongs to the family *Ebenaceae*, which includes hardwood ebony, *Diospyros ebenum*, native to Sri Lanka and Southeast Asia (Table 19.6). There are approximately 200 species in the *Diospyros* genus. Other related plants include *Diospyros virginiana*, American persimmon, and *Diospyros lotus*, poor man's persimmon or date plum. The latter species produce inferior fruit and may be used as rootstocks for *D. kaki*.

The persimmon, which is also called kaki (Japanese), shi zi (Chinese) or oriental persimmon, has a long tradition of use in the Orient. The fruit is particularly appreciated in China, Korea and Japan. The fruit can be eaten fresh or it can be dried. The fruit and parts of the plant have been used for several millennia in traditional Chinese medicine. Japanese craftspeople have used the juice from astringent persimmons, mixed with a natural lacquer, to protect wooden objects. The deciduous tree is also prized for its beauty, especially the autumn leaf colourings.

The persimmon originated in China but has long been grown in Korea and Japan. Much of the variation in plant material used for fruit production is found in Japan, and it is generally held that the modern-day, non-astringent fruiting forms (astringency is discussed later) originated in Japan. The opening up of Japan to the west includes the record of Commodore Perry returning to the USA from Japan in 1885 with persimmon trees. Commercial production began in areas outside of the Orient over the next 100 years, with plantings in areas such as California, Italy, Israel, Australia and New Zealand.

The persimmon grows in areas where summers are warm and winters provide sufficient chilling to overcome a short dormancy requirement. Most consumption of persimmon occurs in Asian countries. Exports from southern hemisphere countries, such as Australia and New Zealand, to the northern hemisphere have provided a continuity of supply for consumers in traditional consumption areas. Whilst people of Chinese, Korean and Japanese ethnic origin are familiar with persimmon, consumption by people from other ethnic backgrounds is slowly increasing. The main producing countries continue to be China, Japan and Korea. Outside the Orient the countries noted with commercial production include Spain, Turkey, Israel, Brazil, Australia and New Zealand. Global production was estimated to be 3.3 Mt in 2007.

Key points

Fruit astringency

Persimmon may have cells containing soluble tannin in the fruit flesh. When the fruit is eaten, the tannin-containing cells rupture, releasing the soluble tannins. The tannin effect in the mouth is to make the mouth feel very dry. An explanation for this effect is that the soluble tannins bind to salivary proteins and mucopolysaccharides. This action prevents saliva from coating and lubricating the oral tissues – hence the dryness sensation. If the tannin in the fruit flesh cells is made insoluble, for example through coagulation, the soluble tannin effect of drying the mouth tissues is avoided. This is the basis for many of the astringency removal treatments used to make harvested 'astringent' fruit non-astringent. The breakdown of tannin cells and the reduction in soluble tannin in the fruit can also occur naturally – for example when the fruits are allowed to soften with exposure to low and freezing winter temperatures.

Fruits that are deemed 'non-astringent' when harvested may have some tannin cells, with soluble tannin present, but the total mass of tannin cells to fruit cells is such that the tannin effect in the mouth may be minimal when the fruit is eaten. Research suggests that when the soluble tannin content (STC) is below 0.3%, the astringency would not be detected and the fruit would be considered to be non-astringent. Astringent fruit can become non-astringent if allowed to soften and become overripe. However, soft fruits are difficult to handle, transport and keep for more than a few days. Ethylene will promote fruit softening. In pollination-variant fruits the flesh condition around any seeds found in the fruit is affected. Pollination-variant fruits with seeds have a dark orange-red–brown flesh surrounding the seed. Seedless fruit have flesh of a different colour (pale flesh) around the place where the seed would have been found.

M. Morley-Bunker

Table 19.6. Persimmons.

Botanical, anatomical, and physiological aspects	Climatic, geographic, soil and water requirements	Cultural aspects
Common name Persimmon Oriental persimmon Kaki Sharon fruit	*Temperature needs* There are two types of persimmons: astringent and non-astringent. The astringent types will probably grow in most fruit-growing areas where grapes are now grown. The non-astringent ones are expected to succeed in somewhat warmer conditions	*Propagation* Budded in late summer on suitable rootstocks. Persimmons do not like root disturbance – great care needed when transplanting. Success has also been reported with softwood or hardwood cuttings in a hot bin
Botanical name *Diospyros kaki*		*Rootstocks* Seedlings of *D. kaki* are used for almost all persimmons. In some colder parts of Japan, *D. lotus* is occasionally selected. Using interstocks of the low-vigour Korean persimmon cultivar Shakokushi to reduce tree size and increase precocity has recently produced successful results in Japan
Botanical name of related and useful species *Diospyros lotus* *Diospyros virginiana*	*Frost tolerance* Young growth in spring is frost tender, and early frosts in autumn may cause damage to fruit	*Spacing* The common spacing in Japan is 6 m × 6 m, although trees may be double planted in the row at 3 m and later thinned out when branches begin to intertwine. Some cultivars may be planted permanently at closer distances
Type of plant and size Deciduous tree 6 m or more in size if unpruned, round and bushy	*Water needs* Trees must not be allowed to dry out during the growing season	*Training and pruning* See under Key points
Sexuality Trees of most commercial cultivars are dioecious and bear only female flowers; some are hermaphrodite and these are often used as pollinators. Persimmon can bear parthenocarpic fruit, in the absence of pollination, but such fruits tend to be smaller	*Tolerance to wet soils* Grow in wet areas in Japan but soil must not become waterlogged	*Thinning* Fruit should be thinned to allow remaining fruit to develop to full size. Biennial bearing is a major problem, and early fruit thinning or even flower thinning by hand are sometimes used in the 'on' year to promote flower initiation for the following season
	Drought tolerance Can survive drought quite well but production will be limited and quality poor	*Time to first harvest* Three years in good conditions
Flower buds Persimmons flower laterally and crop on wood produced in the current season. Buds containing flowering shoots begin development early in the season prior to budburst and flowering	*Humidity tolerance* Good	*Time to full production* Probably 7–10 years
	Wind tolerance In windy areas persimmons need good shelter, since tree growth will be inadequate and fruit rub will seriously detract from its value	*Expected yields* 3 years: 2 t/ha 6 years: 10 t/ha 9 years: 20 t/ha
Growth of fruit Has double-sigmoid growth curve		*Method of harvest and storage* See under Key points

Continued

Table 19.6. Continued.

Botanical, anatomical, and physiological aspects	Climatic, geographic, soil and water requirements	Cultural aspects
Time of flowering Late spring, early summer *Time of maturity* Late summer, autumn	*Edaphic features* Flat land preferred; sun-facing slopes may be used in marginal areas *Soil needs* The tree is best when planted on deep soils of moderate to good fertility *Nutrient requirements* See under Key points	*Main pathogens* The range of pests and diseases will vary according to the district in which they are grown. Some general comments are made under Key points

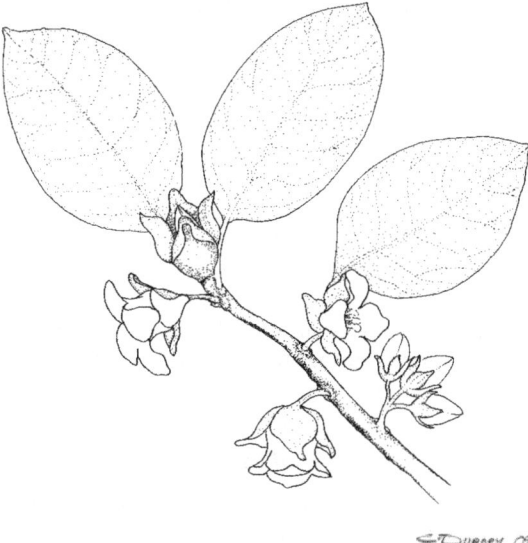

Fig. 19.5. Persimmon – summer shoot, bearing flowers: this shoot is produced from a lateral bud on a branch formed in the previous season.

In fruits with only a few seeds, the dark flesh condition is only found around the seeds that are present.

Persimmon cultivars and astringency groups

It is unfortunately the case that some named cultivars have sometimes been mislabelled, especially when cultivars are introduced to new areas and, for example, have been taken from Asian countries to Europe and America. This has created some problems for producers. New DNA identification procedures can be used to check genetic profiles and the correct naming of cultivars. Additionally, local budwood selections can sometimes produce plants which have different features when bud mutations are selected. New breeding work, especially using micropropagation technology, has opened the possibility of embryo rescue for special crosses and growing plants from chimaeral tissues, with the prospect of expanding the range of persimmon features and persimmon cultivars.

Persimmon cultivars are divided into groups on the basis of the level of fruit astringency present at harvest. The degree of astringency can be related to different patterns in fruit cellular and tissue development and also the effect of the presence of seeds on fruit growth and development. Most descriptions of persimmons list four types of fruits:

1. Pollination-constant, non-astringent (PCNA).
2. Pollination-variant, non-astringent (PVNA).
3. Pollination-variant, astringent (PVA).
4. Pollination-constant, astringent (PCA).

GROUP 1. PCNA – POLLINATION CONSTANT, NON-ASTRINGENT These non-astringent persimmons have the same flesh colour whether they develop with or without seeds. The basis for there being little or no discernible astringency has been explained by the impact of two processes. First, it has been observed that there is relatively little tannin cell development after the first stages of fruit development. The low incidence of tannin cell development means that there is a dilution of tannin content in the fruit flesh at harvest. Second, there is some tannin coagulation late in the growing season, which might be caused by natural oxidation. Temperatures during the growth and development of the fruit seem to influence the extent of astringency in this group. Fruit from locations with high summer temperatures have an almost complete lack of astringency, whereas those from regions with cool summer temperatures may have some noticeable, though residual, astringency.

Some of the notable Japanese-originating cultivars in Group 1 are considered to be:

Early ripening; early autumn

- Izu
- Ichikikei Jiro
- Maekawa Jiro.

These are all weaker trees and may be planted closer in the row than normal. Quality and storage ability are inferior to that of later persimmons.

Mid-season – late ripening; late autumn

- Fuyu – the major world and Japanese cultivar. Fuyu fruit quality is recognized as being good and of acceptable storage potential. Fuyu yields are generally good but the fruit should have adequate pollination provision to ensure good set and growth.
- Matsumoto Wase Fuyu – very similar to Fuyu but seasonally a little earlier. The selection may suit some situations where Fuyu is not suitable.

- Jiro – a major Japanese cultivar but is now being replaced by Maekawa Jiro.
- Suraga – high-quality fruit; 2 weeks later than Fuyu; needs mild climate.

Recent reports have identified PCNA fruit in China and in particular a selection named Luotiantianshi.

GROUP 2. PVNA – POLLINATION VARIANT, NON-ASTRINGENT Many persimmon cultivars are able to produce some seedless (parthenocarpic) fruit. In the PVNA group, seedless fruits are astringent and the flesh is light in colour. However, when there is adequate pollination and fertilization, leading to the production of a full complement of seeds, the fruit is non-astringent. The flesh around the seeds will be a dark reddish-brown but the flesh is relatively firm, even when ripe. In this group, the most promising cultivar is Nishimura Wase, which is an early persimmon, ripening probably about mid-March.

GROUP 3. PVA – POLLINATION VARIANT, ASTRINGENT The cultivar Rojo Brillante, selected by growers in Spain, has become popular with Spanish growers and is sold to both local markets and those within the European Union. Growers have perfected an astringency removal process to control fruit condition and allow commercial handling.

GROUP 4. PCA – POLLINATION CONSTANT, ASTRINGENT These fruits remain astringent or bitter until they soften, at which stage they are difficult to transport. They need to be treated to remove astringency. A notable example of this group is the Japanese cultivar Hiratanenashi. It does not need to be pollinated to produce good crops and is of very fine quality. Some Japanese prefer the texture of astringent to non-astringent cultivars, once astringency has been removed. Hiratanenashi ripens at a similar time to Fuyu. A strain called Compact Hiratanenashi has similar characteristics but the smaller size means higher plant densities are possible. Two earlier-maturing bud sports (mutations) of this cultivar have recently been released and are worth evaluation; they are Tonewase and Sugitawase. Other cultivated PCA cultivars include Hachiya, Atago and Yokono.

Pollination

Apart from a few cultivars, such as Hiratanenashi, most persimmon cultivars benefit from pollination and fertilization, either to reduce astringency, as in the pollination-variant Nashimura Wase, or to reduce fruit drop and increase fruit size, as in most cultivars, but especially Fuyu. It may be necessary to include one pollinator to every nine trees of the cultivar planted. Pollinators recommended are: Zenjimaru, Akagaki, Gailey and Omiyawase.

Although fruit drop is reduced by good pollination, good pruning to allow light penetration, balanced nutrition and cincturing will also help. Cincturing (girdling) entails the removal of a strip of bark 4 mm wide around the trunk of the tree. This is done during the flowering period: late spring to summer.

Removal of astringency

There are many specific protocols that have been developed for persimmon astringency removal. Exposure to carbon dioxide-saturated environments or alcohol-saturated environments form the basis for most of the protocols. Fruit can be treated on the tree or in boxes/containers after harvest. Some management of prevailing temperature and gaseous composition may be incorporated in the methods used. The main methods are as follows:

CARBON DIOXIDE TREATMENT This method is fairly exacting. Different recommendations are found in various Japanese districts, one of which is to have a sealed container with 90–96% CO_2 and 3% O_2 at 22°C. Fruits are placed in these containers for 12 h.

ALCOHOL TREATMENT AFTER HARVEST Fruits are picked before they are completely mature, then graded and packed in cardboard boxes. Then 35% ethanol (ethyl alcohol) is sprayed over the top layer of the fruit (180 ml/15 kg) and a cardboard sheet is placed over this. The box is sealed and the fruit takes 3–4 days to become non-astringent.

ALCOHOL TREATMENT ON THE TREE Individual 'Hiratenenashi' fruit at the stage of early colouring are enclosed in polyethylene bags (0.03 mm thick, 10 × 14.5 cm) containing approximately 1 ml of 40% ethyl alcohol. The bags are left in place for about 3 days. Bags are then cut in half to drain the alcohol and left in position to indicate fruits that have been treated. Treated fruits are left on the tree until mature. The flesh colour with this treatment turns brown, but this is not considered to be a disadvantage.

It should be remembered that treated fruit does not store as well as non-treated fruit. Of the treatments given, CO_2-treated persimmons keep best, on-tree alcohol the next and off-tree alcohol store most poorly. However, alcohol-treated fruits are said to have better quality.

Pruning and training

In Japan, the open-centre or vase tree is most common, although a modified centre-leader tree is also used. It is very important to build up a strong framework of branches in the early years, since the wood tends to be brittle and easily breaks under a heavy load. Securely held fruits are also less likely to be damaged by wind rub. Narrow branch crotch angles should be avoided. Heavy pruning is not necessary but good light penetration should be encouraged, since this assists flower development, fruit set and the growth of good-quality and well-coloured fruit. Problems with limb breakage and wind rub on fruit could be overcome by using permanent support structures such as the Lincoln canopy, the Y or Tatura trellis, or an overhead pergola. Structures that are easy to enclose with bird netting are convenient.

Main pathogens

Persimmon pest and disease problems vary with the region in which the crop is grown. The following generically listed pests could be expected to cause problems: scale insects (various), mealybug, mites, tree borers, fruit flies and fruit-eating bats. Amongst the disease organisms that might be a problem are crown gall (*Agrobacterium tumefaciens*), anthracnose (*Colletotrichum* spp.), leaf-spot-causing organisms such as cercospora leaf spot, white root rot (*Dematophora* spp.), *Cephalosporium* wilt. A number of physiological problems could be important: dehiscence, in which a space or cavity appears beneath the calyx, is encouraged by high nitrogen applications, light crops and over-sized fruit (do not over-thin), and high autumn temperatures. Any fruit marking should be minimized. Marked skin may blacken and make the fruit appearance unattractive. It is important to avoid branch rub and any other potential causes of skin marking, such as thrip damage and spray damage. High humidity, and high nitrogen and potassium may accentuate any skin-marking problems. Trees should be pruned to keep them open to reduce

humidity, and total weed suppression may also help. Birds are a serious problem and many plantings need to be enclosed in bird netting.

Method of harvest

Fruits are harvested when they have attained a deep yellow to red colour but are still firm. A minimum of 15% soluble solids is required at picking. Non-astringent fruits need to weigh 200 g but astringent types are often slightly smaller (160–180 g). Fruits are clipped, often twice (like lemons): once to remove from the tree, the second time to shorten the stem piece to 1–2 mm. Picking is usually done two to three times over a 1-month period for main cultivars.

Storage and processing

Astringent persimmons store 4 months or longer at −1 to 0°C. The non-astringent Fuyu are stored at 5°C but will keep up to 5 months if fruits are individually sealed in polythene bags, 0.06 mm thick, at 0°C. Historically, persimmon fruits were dried and were a significant sugar source in eastern diets. Persimmons can be used to produce fruit leathers, jams and jellies. Persimmon has been used in baked products and as a flavouring for dairy and frozen dairy products. Medical awareness of persimmon vitamin, mineral, antioxidant, fibre and compositional profile could be used to promote persimmon consumption.

Feijoas

Few countries have commercial production areas devoted to growing feijoas (*Acca sellowiana*) (Table 19.7). A related fruit plant, the guava (*Psidium guajava*), is internationally more important. Both are part of the botanical family *Myrtaceae*, which is found largely in the subtropics and tropics, with many species native to Australia.

The feijoa is a native of South America, probably originating in southern Brazil at an altitude of 1000 m above sea level. It is grown on a limited commercial scale in California, Uruguay and New Zealand. There is some interest in feijoa as a minor crop in several countries, including what was the USSR, Japan, China and Australia.

The ripe feijoa fruit, when cut open, has an edible creamy flesh and the locules are filled with a clear, gelatinous pulp. If under-ripe, the jellied

Table 19.7. Feijoas.

Botanical, anatomical, and physiological aspects	Climatic, geographic, soil and water requirements	Cultural aspects
Common name Feijoa Pineapple guava	*Temperature needs* Although often considered a subtropical fruit, feijoa will tolerate winter frosts. Bigger fruit and larger crops are found in warmer temperate and subtropical climates	*Propagation* Seed may be sown into boxes or nursery beds. Cuttings three nodes long of semi-ripe wood taken in autumn may be rooted in mist beds with bottom heat 21°C. Treat with IBA 2200 ppm. Use well-drained media. Layering may also be used
Botanical name *Acca sellowiana* (syn. *Feijoa sellowiana*)	*Frost tolerance* May survive temperature as low as −11°C. Tree is relatively frost tolerant. Late-autumn frost may affect internal fruit quality	*Rootstocks* Seedlings may be used as rootstocks for proven scion clones. Whip graft on to 1- year-stocks. Superior rootstocks have not been selected
Botanical name of related and useful species Belongs to *Myrtaceae* family, which includes *Psidium guajava* (guava), *Psidium cattleianum* (Cattley or strawberry guava) and *Myrciaria cauliflora* (jaboticaba). Family also includes eucalyptus	*Water needs* Irrigation preferable in dry months when fruit is on tree	*Spacing* 2–4 m between plants, depending on whether hedgerow or individual trees wanted. 4–5 m between rows
	Tolerance to wet soils Grows in climates with rainy seasons. There are indications that it tolerates monthly rainfall of at least 125 mm, providing good drainage is present	*Training and pruning* Remove suckers and low branches (especially from grafted plants). Pruning may be practised to restrict tree spread, allow good spray penetration and prevent undue height and shoot length. However, pruning is generally kept to a minimum
Type of plant and size Small tree or shrub; evergreen; 2–5 m tall. Leaves 3–7 cm long; dark green top; white, felted underside. Solitary axillary flowers showy. Fruit is ovoid berry 4–7 cm long, dark green skin, cream flesh, gelatinous pulp	*Drought tolerance* Plant has drought tolerance characters. Some fruit and leaf fall is probable with drought conditions	*Thinning* Not normally done
		Tillage Shallow rooting habit means deep tillage must be avoided; may use grass between rows. Weed-free areas may be maintained with herbicide and/or mulch
Sexuality Hermaphrodite flowers but self-incompatibility in some clones	*Humidity tolerance* Good	*Time to first harvest* 2–4 years after planting
Pollination Insect and bird pollinated		*Time to full production* 10 years

M. Morley-Bunker

Flower buds
Flower buds formed in leaf axils at base of current season's growth. These flower in spring and summer of the following year

Growth of fruit
No information available

Time of bud burst
Evergreen; flush of growth occurs in spring

Time of flowering
Late spring, early summer; later than most fruits

Time of maturity
Matures 20–26 weeks after flowering, i.e. early winter. Tree will contain fruit of varying maturities

Wind tolerance
Wind will cause fruit bruising and marking. Brittle wood may break. Shelter is desirable

Edaphic features
Flat land preferred

Soil needs
Grows on wide soil range provided drainage is satisfactory. For adequate soil moisture retention, and for producing good fruit size, loam soils would be preferable to sandy soils

Nutrient requirements
Split applications in late winter and at flowering. Total nutrient (per ha): young trees: 25–30 kg N; 40 kg P; 20 kg K

mature bearing trees: 120 kg N; 80 kg P; 100 kg K. Reduce these rates if regular soil tests indicate a surplus of any element

Expected yields
2–4 years: 4–5 t/ha
10 years: 20–25 t/ha

Normal productive life
Probably 30–40 years

Method of harvest
Often picked after natural drop but this is not desirable. Best to anticipate drop and pick from tree. Fruit detaches easily when mature. Some skin colour change to a slightly lighter green occurs and fruit softens slightly

Storage
Will store at 0°C for 4–5 weeks

Main pathogens
Chinese wax scale, greedy scale, leafroller caterpillar

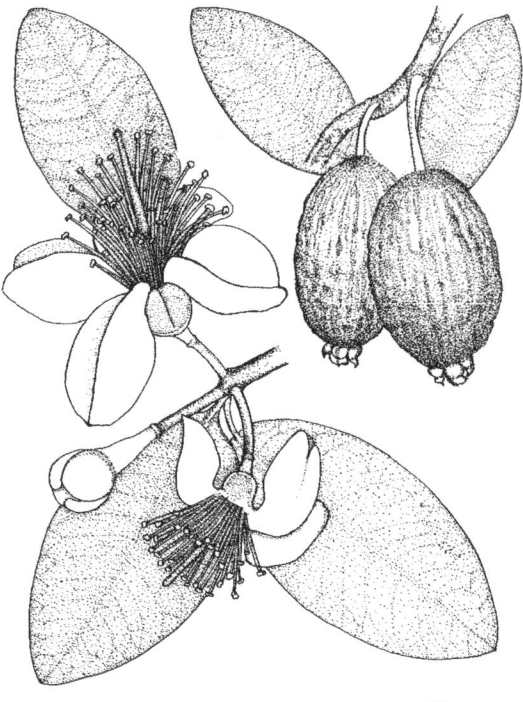

Fig. 19.6. Feijoa, young shoot, fruit and flowers.

section is half-white/half-clear. If overripe, the flesh and jellied section will show signs of browning. The feijoa may be eaten as a dessert fruit, mixed with other fruit in a salad, stewed or canned, or used for jellies and juice. A study of feijoa use in southern Brazil reported that fruit was used for medicinal preparations, candy and confectionery, and making fresh juice and alcoholic beverages. The plant was also used for firewood, tools, poles and handicrafts.

The feijoa has some value as an ornamental plant. The white leaf undersides and shoots contrast with the dark green upper leaf surface and it has attractive red flowers. The plant has sometimes been used as a wind-resistant hedge.

Key points

Selection of plant material and propagation

Seedling trees do not come true to type. To obtain fruit with desirable characteristics requires either grafting seedling trees to superior selections or using rooted cuttings taken from superior trees. Cleft and whip grafting of seedling stocks, once

they have reached pencil thickness, has been suggested. Graft union formation may be assisted if the plants are kept in a warm (25°C) mist/fog environment. The rooting of woody material has been reported as difficult and research has shown softwood cuttings, taken at the appropriate time, might be more successful.

There are named selections with a range of fruiting and tree characters. Unfortunately, at some time in the past, some plants were incorrectly named. As a consequence some older named selections may not have uniform characters typical of the named selection. Growers should ensure that the provenance has been tested and is true to its name.

Cultivar selection has been for locality adaptability, fruit size, fruit shape, fruit internal quality, fruit firmness, bruising resistance, fruit keeping quality, self-compatibility, reliable cropping and seasonal timing. Needless to say, plants selected for the production of fruit for, say, canning, may not produce fruit suited to dessert use.

Named selections have been made in several countries, including Australia, the USA and New Zealand. Coolidge and Choiceana originate from Australia; Edenvale Supreme, Nazemetz and Trask are named cultivars listed in California. Mammoth and Triumph were early selections in New Zealand. Growers and plant scientists in New Zealand have continued selecting new cultivars to complement or replace older cultivars. Apollo and Gemini were selected in the 1980s; later introductions included Unique, Opal Star, Pounamu, Kakapo and Wiki Tu. New releases include Kakariki, Kaiteri and Anatoki. It has been suggested that the existing cultivars come from a narrow genetic base. This would indicate that there is a need for some genetic conservation of indigenous plant material in South America.

Harvesting and storage of feijoas

Fruit may take from 5 to 7 months from flowering to reach maturity and ripeness, depending on seasonal temperatures. Traditionally, feijoas have been collected after falling to the ground. Nevertheless they are best picked from the tree. There is only a slight change in the colour of fruit as it approaches ripeness. The colour change is generally regarded as too subtle as an indicator for pickers to use as a harvest guide. The method presently used to determine maturity is to apply a minimum force to detach the fruit. The fruit should detach if it is mature. Nets to catch fallen ripe fruit have been tried, but they are expensive and make movement in the orchard difficult. Net inspection every 1 or 2 days is required. Fruit ripen over several weeks.

It is normally considered that tree-ripe fruit has the best eating quality, but firm, mature fruit can ripen off the tree. The fruit will soften and there will be some continued internal development. A ripe cut fruit should have a creamy flesh and locules with a clear gelatinous pulp. The ratio of the translucent part of the fruit to the overall cross-sectional area can be used as a measure of fruit maturity. Research into the acoustic properties of feijoa fruit may lead to a practical, non-destructive measure of internal fruit quality.

Mature fruit can be held at temperatures of between 0 and 4°C for 4–7 weeks, with a further week's shelf life at room temperature – the length of storage time is influenced by the cultivar. Controlled atmosphere storage has been shown to prolong storage life by a few weeks.

Fruit set

Fruit set in feijoa can be variable. Factors that have been implicated in poor fruit set include pollen incompatibility and a lack of effective pollinating agents. Some scientists have suggested that birds can act as pollination agents for feijoa. It seems likely that insects may also be involved. Trials with hand pollination (pollen from Gemini, Triumph or Mammoth) of the self-fertile cultivar Apollo have demonstrated improved fruit set, fruit quality and seed number. There is also evidence of self-incompatibility in some cultivars in some situations. The cultivars Apollo and Unique are reputed to be self-fertile. However, cross-pollination provision is a sensible approach.

Figs (*Ficus carica*)

The common cultivated fig originated in western Asia. It is one of the most ancient fruit crops, with evidence of cultivation and use at various Neolithic, late Neolithic and Bronze Age sites in the Mediterranean basin. Most of the world's production still occurs in and around the Mediterranean basin, the major producers being Turkey, Egypt and Iran (Table 1.2). Italy, Portugal, Spain and Greece have historically been important European producers. Figs are also grown in an area stretching eastwards from the Balkans and Turkey into Iran and India. Figs are grown in North Africa and Middle Eastern countries, where the ability to tolerate low rainfall and drought conditions makes the tree a valuable asset. Spanish missionaries were responsible for introducing the common edible fig into California, and figs are now grown in the southern, drier areas of the USA. In the southern hemisphere, Argentina and Australia have limited production.

The fig has a history that includes religious associations. It is cited in the Bible (Genesis 3:7), when the first man and woman, Adam and Eve, cover themselves with fig leaves. The fig is one of the two sacred trees of Islam. Fig trees also have a pivotal presence in Buddhism, Hinduism and Jainism. Siddhartha Gautama, the Supreme Buddha, is traditionally held to have found bodhi (enlightenment) while meditating under a sacred fig (*Ficus religiosa*).

The number of species (about 750) and the range of plant habit in the genus *Ficus* is large (Table 19.8). Whereas the common fig is a deciduous temperate tree, many other species are subtropical and tropical evergreen plants, ranging in size and form from large trees, sometimes with aerial roots, to small trees, shrubs and climbers. *Ficus elastica* (the rubber tree) and the weeping fig (*Ficus benjamina*) are used as houseplants in temperate regions. The creeping fig (*Ficus pumila*) is a vine whose small, hard leaves form a dense carpet of foliage over rocks or garden walls. Despite the *Ficus* genera having a broad range of plants there are several distinguishing botanical features. In particular, *Ficus* species plants have a white to yellowish sap (latex). Tissue wounding normally leads to the exudation of the latex, sometimes a copious exudate.

Many *Ficus* genera plants are gynodioecious (have two sexual forms). The plants may have two flower-bearing structures – one is termed the caprifig and has staminate (male) flowers and short-styled pistillate (female) flowers; the other, the fig, only bears long-styled pistillate flowers. The structure typically recognized as the fig 'fruit' is a specially adapted type of inflorescence (an arrangement of multiple flowers). This structure is botanically termed a syconium. On examination, the structure is found to be an involuted and nearly closed flower receptacle, with many small flowers arranged on the inner surface of the 'fruit'. The true, botanical flowers of the fig are not visible unless the

Table 19.8. Fig.

Botanical, anatomical and physiological aspects	Climatic, geographic, soil and water requirements	Cultural aspects
Common name Fig, common fig, edible fig *Botanical name* *Ficus carica* *Botanical name of related useful species* *Ficus sycomorus* – sycomore fig, *Ficus religiosa* – sacred fig, *Ficus racemosa* – cluster fig *Ficus microcarpa* – Chinese banyan *Ficus elastica* – Indian rubber plant *Ficus benghalensis* – Indian banyan *Ficus benjamina* – weeping fig, Benjamin's fig *Morus* spp. – mulberries *Artocarpus altilis* – breadfruit *Artocarpus heterophyllus* – jackfruit *Type of plant and size* Deciduous or partially deciduous tree. Grows 6–10 m tall. Has smooth grey bark. Leaves are 12–25 cm long and lobed. Weeps milky latex exudate from cut or wounded tissues *Sexuality* The organ thought of as the fig 'fruit' is a specially adapted type of inflorescence. The urn like structure is termed a syconium. The male and female flowers may be found inside the syconium. Various combinations of flower presence and presentation are possible	*Temperature needs* Figs are adaptable, however they are well adapted to a Mediterranean type climate (wet winters, dry summers) with average monthly temperatures of approximately 20–25°C between May and October (northern hemisphere) *Frost tolerance* Tolerant of freezing temperatures (upto –15°C when dormant) but susceptible to frosts once actively growing *Water needs* Will grow satisfactorily in locations with a total yearly rainfall of 500–550 mm *Water tolerance* Does not tolerate excessive rainfall. Poorly adapted to soils with poor drainage conditions *Humidity tolerance* 40–45% humidity for the drying period (Northern Hemisphere - between July and September) *Wind tolerance* Not tolerant – subject to shoot and branch breakage *Edaphic features* Prefers sun exposed sites with wind shelter and low spring frost risk	*Propagation* Fig trees can be raised from seed. Ground or air layering is possible. Figs are most commonly propagated by hardwood cuttings (mature wood 2 to 3 years of age). Micro-propagation of apical tips has been reported. Grafting over stocks of existing trees can be achieved shield and patch budding and with cleft grafts *Rootstocks* Not normally used but may be useful when nematode and other soil problems present *Spacing* 5–6 m between row spacing should be sufficient. Densities within row depend on pruning and training regimes and can be very variable ranging from 0.5–4 metres *Training and pruning* Prune to maintain a balance between new and old wood. Selectively prune to encourage strong new shoot growth and remove dead, diseased, damaged and low vigour shoots. Prune to produce trees with low branch density and for good light penetration of the canopy. Cut back long branches to the desired length. Growers may follow: (i) open centre tree (vase) training systems; (ii) regrowth systems where trees are pruned to ground level and shoots regrown each growing season; or (iii) espalier training against a wall. Pruning may be practised in late summer and/or in early spring. Care should be taken to protect late summer pruned trees from winter damage. Spring pruning can encourage vigorous spring growth and larger fruit. Root pruning is practised in some situations where vegetative vigour is promoted at the expense of flowering and fruiting. Girdling has also been used to influence the balance between flowering and vegetative growth *Thinning* Not normally practised

M. Morley-Bunker

Pollination
See text. Wasps (in particular fig wasp *Blastophaga psenes*), may enter the synconium to pollinate the flowers and lay their eggs. Smryna figs in particular require wasp vistation. Not all fig flowers need pollination for the 'fruit' structure to grow. Unpollinated fruits are 'parthenocarpic' fruits

Flower buds
See text. Usually two sets of flower buds – breba flower buds overwinter and become apparent in early spring. Main crop flowers are produced in the leaf axils of current season shoot growth.

Growth of fruit
Double sigmoid growth curve. Fruit is ethylene responsive in final maturation stage. Fruit has a respiratory climacteric as ripening and fruit softening approaches.

Time of bud burst
Growth resumes in spring (northern hemisphere – February–March)

Time of flowering
Main crop: May–July (northern hemisphere). Breba crop (see text) in March–April (northern hemisphere)

Time of fruit maturity
Main crop figs ripen from August to October (Northern Hemisphere). Ethephon may hasten ripening. First crop or breba figs ripening occurs earlier (see text)

Soil needs
Should be free draining. Rooting can be extensive – and promotive of vegetative vigour. Plants can be grown in large containers where some root restriction will occur. Prefers soils that have a pH between 6–7. Will tolerate some alkalinity

Nutrient requirements
Trees are reputed to not need fertilization every year. Fruit growth may benefit from potassium containing fertilisers. Fertilizers high in nitrogen, will promote green leafy growth which may reduce flower bud development. Nitrogen dressings to maintain shoot growth can be given in split applications - avoid fertilizing late in the growing season and delaying hardening for winter. An annual total dressing of between 25–60 units of nitrogen (N), 20–50 units of phosphate (P_2O_5) and 50–100 units of potash (K_2O) per hectare – depending on climate, soil, irrigation, plant vigour and yields – may be satisfactory

Tillage
Minimal soil disturbance so as not to disturb roots and potentially stimulate suckering. Soil movement on sloping sites should be minimised. Bare soils may assist yields in arid regions. Bare soil also assists mechanical sweeping of fallen fruit. Cover crops could assist in reducing vegetative vigour

Time to first harvest
Some fruit should be produced in the second growing season

Time to full production
Trees may reach full commercial yields in about 5 years

Expected yield
Yields of between 6 and 15 t/ha are achievable

Normal productive life
Orchards should remain productive for 15–20 years, although trees may be long-lived

Method of harvest
Table fruit should be cut or twisted and snapped from the tree

Storage
Fresh fruit have a short storage life. Refrigerate between 0–4°C. Shelf life may be no greater than 8 days. Dried fruit (using solar or hot air technology) can be kept for several months, especially in dry refrigerated conditions

Main pathogens
Root diseases include *Rosellinia necatrix* and *Armillaria mellea*. *Botrytis cinerea* and *Alternaria* can affect both foliage and fruit. Fig trees pests in Portugal include two fly species – Mediterranean fruit fly (*Ceratitis capitata*) and *Lonchaea aristella*. Scale insects can be problematic e.g. fig wax scale, *Ceroplastes rusci*. Root-knot and plant-parasitic nematodes have been shown to affect figs. Fig leaf miner (*Eutromula nemorana*) is an important pest in southern Portugal

fig structure is cut open. There is an opening (ostiole) into the structure and a means of passage into the internal flowers. It is through this opening that tiny wasps, know as fig wasps, may enter to pollinate the flowers and lay their eggs. Not all figs need to be pollinated in this manner for the edible 'fruit' structure to grow. These unpollinated fruits are therefore 'parthenocarpic'.

There are, in practice, three main types of cropping figs: those that will set crops without pollination, those that require pollination (the Smyrna fig) and, finally, the group that will set fruit by either mechanism (the San Pedro fig). The Smyrna group of figs produces figs with wasps as the pollination agent. Calimyrna is California's main Smyrna cultivar. One type of fig that will set seedless parthenocarpic fruit is the Adriatic fig. There are several well-known cultivars of Adriatic fig including: Brown Turkey, Brunswick, Kadota, Mission and White Adriatic. Many countries producing figs have their own named selections. Sometimes the selections may be very localized in distribution, indicating adaptation to local environmental, agronomic and agroeconomic conditions. The parthenocarpic figs are probably the most cultivated type of fig.

The plant is propagated by 20 cm cuttings from wood of 1–3 years old, taken during the dormant season. The plants may be set out at 5–8 m squares, and fruiting may begin after about 3 years.

Key points

Growing conditions and plant phenology

It is commonly held that the fig has a short chilling requirement of up to 300 h below 7°C. It will withstand a little frost when leafless and dormant but is considerably less cold tolerant than most temperate fruit. A frost of –5 to –10°C may kill the plant down to ground level. Some home growers of figs may grow the plants in containers, and when dormant, move the plants under cover and keep the plants dry until dormancy appears to be over.

The ideal environmental conditions for fig production are a warm, semi-arid climate with irrigation. A Mediterranean climate is suitable. High rainfall areas will have problems with fruit splitting. Extreme drought will cause fruit drop. Shelter is advisable as the branches break quite easily. Pruning can assist in producing plants with a strong framework and selecting the more robust shoots. Figs are

not demanding of soil conditions and need sufficient depth and soil drainage. The pH should be between 6.0 and 6.5. The fig has good salt and drought tolerance.

The plant resumes active growth in spring. Water and fertilizers can be given to promote new shoot growth. Irrigation may be necessary to promote and maintain vegetative growth, especially if spring soil moisture is low. Some selective spring pruning with shoot selection can also be undertaken to promote new growth.

The fig can carry two crops in a year. Good fruit growth and development will be promoted by an accompanying period of several dry, warm months. The first crop, called the breba crop, is produced from the flower buds initiated in the preceding late summer. They appear as flowers and young fruit near the end of the overwintering shoots in early spring. Typically this fruit is small and does not grow into succulent, well-flavoured fruit. In cold climates the breba crop is often destroyed by spring frosts. In the southern hemisphere these fruits can ripen in November–January, depending on the site and cultivar selected. If the breba crop has some value to the grower, then he or she may remove the terminal buds at spring bud burst, with the aim of limiting early vegetative growth and promoting fruit growth. The second or main crop is produced from flowers borne in the axils of the current season's shoot growth. This crop ripens in late summer and normally produces the better-quality fruit.

Pruning fig trees

Pruning should be used to maintain a balance between new and old wood and provide an appropriate tree structure suited to harvesting. Pruning should include the cutting back of long branches to the desired length. Controlling branch density to allow light penetration will encourage better fruiting and easier management. Inward-growing branches, water sprouts and crossing branches should be removed. For some growers this is achieved using open-centre tree (vase) training systems. However, care is needed to not overexpose the main frame of the tree and create a situation where sunburn injury can occur. Some growing systems do not retain a winter tree structure and a breba crop but instead regrow the fig from the ground in spring, after pruning it down to ground level as winter approaches.

M. Morley-Bunker

In cold climates the fig may be trained against a sun-facing wall as an espalier, and sucker growth is removed from the tree base. Sucker removal may be needed with other fig cropping systems as well. In cool growing areas the breba crop is usually of poor quality; consequently any removal of the immature overwintering breba crop, by severely pruning the tree from late summer onwards, is not considered a waste of potential crop. Care should be taken to protect plants during winter from extreme frosts killing pruned trees. Some growing systems may prune fig trees in spring. Hard spring pruning encourages vigorous spring growth and favours the production of larger-sized main crop fruit in late summer. Spring pruning aims to prune selectively to encourage strong new shoot growth and to remove dead, diseased, damaged and low-vigour shoots. If a breba crop is present but deemed to be of no value, the shoots with fruit can be removed.

Fruit harvest and fruit drying

In the southern hemisphere the breba crop may ripen between November and January, while the main crop can ripen from March to May. The fig should be allowed to ripen fully before harvesting. However, fruit damage by birds may make protection an absolute necessity as the fruit ripens. Most figs intended for fresh consumption are not in a condition to detach readily from the branch. Therefore the fruit must be cut or twisted and snapped from the tree. Some fig production systems allow the fruit to partially dry on the tree before harvesting and then allow the fruit to fall on the ground, where it can be swept up with mechanical sweepers.

The skin of the fig is thin and tender and begins to change in colour as it ripens. The fresh fruit must be handled very carefully when hand harvesting, preferably with gloves. The gloves also reduce the exposure to any irritating latex. The picked fruit should be placed in single-layer trays. The fruit needs to be dried for 1 or 2 days to extract out any latex before being taken to a fruit handling centre for sorting, packing or processing.

Depending on the cultivar, the fruit skin may change from green to brown and purple and the inside of the fruit should be sweet and soft. The fleshy fruit wall may be whitish, pale yellow or amber, while other cultivars may have inside colours of pink, rose, red or purple. The whole fruit should be juicy and sweet when ripe. It will be gummy with latex if unripe. There may be some seeds if the crop has been pollinated. The seeds may range from large to minute and range in number from 30 to 1600 per fruit.

Examples of the variation in fruit colour are the Mission fig, which, when ripe, is a dark purple-black. Adriatic (White Adriatic) is green to yellowish-green, with a red pulp. Genoa (White Genoa) has a downy skin that is greenish-yellow and an internal flesh that is greenish-white near the skin and mostly amber tinged with red in the centre. Brown Turkey can be variable but should have a slightly bronze-coloured skin and pinkish or light amber flesh.

Fresh figs do not keep well. They should be refrigerated to extend their shelf life. Many figs are high in sugar (22–26% TSS) and will dry in the sun or dehydrators. Sulphuring may be required to prevent any rotting during drying. Dried figs can be stored for 6–8 months.

Olives

The olive, *Olea europaea*, has been a significant tree crop in the Mediterranean region for several millennia, providing ancient civilizations with valuable food and oil. The area in which the olive originated and was domesticated has been subject to debate, but the area between the coast of Syria and Israel through to north and west Iraq is the most likely area of origin. The olive has a long historical association with the people of Israel and Palestine.

Recent reappraisals of the value of different forms of oil in the human diet have once again highlighted the value of the olive and olive oil. Many studies have shown that olive oil can protect against cardiovascular disease, as opposed to the effect produced by many animal-sourced fats. The term 'Mediterranean diet' refers to a diet that relies on olive oil for much of the oil component of the diet, together with traditional Mediterranean food ingredients and culinary practices. This diet is associated with coastal regions of southern Italy, Crete and coastal Greece. Studies of human health and longevity for people in this region have suggested that if the diet and lifestyle of this region were adopted in other countries there would be a reduction in the incidence of several health conditions linked with early mortality, such as cardiovascular failure and type 2 diabetes.

Fig. 19.7. Olive flowering with axillary inflorescences arising from nodes on 1-year-old shoots.

Olives are used as table olives or for oil production. Mediterranean countries that are significant producers of table olives and oil include Italy, Greece, Spain, Portugal, Turkey and Tunisia. However, production has spread to other world areas – North and South America and Oceania. Countries like the USA, Argentina, Mexico and Australia all produce olives to a greater or lesser degree. Olives are grown at latitudes between 30 and 45° north or south, although this generalization does not take account of the influences of altitude, maritime and continental land mass on local climate (Table 19.9).

Key points

Plant material

There are more than 1000 olive cultivars, with over 3000 synonyms. Italy and Spain predominate as the originating home of many olive cultivars. Many of the cultivars now grown in areas other than the Mediterranean are of Mediterranean origin; for example, the four most prominent cultivars in the Californian olive industry are Manzanillo, Mission, Sevillano and Ascolano.

Over 800 cultivars are typically identified with oil production. Prominent among these are Picual, Souri, Cornicaba, Frantoio, Leccino, Koroneiki, Sourani and Chemlali. Amongst those cultivars typically used for table olives are Manzanillo, Sevillano, Calamato, Ascolano, Hojiblanca, Kadesh, Domat, Gemlik, Conservolea, Zitoun, Picholine, Merhavia, Leccino, Frantoio and Barouni. There are over 100 cultivars used just to produce table olives, while another 100 or more olive cultivars may be used as dual-purpose olives – for either oil or for table olives.

Some olives are propagated from softwood and semi-hardwood cuttings. Cultivars vary in their readiness to root from cuttings. The rooting of cuttings is assisted by the synthetic auxin indole-3-butyric acid (IBA), and being placed under mist. Cutting-raised plants are clonal plants. When the desired cultivar is hard to root from a cutting, it may be budded or grafted on to seedlings or plants raised from cuttings. There are some plants that have been selected as suitable for use as rootstocks.

Pruning and training

The traditional image of an olive grove is one where the trees are widely spaced on dry, hilly, rock-strewn terrain. The wide spacing was considered necessary to achieve growth and maintain productivity where both nutrients and water were often limiting. Some mature groves with full-grown trees may also be grazed by livestock where the climate allows for the growth of a ground cover crop. In these circumstances tree numbers will be as low as 100 trees/ha at 10 m square spacing.

Wide spacings are unavoidable when trees are allowed to develop to full size with a large canopy spread. Large, full-grown trees, with three to five trunks and large scaffold branches, are pruned to maintain a canopy composed largely of previous season shoots. The olive flowers on 1-year-old shoots; the most productive flowering shoots are those that are 20–30 cm long. Both the shorter and the vigorous long shoots tend to flower poorly – or not at all.

M. Morley-Bunker

Table 19.9. Olives.

Botanical, anatomical and physiological aspects	Climatic, geographic, soil and water requirements	Cultural aspects
Common name Olive	*Temperature needs* Olives need long warm/hot summers but some winter chilling. Many productive areas have summer month average temperatures of 20–25°C. Floral bud development in winter and early spring appears to require fluctuating day and night temperatures between 1.5 and 15.5°C	*Propagation* Plants raised from seed are not true to type but may be used as rootstocks. Hardwood and softwood cuttings may be taken; softwood cuttings should be rooted under mist and will benefit from rooting hormone treatment. Some cultivars are hard to root. Grafting and budding can be practised and may be used to work over easily rooted cuttings to the required cultivar
Botanical name *Olea europaea*		
Botanical name of related and useful species Wild olive – *Olea africana*	*Frost tolerance* There is some cultivar variation in tolerance to freezing and frost damage. Winter acclimation may also vary the tolerance to freezing conditions. Leaf damage may be expected with exposure to temperatures below –9°C. Damage to green fruit occurs at –2.5°C	*Rootstocks* Clonal rootstocks have been selected in some production areas. Seedlings may also be used as stocks. Some cultivars may be grown on their own roots. Rootstock work suggests some low-vigour and compact-growth plants could be selected and become available if intensive, high-density planting trends are to be followed
Type of plant and size Evergreen tree. May grow to 15 m height and 10 m width. Tree training, plant density and age will influence tree size. Plant develops a long tap root with a shallow fibrous root system. Branching is dense. Leaves, 4–5 cm long, are borne in pairs, with one surface green and the lower surface whitish-green		*Spacing* Wide spacings (appropriate for infertile soils) may be 10 × 10 m. Spacing regimes using densities of 6 m between rows and 4 m between plants are common. Hedgerow plantings of 3.5 m between rows and 1.5 m between plants have been made, but highly intensive plantings have not yet proved especially advantageous
	Water needs Some production areas have rainfall of 200 mm per year but annual rainfall should be at least 500 mm. Some estimates suggest a minimum requirement of 2000 m³/ha/year. Water need is most critical at flowering and fruit set and later for fruit size	*Training and pruning* Systems to complement planting density fall into two main types: open vase shapes for low-density plantings and centre-leader (monoconical) configurations for higher-density plantings. Pruning should ensure good light penetration while removing dead, diseased and interfering growth. Sufficient new growth must be generated annually to ensure fruiting in the following season (see sexuality)
Sexuality Perfect and imperfect flowers on each flower inflorescence. Most of the flowers in the inflorescence are imperfect and functionally male or staminate. Flowering only occurs on previous season's shoot growth. Older wood/shoots do not have inflorescences	*Tolerance to wet soils* Not tolerant	*Thinning* Heavy fruit set will require fruit thinning. Thinning and/or NAA sprays should be used as early as practicable and preferably within 8 weeks of bloom. Heavy fruit set will induce biennial bearing and fruits will be small and may not mature properly on overladen trees.
	Drought tolerance Very tolerant	
Pollination Wind pollinated. Some cultivars set fruit best if grown with other cultivars nearby		*Tillage* Groves may be kept free of grass and weed in low rainfall areas. If ground cover is likely to encourage cool, wet spring soil conditions, it should be removed

Flower buds

Flowering is affected by previous year treatment and conditions, suggesting that floral induction occurs in the previous season about 7–8 weeks after bloom. Winter chilling is required for optimum flower development and emergence.
15–20 flowers are borne in the leaf axils of the previous season's shoot growth

Growth of fruit

The development of one fertilized ovule leads to ovary growth in the first 6 weeks following bloom. Strong ovary growth is necessary for successful fruiting. Fruit abscission occurs in the first 6 weeks where ovaries are not fertilized or are not competitive for assimilates. A double-sigmoid growth curve takes the fruit to final fruit size

Time of bud burst

Evergreen with vegetative growth episodes. One episode occurs about 6–8 weeks before flowering and produces new seasonal growth. Further episodes in the season tend to produce lateral shoot growth

Time of flowering

Exact timing is dependent on climate and cultivar.
Northern hemisphere between May and June.
Southern hemisphere between late October and early December

Time of maturity

Exact timing is dependent on climate, cultivar and intended product use.
Northern hemisphere between October and December.
Southern hemisphere between April and July

Humidity tolerance

Poor tolerance if conditions lead to wet soil conditions

Wind tolerance

Normally tolerant but hot drying winds at flowering and early fruit set will adversely affect crop yields

Edaphic features

Historically olives have frequently been planted on hillsides and sloping land but flat land is suitable

Soil needs

Must be grown on well-drained soils. Soils may vary considerably from low to high fertility, although high-fertility soils are most likely to give best yields. Olives will tolerate saline soils and a pH of up to 8.5

Nutrient requirements

The wide range of soils used makes precise recommendations difficult. Many growers use soil and plant tissue analysis to determine need.
FAO suggests annual need of
200–275 kg N/ha
160–210 kg K/ha
55–75 kg P/ha

Time to first harvest

Assuming minimal pruning, trees can bear two seasons after planting

Time to full production

Trees may approach full production as early as 6 years from planting

Expected yields

Year 2: 1 t/ha
Year 3: 5 t/ha
Year 6: 10 t/ha
Olives are susceptible to biennial bearing and may bear up to 20 t/ha in an 'on' year

Normal productive life

Bearing trees may be very long lived. Economic crops may be obtained from groves of 80–100 years old

Method of harvest

Fruit may be hand picked or machine harvested. Olives intended for table olives can be bruised if handled roughly or shaken on to hard ground

Storage

Fruit is normally processed, and such fruit and fruit products may have a long life. Fruit intended for oil pressing may be kept for up to 45 days at 5°C

Main pathogens

Pests include olive fly (Dacus oleae), olive moth (Prays oleae), black scale (Saissetia oleae) and olive fruit fly (Bactrocera oleae).
Diseases include verticillium wilt and peacock spot (Cycloconium oleaginum)

M. Morley-Bunker

Competitive pressures have caused olive growers to look for efficiencies in production. One approach has been to plant trees more intensively. These systems aim for early development of the tree canopy, early onset of production and rapid rise to full production. This can be achieved with single-trunk, centre-leader trees (sometimes called monoconical trees). The trees are allowed to develop with minimal pruning except for any removal of suckers and upright shoots that would be competitive with the central shoot. Lateral growth is allowed to develop, spiralling outwards from the main trunk. Low branches that impede orchard operations are also removed. The less pruning performed in the early years of establishment the better. The objective should be the production of a leafy pyramidal canopy without any growth check. Once the allotted tree space is occupied, pruning should aim to renew lateral growth and maintain canopy porosity for light interception. To minimize the costs of production there have been investigations into using mechanical pruning and pruning in alternate years.

Plantation density for monoconical trees spaced 6 m between rows and 4 m between trees is 420 trees/ha. Some 'hedge' systems have used higher populations – some with spacings of as little as 1.5 × 3.5 m. It may be that the selection of olive cultivars needs to be focused on finding cultivars that are optimally suited to the adopted pruning, training and harvesting approaches for the intensive system practices being used.

Flowering, pollination and fruit set

The number of flowers produced by an olive tree is considerable; mature trees may have more than 500,000 flowers. Axillary buds, at every node, on 1-year-old wood are capable of bearing an inflorescence, and each inflorescence contains between 15 and 20 small, whitish-green flowers. Only a small proportion of the flowers are required to produce the fruit considered necessary for a full crop. It has been suggested, taking into account that many flowers in the inflorescence are imperfect and therefore not capable of developing a fruit, that only 1.2% of the entire flower population need be involved to ensure a good commercial crop. Indeed fruit set levels of one fruit per inflorescence are not necessarily regarded as low, and fruit thinning treatments may be required to adjust fruit numbers to less than this.

The aim of fruit thinning is to improve fruit size and quality and to ensure biennial-bearing patterns are not induced by over-heavy cropping. Chemical thinning agents have been used commercially; hand thinning is not an attractive option because there is a limited period in which thinning is best performed (within 4–6 weeks after bloom) and because of the high cost. NAA (about 150 ppm applied 12–18 days after bloom) has been used in California for many years. Other compounds that have been used for fruit thinning include urea (6% and 20 days after bloom).

On some occasions fruit set is too low to achieve a good crop. This is most common when cultivars require cross-pollination and when environmental conditions, in particular temperature, adversely affect the growth of the pollen tube required to effect fertilization. Amongst important cultivars known to be partially or completely self-infertile is Manzanillo. One study showed that when this cultivar was interplanted with Sevillano, fruit set was quadrupled.

Treatments to improve fruit set have been investigated, including pollinating with stored pollen, use of chemicals such as benzyladenine at flowering, and removing apical growth at flowering. Natural post-bloom fruit drop can be heavy and occurs within a period of 4–45 days after bloom.

Harvesting and postharvest processes

The olive is either hand or machine harvested. Hand harvesting individual fruit may be considered too costly and other manual means are used instead. Fruit may be removed by raking through the foliage with a wooden rake or striking the foliage, causing the fruits to fall into nets. Nets are also spread around the tree when machine harvesting is used. Machine harvesting typically involves shaking the trunk or main limbs. Access for the machine arms to reach and attach to the tree must be provided.

The shaking energy that must be conducted through the trunk needs to be particularly high to dislodge a small fruit like an olive. The force required to separate an olive from the tree (fruit removal force) can be measured and used as an indicator of olive maturity and when to commence harvest. Ethephon can be used to promote olive fruit loosening (2000 ppm during ripening) but caution is required as excessive leaf abscission may also result. Machine harvesting can result in about 80% of the fruit being removed.

Table olives are harvested when sufficiently ripe to produce the desired product, be it green olives, olives turning colour, naturally black olives or olives that become black following oxidation in an alkaline medium. Machine-harvested fruit to be treated for use as table olives should begin the preparation process within hours of being harvested. Otherwise product quality will be severely compromised. Bruising, which manifests itself as brown spots on the fruit, is one problem that will become apparent on fruit that is not quickly processed. The processing of olives for table use involves the removal of some or all of the natural fruit bitterness. Once this occurs, many table olive products can be generated, ranging from whole fruit (sometimes pitted and stuffed) to olive paste.

The timing of harvesting depends on the intended product. Green table olives are harvested first. Naturally black table olives and olives to be pressed for oil are harvested last. There are several approaches to determining the time of harvest.

1. Using visual signs such as the colour of the olive, including the outside colour of the fruit as well as the colour of the flesh; lenticels become less visible as the fruit matures and ripens.
2. Sample fruit may be assessed for oil content, as both a percentage of fresh or dry weight and composition, i.e. oil content in relation to acidity loss.
3. Fruit removal force may be measured.
4. Fruit can be assessed for the presence of white juice exudate when squeezed and how readily the stone is freed from the flesh when transversely cut and then twisted.

In practice, several indicators are available to determine harvest maturity, but growers of olives, especially growers of table olives, may decide that the best time to harvest is when the best size grade is forecast. This is because the price paid for fruit is mainly determined by fruit size or weight. Lastly, since harvest time occurs in late autumn, as temperatures fall, the time of harvesting may be determined by the threat of frost or freezes.

Since the range of table olives extends from green to naturally black, a skin and pulp colour index (CI) has been developed to describe olives. The scale begins at zero. A zero score defines olives as having 'lime green skin', while seven defines the olives having 'black skin and entirely coloured pulp'. It may take about 6 weeks for an olive fruit to develop from a CI score of one (yellow green) to a score of seven. Green table olives are harvested when they are at a CI of one. Fruits that are to be processed as black olives may be harvested before they have substantially turned in colour. The processing of the fruit may turn the fruit black.

As a general principle the processing of table olives involves a lye treatment (1.3–2.6% sodium hydroxide), although natural black olives need not be treated with lye. The lye removes most of the bitterness component in olives – a glucoside, oleuropein. Green olives have more bitterness than black olives. The lye is allowed to penetrate nearly to the stone, after which the olives are washed (the number of washings and the duration of washings is variable) and then brined. The brine is a solution of 5–6% sodium chloride. While the olives are in brine they undergo fermentation. After fermentation they may be further processed (e.g. have the pit removed and be stuffed) and packed.

There are variations in this processing procedure. For example, the timing of the events can differ with olive cultivar, fruit condition, required end product and general conditions during processing. One processing procedure that is somewhat different to that described above is the production of Greek-style Kalamata black olives. This product involves the use of black olives washed in water or brine to remove some of the bitterness; they are then immersed in wine vinegar before canning in olive oil.

The production of olive oil requires a press and most olive presses are either hydraulic or centrifugal machines. The olives, including the stone, are ground before pressing. The grinding breaks the oil cells. If a hydraulic press is used the paste is spread on nylon mats, layered in the press and pressure is applied. If the temperature at which the pressing takes place is no greater than room temperature, that is the olives are pressed cold, then the oil is termed virgin oil. Under these conditions the flavour and the organoleptic properties of the oil should be unaffected by the extraction method. The pressed fluid is allowed to stand and the oil decanted from the top of the fluid. Cleanliness, such as the use of stainless steel vats, is important if the quality of the oil is to be assured. High-quality oils should have a low acidic content (less than 1%) and be free of taste flaws.

Further Reading

Bellini, E. and Giordani, E. (eds) (2009) IV International Symposium on Persimmon. *Acta Horticulturae* 833.

Collins, R. (ed.) (2003) II International Persimmon Symposium. *Acta Horticulturae* 601.

Leitao, J. and Neves, M.A. (eds) (2008) III International Symposium on Fig. *Acta Horticulturae* 798.

López Corrales, M. and Bernalte García, M.J. (eds) (2003) II International Symposium on Fig. *Acta Horticulturae* 605.

Özkaya, M.T., Lavee, S. and Fergus, L. (eds) (2008) V International Symposium on Olive Growing. *Acta Horticulturae* 791.

Park, Y.M. and Kang, S.M. (eds) (2005) III International Symposium on Persimmon. *Acta Horticulturae* 685.

Vitagliano, C. and Martelli, G.P. (eds) (2002) IV International Symposium on Olive Growing. *Acta Horticulturae* 586.

20 Miscellaneous Fruit Crops

Michael Morley-Bunker

This chapter examines fruits that have been termed 'miscellaneous fruits'. They tend to be less well known and therefore less produced and traded. In Chapter 1 data was presented for fruit crops that are widely traded (Table 1.1). There are many fruit crops not listed in this table. Some authorities use terms like major and minor crops and set numerical ranges for either production or trade to distinguish between major and minor fruit crops. Although 10 Mt might be a rather arbitary value, it can be used to designate the boundary between 'major' and 'minor' fruit crops.

A chapter on miscellaneous fruits can make readers aware of future opportunities for further development. An example of this is the pomengranate. It is going through a rapid increase in production and increase in sale volume. Reasons advanced for the rise in pomengranate production include new technologies in fruit processing and fruit storage and the renewed interest in the health-promoting effects associated with the fruit. Modern chemical analysis of fruit crops, including the pomengranate, is providing increased information about the presence of bioactive phytochemicals in the fruit (and the foods) that we eat.

The term 'miscellaneous' can be taken to mean lesser-known fruit-bearing plants. Table 20.1 presents a few plants from several botanical families and from different places of origin as examples of miscellaneous fruits. By implication, it is suggested that these fruit-bearing plants could be developed to become globally important. It is also possible that other plants could be candidate fruit crop plants. As was pointed out in Chapter 1, our knowledge and appreciation of the world's flora may be limiting the number of fruit crops that could be grown for sustenance, pleasure and health.

Babacos (*Carica pentagonia*)

There are a number of cool-tolerant members of the papaya family, including mountain papaya (*Carica candamarcensis*) and oak-leaved papaya (*Carica quercifolia*). This group of plants has been termed 'highland papayas'. The babaco was considered to be a distinct species (*Carica pentagonia*), but, as the fruit is seedless, it probably originated as an interspecific hybrid, maintained by cultivation in South American countries such as Ecuador.

The plant is a slender-stemmed, herbaceous, tree-like plant, reaching 2 m in height. Flowering occurs in the leaf axils on the newly formed trunk. Fruit is set parthenocarpically and the five-sided fruit may grow to in excess of a kilogram in weight, being about 30 cm in length and 10 cm across. Fruit size is determined by its position on the stem and stem vigour. Smaller-sized fruit can be obtained by growing the plant with two or three main stems. The fruits hang against the trunk and take about 9–10 months to show yellow coloration, as they begin to ripen.

Full fruit size is reached about 2 months prior to ripening. The first fruits set are usually the largest and ripen first. As the plant loses vigour, the stem diameter decreases, together with fruit size. One axillary shoot at the base of the plant may be chosen and used to renew the main trunk. Axillary growth may be stimulated by heading back the main stem above the last-formed fruit shortly before harvesting is complete. Axillary growth should occur at the top as well as the tree base. After harvesting in early summer the main trunk is cut back to the low-borne axillary shoot. There is some loss of leaves during the winter months and the growth of the main trunk is very slow compared with the phase directly after fruit harvesting.

Table 20.1. A short list of lesser-known fruiting plants.

Common name	Scientific name	Family	Habit	Area of origin
Babaco	*Carica pentagonia*	*Caricaceae*	Herbaceous tree	South America
Cape gooseberry	*Physalis peruviana*	*Solanaceae*	Herbaceous perennial	Tropical America
Capulin cherry[a]	*Prunus salicfolia*	*Rosaceae*	Tree	South America
Carob[a]	*Ceratonia siliqua*	*Leguminosae*	Tree	Greece, Turkey
Cattley guava	*Psidium littorale* syn. *P. cattleianum*	*Myrtaceae*	Shrub or low tree	Brazil
Cherimoya	*Annona cherimola*	*Annonaceae*	Shrub or low tree	South America
Elderberry	*Sambucus nigra*	*Caprifoliaceae*	Shrub or low tree	Europe
Jaboticaba[a]	*Myrciaria caulifJora*	*Myrtaceae*	Tree	Brazil
Jujube	*Ziziphus jujuba*	*Rhamnaceae*	Shrub or low tree	South America
Longan[a]	*Euphoria longana*	*Sapindaceae*	Tree	South-east Asia
Loquat	*Eriobotryra japonica*	*Rosaceae* subfamily *Pomoideae*	Tree	China
Lucuma[a]	*Pouteria lucuma*	*Sapotaceae*	Tree	South America
Medlar	*Mespilus germanica*	*Rosaceae* subfamily *Pomoideae*	Tree	Europe
Mulberry (black)	*Morus nigra*	*Moraceae*	Tree	West Asia
Naranjilla	*Solanum quitoense*	*Solanaceae*	Herbaceous perennial	Ecuador
Northern pawpaw	*Asimina triloba*	*Annonaceae*	Shrub or low tree	North America
Pepino	*Solanum muricatum*	*Solanaceae*	Low bush, herbaceous perennial	Peru
Pomegranate	*Punica granatum*	*Punicaceae*	Shrub	Iran
Prickly pear[a]	*Opuntia ficus-indica*	*Cactaceae*	Spiny large shrub	Mexico
Quince	*Cydonia oblonga*	*Rosaceae* subfamily *Pomoideae*	Tree	Central Asia
Star fruit, carambola[a]	*Averrhoa carambola*	*Oxalidaceae*	Tree	South-east Asia
Tamarind[a]	*Tamarindus indica*	*Leguminosae*	Tree	Dry African tropics
Tomatillo	*Physalis ixocarpa*	*Solanaceae*	Low bush, herbaceous perennial	Tropical America
Ugni[a]	*Ugni molinae*	*Myrtaceae*	Low-growing shrub	Chile
White sapote	*Casimiroa edulis*	*Rutaceae*	Tree	Central America

[a]Not described in text.

The stem, fruit stalk and fruit skin produce latex when wounded. This can be unsightly on the fruit. Fruit is picked when showing yellow, by cutting with a short stalk length attached. It may be kept at 6°C and is ready for consumption when fully yellow. High soluble solids levels may enhance eating quality, but it is not certain whether these are promoted by judicious clonal selection or by environmental conditions. High soluble solids levels probably result as a consequence of both factors. The fruit is juicy and slightly acid, with a very mild papaya flavour.

The plant is propagated by rooting cuttings from apical sections of axillary shoots. It is also possible to root stem sections. The plant should be planted in a frost-free situation with good wind protection. Regular water and good moisture-holding soils are desirable, but the plant will not tolerate waterlogging and wet feet. Although the babaco tolerates cooler subtropical/warm temperate climates, warm microclimate conditions will help ensure good fruit development, maturity and ripening. There is a general trend to grow babacos under protection in glasshouses and shade houses in warm temperate areas.

Cape Gooseberries (*Physalis peruviana*)

In the solanaceous family a group of plants are distinguished as having papery 'husks' around the fruit. One plant in this group is the Cape gooseberry, sometimes also called goldenberry, *Physalis peruviana*. The plant originates from the Andean region in South America. Cape gooseberry is frost sensitive and grown as an annual crop in temperate regions and a perennial in frost-free regions. It is related to the tomato and the tomatillo (see later) and it can be grown in similar soils and climate to the tomato. However, the Cape gooseberry requires a longer frost-free season to mature its fruit. The plant is a branching annual or perennial plant. Very fertile conditions will promote vegetative growth at the expense of reproductive growth, which will, in turn, delay fruiting. The plant may be raised from seed or from cuttings. The fruit is contained within a papery husk. Fruit is normally hand harvested when the colour and size indicate the fruit is mature. The fruit progresses from a green to golden-orange colour as it matures. The fruit has a climacteric respiration peak together with an ethylene climacteric. Fruit size is variable, 1–2 cm in diameter. The Cape gooseberry can be eaten fresh and it can be processed for use in pies, jams or made into juice. The acid:sugar balance may need adjustment with extra sugar when the product is processed. Chemical analyses show that the fruit has a significant vitamin A and C content as well as antioxidant potential. The plant has been used as a herbal medicine. The Cape gooseberry or goldenberry has received occasional attention in various countries over the last 100 years – usually as a potential new crop for various projects. Commercial production has been associated at various times with South Africa, Australia and Colombia.

Cattley Guavas, syn. Strawberry Guavas (*Psidium littorale* syn. *Psidium cattleianum*)

There are many guava-like plants. The most usual commercially grown guava is *Psidium guajava*. The most hardy form is the Cattley guava, which may withstand and recover from light frost and grow in warm temperate and cool subtropical climates. The plant is native to Brazil. There are a number of synonyms, including cherry guava, Chinese guava, purple guava and strawberry guava.

The shrub grows rather slowly, up to 3–5 m in height. Flowers are white and the fruit is about 2–3 cm in diameter – small compared to *P. guajava*. There are purple, coppery-red and yellow fruiting plant forms. The fruit has a white flesh, may be

rather acid and it is usually recommended for jams or jellies rather than as a fresh dessert fruit. The main fruiting season is during the winter. Fruiting can occur soon after planting, although full bearing may not be achieved for up to 9 years.

The plant may be raised from seed and is reputed to be difficult to propagate vegetatively. It may be planted in winter and managed in a similar fashion to feijoas.

Cherimoyas (*Annona cherimola*)

There are several fruit-producing plants in the *Annonaceae* family, including *Annona atemoya*, *Annona squamosa* (sweetsop), *Annona cherimola* (cherimop) and *Annona muricata* (soursop). The cherimoya is most suited to the cooler regions of the subtropics. The plant is considered to be an Andean fruit and southern Ecuador is considered to be the centre of origin. The cherimoya has been introduced to many countries, including California, Chile, India, Spain, Australia and New Zealand.

The tree tends to look untidy. It is deciduous and may bear fruit after producing rather inconspicuous axillary flowers on both young growth and old wood from early to midsummer. The time of flowering can be influenced by pruning regimes. Flower initiation takes place nearly a year beforehand, and flower appearance is dependent on bud activation and growth.

Pollination is a problem, with different times of activity (dichogamy) for the pistil (female) followed by the stamens (male). Cherimoya is not pollinated by bees and is visited by a range of spiders and insects. In several growing areas hand pollination is the recommended practice to secure satisfactory yields. Pollen can be collected a day beforehand and kept overnight, dry, at between 7 and 10°C. Fruit set can be obtained with pollen from the same cultivar. Chemical stimulation of parthenocarpic fruit has been tried using GA_3. Cooler temperatures may encourage a longer receptive period for the female phase, which may benefit pollination.

The fruit is a syncarpium, or multiple ovary. The skin is composed of overlapping, scale-like carpel walls, which may be slightly raised or depressed from the fruit surface. The fruit covering is fairly thin and the flesh is soft when the fruit is ripe. It should be symmetrical to be of attractive appearance.

The flesh has a pleasant blend of sweetness and mild acidity, with the consistency of baked custard.

M. Morley-Bunker

The latter characteristic is the reason why the fruit of cherimoya and other *Annona* species are sometimes called custard apples. When ripe, the skin may be brownish or yellow-green. Harvesting should commence when the skin is beginning to change from green so that fruit handling is less likely to damage the soft fruit. The fruit has a 3-week storage life at 10°C but is sensitive to lower temperatures.

The tree is sensitive to frost, and branches are susceptible to breakage by wind. The natural growth habits of cherimoya make it necessary for producers to prune throughout the life of the tree. Growers prune to shorten branches and make the tree more compact. Weaker branches tend to carry more flower buds than vigorous branches. Trees may be trained as vase- or pyrimidal-shaped trees to standardize tree size and management required for the trees. The plant can be grown from seeds but propagation is normally by budding or grafting selected clonal material onto seedlings. Named specific rootstock selections which will advance cherimoya production are still to be identified. Producing cultivars include Madeira, Smoothy and Spanish, while, in Equador, named cultivated selections include Fino de Jete, Bays, White, Bronceada and Concha Lisa.

Elderberries (*Sambucus nigra* and *Sambucus canadensis*)

Elderberry plants are deciduous and usually much-branched shrubs rather than trees. *Sambucus canadensis* is the American elderberry and *Sambucus nigra* is the European elderberry, commonly found in hedgerows and waste ground. Both are quite similar in growth habit and cultural needs.

The European elderberry has had many traditional uses other than providing fruit. These range from providing herbal medicines to the use of hollow young stems for pipes. Large white flower panicles are produced in late spring and the fruit is ready by mid- to late summer. The fruit can be harvested in clusters stripped from the stalks. It does not last once it has been stripped. The fruit is normally processed into sauces, jams and jellies or it can be pressed for juice and fermented to make an alcoholic beverage. The flowers can also be used for flavouring stewed fruit, milk and cold drinks and for winemaking.

Current interest in the chemical properties of berry fruit generally has included some examination of elderberry fruit components such as soluble solids, vitamin C, total antioxidant capacity (FRAP assay), total phenolics and total anthocyanins. European elderberry was pronounced very rich in terms of health components.

The plant is capable of making rapid growth, with long new shoots or cane-like growth. It may be propagated from hardwood or softwood cuttings or suckers. Virus problems emphasize the need for virus-free propagation material. The flowers are borne terminally on new growth and are partially self-fertile. The planting of two or more selected cultivars would be advisable. The plants should be pruned to maintain the production of new growth for flowering. This may entail annual selective pruning but some systems of production cut the plant down to the ground either every year or every second year.

Jujube (*Ziziphus jujuba*)

The *Rhamnaceae* family includes the genus *Ziziphus*, which has approximately 170 species native to the tropics and subtropics. Two species are noted for producing edible fruit: the common jujube or Chinese date, *Ziziphus jujuba* (called Zao or Hongzao in Chinese), and the Indian jujube, *Ziziphus mauritiana*. China is one of the origin and distribution centres of *Ziziphus*. There are at least 14 species of genus *Ziziphus* in China, including sour jujube or acid jujube *Ziziphus acidojujuba*. The Chinese jujube is a deciduous tree, derived from sour jujube, and is distributed mostly in temperate areas in China. Indian jujube (commonly called ber) is evergreen and is mainly distributed in tropical and subtropical areas.

The Chinese jujube has a long history of over 7000 years. The Chinese date has a fruit between a cherry and plum in size, which can become very sweet. The fruit can be allowed to ripen and lose moisture, turning a dark red or black, becoming more date-like in appearance. Production of the Chinese jujube in China has increased rapidly in the latter part of the 20th century. In 2003 there were over 1,000,000 ha in production and about 1.5 Mt of fresh jujube produced. The major commercial production provinces were those with arid and semi-arid temperate climates in the north of China. These areas tend to experience cool winters and hot summers. When dormant the tree can withstand winter temperatures as low as –30°C. The tree usually comes into leaf late and consequently normally avoids frost and freeze damage. The plant is quite tolerant of a wide range of moisture conditions including severe drought, but fruit splitting can be a problem and cause saleable fruit yield loss (Table 20.2).

The *Ziziphus jujuba* tree is small and has small, glossy leaves. The root system is extensive in area and soil volume but is, compared to other fruit trees, quite sparse. Surface-zone roots are prone to suckering. The new shoots tend to zigzag in direction. Some plant selections have spines, usually paired, while others are spineless. The bearing branches, usually about 15 cm long, are typically thin and may weep with the weight of fruit. The flowers are borne in the leaf axils of the current season's shoots. There may be one flower or more, clustered on a short branch. The tree blossoming period may extend for about 3 weeks, although individual flowers flower over a short period of time and may have distinctive patterns of opening in a day. For example, some cultivars have been observed to only open in afternoons and others in the morning. Chinese jujube may have abundant flowering and low fruit set. Cross-pollination is advisable, even with self-fertile clones. Fruit drop immediately after and during fruit growth development can also reduce yields. Unfertilized fruits and fruits with poor embryo development drop readily. The fruit is a drupe, and pit or stone hardening usually begins about 40–50 days after flowering. Maturity is about another 40–60 days after pit hardening. Cultivars may have fruits with different shapes, with some round, oblong and oval, and others with distinctive and unusual shapes, for example those that have a long capsicum-like shape. Fruit size is also variable between 1.5 and 6 cm long and 1.5–4 cm in diameter.

The Chinese have as many as 700 cultivars of Chinese jujube. They include Jinsxiaozao, Pozao (Fupingdazao), Yuanlingzao, Changhongzao, Muzao and Bianhesuan as leading cultivars for the production of dried fruit. Zanhuangdazao and Huizao are listed as multi-purpose cultivars. American listed cultivars include Li, Lang, Tigertooth (syn. Silverhill), Sherwood and So. The plants are grown from suckers or from seed. Seedlings should be grafted over to a selected cultivar to ensure reliable fruiting.

Key points

Propagation

Selection of superior and adapted cultivars is an ongoing process in China and still to be addressed in other parts of the world. The rooting of air layers, hardwood cuttings and softwood cuttings have all been reported, with average success rates. Much more common is the use of root and basal suckers. Chinese workers have, in particular, worked with plant material *in vitro* – both for propagating pieces and plants and for purposes like embryo rescue and genetic transformation.

Superior scion selections can be grafted on to seedlings for transplanting into orchards. Seedling jujube or seedling sour jujube (*Z. acidojujuba*) are the most common rootstocks used. Mature stones and kernels are stratified before germination. Treated seeds can be sown in nursery rows 10–15 cm apart in spring. Removing the stone and sowing the kernels (after stratification) usually ensures the new emerging seedlings are more vigorous. Grafting should be possible 1 year later. Cleft or whip grafting is usually successful. The best time for grafting is when bud break takes place. Scions from 1- to 2-year-old extension shoots can be taken from dormant wood and held in refrigeration until the stocks are ready about 1 or 2 months later. It is recommended that grafts be wrapped or waxed after grafting and until there is sign of the formation of a graft union. New plants are transplanted into the orchard position at about bud burst, usually 1 year after being grafted. New plants should be given protection from dry, windy conditions until they are established.

Training, pruning and crop load

A number of different tree configurations can be found in China, including trees that could be termed as central-leader trees and others that could be termed open-centre or vase trees. Trellising (for example, Y trellis) and the introduction of intensive planting systems has also meant, in recent times, the use of cordon, pillar, espalier, hedgerow and spindle trees.

Growers usually aim to have trees with six to eight primary branches at a height no greater than 3–5 m, well spaced in all directions. To obtain the desired sprouting of lateral main buds, the

Table 20.2. Jujube.

Botanical, anatomical, and physiological aspects	Climatic, geographic, soil and water requirements	Cultural aspects
Common name Jujube, red date, Chinese date, Zao, Hongzao	*Temperature needs* Fruits in areas with average annual temp between 5.5 and 22°C. Needs at least 100 days frost free. Summer temps should be in range of 18–35°C	*Propagation* Can propagate with softwood cuttings. Tissue culture protocols have been developed. Suckers commonly used for plants on own roots or for grafting. Seedlings can be grafted. Seed must be stratified before sowing (3–4 months at 2–5°C). Graft stocks when large enough (about 1 year). Grafting time approximately 1–2-year-old extension shoots
Botanical name *Ziziphus jujuba*. Mill. Name is sometimes written as *Zizyphus jujube*. Synonms include *Ziziphus zizyphus, Ziziphus jujube, Ziziphus vulgaris, Ziziphus sativa*	*Frost tolerance* Leaf and flower emergence late – usually avoids frost	
Botanical name of related and useful species Indian jujube or ber (*Ziziphus mauritiana* Lam.). Sour jujube or acid jujube, *Ziziphus acidojujuba*	*Water needs* Tolerates a wide range of rainfall but better suited to low-rainfall regions	*Rootstocks* Use suckers or seedlings. Sour jujube sometimes used as a rootstock
	Tolerance to wet soils Poor	*Spacing* Traditional spacing at 4–5m between trees in row. Higher-density orchards place trees 0.5–3m apart in rows. Distance between rows depends on farming practice. Some Chinese plantings intercrop jujube with other crops
Type of plant and size Small deciduous tree or shrub. Typically 5–10 m tall but sometimes taller; drooping branches, thorny new shoots. Older trees often less spiny. Prone to produce suckers from surface roots	*Drought tolerance* Very good – surviving and producing in areas with annual rainfall as low as 200 mm	*Training and pruning* Trees trained in first 2–5 years to produce tree form – open centre (vase), centre leader, espalier, etc. Prune mature trees to allow good light interception and canopy penetration. Can prune in late winter or summer
	Humidity tolerance Prone to fruit splitting with high humidity and rainfall during fruit maturity	*Thinning* Not normally practised
Sexuality Hermaphrodite	*Wind tolerance* Good	*Tillage* Avoid disturbance of soil and promoting excessive suckering *Time to first harvest* First leaf after planting

Continued

Table 20.2. Continued.

Botanical, anatomical, and physiological aspects	Climatic, geographic, soil and water requirements	Cultural aspects
Pollination Pollen viability often poor and effective for short periods. Pollen agents mostly bees and wasps. Fertilization rates can be low. % fruit set often low. Some reports suggest cross-pollination provides for better fruit set	*Edaphic features* Best placed in positions with good light exposure	*Time to full production* Fifth leaf
Flower buds Borne in axils of current season's shoots. May appear singly or in cymose collection of up to 10 in a cluster	*Soil needs* Tolerant of wide range of soils (composition and depth), including slightly alkaline soils	*Expected yields* Up to 300 kg/year/tree
Growth of fruit Follows double-sigmoid curve with growth suspended during pit hardening	*Nutrient requirements* Yields possible in unfertilized conditions but improved with fertilizers applied after harvest, at bud break and during the rapid fruit growth phase. Nitrogen promotes shoot growth. 100 kg of fruit should need a replenishment of about 1.5 kg N_2, 1 kg phosphate and 1.3 kg potash	*Normal productive life* Trees can be long lived. Up to 50 years
Time of bud burst Late spring (northern hemisphere April)		*Method of harvest* Hand harvesting to ensure good-quality fruit. Some crops are harvested by tree shakers. Ethephon treatment (200–300 ppm 5–7 days before harvest) can help loosen fruit. Fruits can be harvested white mature (use for confectionery), crisp mature (table use) and fully mature (drying)
Time of flowering Late spring and early summer (northern hemisphere May–June)		*Storage* Fresh table fruit has poor storage life – 2–3 days in ambient conditions. Cool storage at 1°C in polyethylene bags may allow storage for up to 100 days. Fully matured fruits are dried to moisture content below 25%
Time of maturity Late summer and autumn – March and April in southern hemisphere. Northern hemisphere – late August–October		*Main pathogens* In China peach fruit moth is a serious pest and jujube witches broom (a phytoplasma) a serious disease. Anthracnose can damage fruit and leaves

M. Morley-Bunker

secondary shoots beneath the heading cut must be removed, because the lateral main bud does not sprout new shoots without strong stimulation. This is the so-called 'one scissors stop, two scissors sprout' in Chinese jujube.

The bearing branches of the Chinese jujube tree require good light exposure. Annual pruning is normally required. Pruning can be undertaken in both the dormant and growing seasons. In environments where winters are severe it is best to leave dormant-season pruning until late winter or early spring to protect new shoots and trees from low-temperature winter injury. Dormant pruning mostly involves removing shoots that are incapable of producing fruit of satisfactory size and quality, including weak, diseased, pest-damaged and crowded shoots. Heading back old branches, drooping branches and excessively long branches is also done.

Summer pruning includes removing useless or crowded sprouted buds and new shoots. It aims to expose the bearing branches to more light. Summer pruning may reduce plant vigour, which can assist in improving fruit set.

Harvesting and postharvest management

Fruit set of Chinese jujube in China occurs from late May to July. The fruit harvest begins in late August and continues until late October. Fruit use should determine the time of harvesting and fruit condition. Fruit that is to be eaten fresh should be at the crisp mature stage. Fruits for drying should be picked when fully mature. Fruits are mostly hand harvested, which requires care to prevent bruising. Fruits that are to be processed or dried can be collected by shaking the tree or beating branches. Ethephon can be used to assist harvesting by shaking. A spray of 200–300 ppm of ethephon can be applied 5–7 days before harvest. After harvest, fruits should be graded according to the degree of maturity and fruit size and then sold immediately or stored at low temperature, dried or processed.

Fresh Chinese jujube fruits do not keep well at room temperature. The shelf life is usually only 2–3 days. Storage life is much better at low temperature, and some selected Chinese jujube cultivars can be kept crisp for more than 100 days if packed in vented polyethylene bags and stored around 0 ± 1°C. The bag helps maintain the appropriate relative humidity and CO_2 concentration. There are

reports that the rates of respiration and ethylene production of the fruit can be reduced by 1-methylcyclopropene (1-MCP).

Chinese jujube is often better known as a dried processed fruit product. Fully matured and ripe fruits can be dried using traditional methods of sun-drying for 2–3 weeks. The fruit can also be dehydrated in specially designed cabinet driers set at temperatures around 60 °C and allowed to run for 36–48 h until the moisture content of the product is reduced to less than 25%.

Loquats (*Eriobotrya japonica*)

The loquat is an evergreen tree in the *Rosaceae* family, *Pomoideae* subfamily. It is indigenous to south-eastern China. The genus *Eriobotrya* has about 15 species, of which the loquat, *Eriobotrya japonica*, is the significant fruit crop species. The loquat is grown commercially in China, Taiwan and Japan and has also attracted interest in the Mediterranean region, particularly Spain and Turkey. In China an estimated 120,000 ha of loquat is in production, with an annual production of 400,000 t.

There are many named cultivars of loquat. Spanish commercial production includes the use of Algerie, Magdal, Golden Nugget (from the USA) and Tanaka (from Japan). Japan also has cultivars Mogi and Nakasakiwase. China has 83 cultivars (or selections) in the Zhejiang Province alone.

Although the loquat tree may only be grown in a few countries as a fruit crop, the tree is grown in many countries as an ornamental tree. The loquat makes an attractive, round-crowned tree, up to 6 m high. It has new shoots and large, green, glossy leaves, felted on the underside. The time of flowering is slightly unusual for fruit crops, occuring during autumn and winter. The fruit ripens in late spring and the following summer. Loquat fruit has a useful position in the marketplace, arriving before most other spring fruit crops.

Flowers occur in stiff-stemmed clusters or flower spikes emerging as new growth from the terminal points of quiescent shoots. There may be three to ten flowers in the cluster (Fig. 20.1). Trees are self-fruitful, although reports have suggested that cross-pollination can improve fruit set values. As the loquat develops, it becomes less hairy and turns from green and cream to yellow and orange. The fruit is soft and bruises easily when fully ripe. Fruit size can vary; unthinned fruit may be 35 mm long.

Fruit thinning can help improve fruit size, and manual fruit and flower thinning have been traditional practices in some countries. Chemical thinning with 60 ppm NAA at the end of full bloom, will thin fruit to four fruits per cluster. The fruit contains one or more large seeds. Reducing the ratio of seed to flesh volume and increasing the overall fruit size makes the fruit more attractive. The loquat fruit taste can be rather sharp, with a fresh apple-like flavour. Total soluble solid values can vary from 8 to 19% in different cultivars and conditions. Consumer tests indicate a general preference for large-size fruit, sweet, well coloured and of good shape.

The fruit can be kept at temperatures below 5°C after harvesting, but there is the possibility of chilling injury; various treatments have been investigated to improve storage and reduce chilling injury. Low-temperature conditioning at 5°C for 6 days has been suggested as facilitating storage at 0°C for about 50 days.

Loquats prefer similar soils to those suiting its close relative, the apple. Deficit irrigation can be used to manage plant vigour, timing of phenological events and fruit condition. The tree can be pruned to produce a clear-stemmed tree with strong scaffold branches beginning at about 1 m high. Pruning is done just before the autumn growth flush. After harvest, the lateral shoots below the fruit cluster can be thinned and the defunct fruit stalk removed. Selected cultivars can be grafted on to a range of rootstocks, although no

Fig. 20.1. Loquat.

general practice has emerged to do this. Loquats can also be propagated from seed.

Medlars (*Mespilus germanica*)

Medlars belong to the *Rosaceae* family and are most closely related to the hawthorns, (*Crataegus* spp.), which are not commonly grown for cropping. There are graft compatibilities with other plants in the *Rosaceae* family; for example, medlars have been grafted on quince (*Cydonia* spp.). Medlars are native to south-west Asia and possibly south-east Europe. Work in Poland suggests a dispersion into north-east Europe as well. The trees are small and bear fruits that resemble small, thick-skinned apples (to which they are related), with a large, open face ringed with the remnants of the flower sepals. The fruit is inedible when harvested in autumn, being hard and very acidic. However, after a few weeks storage and softening, called 'bletting', the fruit can be eaten raw, preserved or made into jelly.

Mulberries (Black) (*Morus nigra*)

The mulberry is an aggregate fruit, ranging from white to red to black in colour. When ripe the fruit is juicy and appears like an elongated bramble fruit. It is attractive to birds. *Morus nigra* is more commonly used for fruit, while *Morus alba* (white mulberry) is better known for its connection with oriental silk production – the leaves are fed to silkworms. Both can be fed to silkworms, although palatability does vary.

Mulberries produce sweet fruit, but they are easily bruised and squashed, so transporting and marketing the fresh fruit is very difficult. Harvesting is also difficult, as the fruit tends to fall easily from the tree, even when still immature. Nevertheless, spreading sheets under the tree, shaking it and sorting the debris is probably the easiest harvesting method. The black mulberry is a very pleasant dessert fruit or can be used for jams, pies, fruit drinks and wine.

White mulberry (*M. alba*) originated in China. The origin of the black mulberry (*M. nigra*) is thought to be south-western Asia, although wild populations exist in Greece, Turkey, the Balkans and forested areas stretching from Europe towards China. Black mulberry has also become semi-naturalized in parts of southern Africa. Although it is convenient to distinguish between *M. alba* and

M. Morley-Bunker

M. nigra on the basis of fruit colour, some *M. alba* selections do produce black fruit.

Black mulberries can be grown in warm temperate climates, the subtropics and the high-altitude tropics. It is the least cold hardy of the mulberries, compared with *M. alba* and *Morus rubra* (American red mulberry). It is deciduous and may require a short chilling period. The plant may be propagated in micro-propagation facilities using shoot tips and leaf tissue. The tree can also be propagated using hardwood cuttings or seeds. Seeds germinate best after stratification. Budding and grafting is possible but with mixed results. The roots dislike disturbance and prefer a light, free-draining soil of reasonable fertility, although the plant may grow on a range of soil types. The tree has scaly bark and may grow to 10 m, but is usually kept pruned to a smaller, open, spreading shape. It can produce quite a dense and shady canopy. Some mulberries exist as male or female trees (dioecious), so both will be required in order to produce fruit. Self-fruitful trees are preferred. Named cultivars in the USA include Black Persian and Shangri La. Mulberries can be long lived, and in London there is a tree named the The Charlton House Mulberry, which is said to have been planted in 1608 for Charles I.

Naranjillas (*Solanum quitoense*)

The naranjilla or lulo originated in South America and is cultivated in the Andes of South and Central America. The main production areas are located in Ecuador, Colombia and Peru. It is a frost-sensitive, erect subshrub, 2–3 m high, and has large dramatic leaves. The leaves have purple veins, especially when young. Some naranjilla forms have spines on the leaves and petioles. Other *Solanum* relatives are also spiny, including the cocona or peach tomato (*Solanum sessiliflorum* syn *Solanum topiro*). The naranjilla flowers are pale lilac, and the globular fruit, about 5 cm in diameter, ripens from green to a bright orange colour, although internally the pulp remains green. The fruit is slightly hairy and is best harvested with gloves. The hairs can be rubbed off ripe fruit. Ripe fruits have a shelf life without refrigeration of 1–2 weeks. The green pulp is versatile and is most used for drinks but can be used for preserves, pies and toppings. The fruits have a high vitamin A and ascorbic acid content. The plant can be propagated from seeds or cuttings and grows rapidly. New plants may begin yielding as early as 6 months. The plant responds to fertilization and may

be pruned to maintain vigour. Naranjilla plants in South and Central America tend to lose vitality through nematode, insect and pathogen incidence. Root knot nematodes have been recognized, in particular, as limiting naranjilla production. Hybridization with other plants has been researched as a way of improving naranjilla production. Crossing *Solanum hirtum* with naranjilla and then backcrossing has introduced some nematode resistance in naranjilla. Other hybridization attempts using cocona (*Solanum sessiliflorum*) have resulted in the development of naranjilla-like crops with similar fruit characters but often better growth features and tolerance of the established pest and pathogen complex in countries like Ecuador.

Northern Pawpaws (*Asimina triloba*)

The term 'pawpaw' incorrectly suggests a papaya relationship, whereas in fact the plant is related to the cherimoya (see earlier). *Asimina triloba* is native to the temperate, wooded eastern areas of North America. It may grow as a gaunt, narrow deciduous tree up to 8 m high but has a tendency to sucker and be more shrub-like. The long, drooping leaves, up to 30 cm long, are deciduous.

Flowering takes place in spring, before the leaves appear, on the previous season's growth. The flowers may be up to 5 cm in diameter and are cross-pollinated by flies or beetles. Bees show no interest in pawpaw flowers. Lack of pollination and fruit set are causes of low yields. Although the pawpaw flowers have both male and female organs, the organs are not functional at the same time. Pollen and ovules from the same tree are self-incompatible, so cross-pollination from unrelated trees is required to set fruit. Hand pollination using a small, soft artist's brush to transfer pollen to the stigma will produce fruit. The pollen is ripe for gathering when the ball of anthers is brownish in colour, loose and friable. Pollen grains should appear as small, beige-coloured particles on the brush hairs. The stigma is receptive when the tips of the pistils are green, glossy and sticky, and the anther ball is firm and greenish to light yellow in colour.

The fruit is about 8 cm long, ellipsoid to oblong in shape and greenish-yellow turning to brown. It is soft and thin skinned, containing several seeds, about 2 cm long, set in aromatic, sweet, yellow pulp. Some descriptions of the taste suggest a mixed flavour of banana, mango and pineapple. The seeds should not be ingested or chewed.

The plants may be grown using seedlings grafted with selected cultivars. Fresh-collected seed needs stratification before sowing. Seedlings may be chip budded or grafted. Nursery plants are best grown in containers, as open-ground plants do not transplant readily. Other forms of propagation have had limited success; this includes softwood cuttings and root cuttings. Named cultivar selections include Sunflower, Overleese and PA-Golden from the USA and Prima 1216 and Prolific from Italy.

Pepinos (*Solanum muricatum*)

The Spanish name pepino dulce, a member of the *Solanaceae* family, may be translated into English as sweet cucumber or melon. Sometimes the plant is called the melon pear. From Peru, where it is thought to have originated, it has been introduced to most of South and Central America, as well as to the USA, Spain and countries in Europe and around the Mediterranean, as well as Australia and New Zealand.

The plant is an herbaceous perennial under frost-free conditions and may grow into a spreading bush about 150 cm wide and 80 cm high. The foliage may be composed of largely simple, entire leaves or compound leaves with three, five or more leaflets. The latter tend to look rather like potato leaves. Flowering occurs on new growth and flow-ers are similar in form to potato flowers. Further shoot growth and flowering will continue in a manner similar to the indeterminate growth and fruiting habit of the tomato. Selection of shoots for flowering and fruiting is essential. This is especially true for plants that may be grown in a glasshouse or protected cropping conditions.

Self-incompatibility may prevent fruit set. Pollination can be hindered by poor pollen release and the position of the stamens relative to the stigmatic surface. Temperatures in excess of 30°C and low humidity conditions contribute to poor pollination. Growth regulators can promote the development of seedless fruit. Seedless (parthenocarpic) fruit set in pepino has established that parthenocarpy has a genetic base, and some plants readily produce parthenocarpic fruit.

The green fruit can be consumed as a vegetable for cooking. When fully ripe it is used as a dessert fruit. The time from flowering, fruit set and ripening is slow relative to the tomato, necessitating a longer growing season without frosts. Different ripening stages have been described, principally on the basis of colour. Colour parameters a* and the hue angle were proposed as reliable in describing ripeness. Ethephon treatment can accelerate colour change and ripening.

Pepino fruit shape and size can be variable, with some selections producing round, melon-like

Fig. 20.2. Pepino fruit.

M. Morley-Bunker

shapes and others producing elongated, tapering shapes. Fruit weight may vary between clones as well as with fruit set. Individual fruit weights of 100 g to 1 kg are possible. The fruit skin colour can range from light cream to yellow and orange. An overblush or striping with shades of purple is normally considered attractive. The internal flesh colour ranges from pale yellow to deep orange. The fruit has a mild, juicy, sweet, melon-like flavour. Soluble solids increase as the fruit reaches the first stages of ripeness but not thereafter. Some selections have a harsh, bitter, even hot flavour, which growers and plant selectors try to minimize.

Clonal selections are maintained by propagating with softwood cuttings. Producing countries have selected cultivars for their particular growing conditions and production systems. Viruses have been identified in pepino. Virus-free plants should be used for propagation. A wide range of pests may attack pepinos, including spider mites, white flies and aphids.

Pomegranates (*Punica granatum*)

The total annual world production of pomegranates is estimated to be around 1.5 Mt, and the four largest producers of pomegranates are Iran, India, China and the USA. The pomegranate has long been cultivated in the Middle East and the Mediterranean regions and is recorded as growing in the Hanging Gardens of Babylon. It probably originated in Central Asia from Iran and Turkmenistan to northern India. It is considered that the culture of pomegranate began in Neolithic times, with the plant valued for its fruit and perceived medical benefits. The worldwide increased demand for pomegranates is thought to be due to clinical trials using fruit extracts and juice that show retardation of some forms of cancer.

The plant usually grows as a bush up to 4 m high, but there are also dwarf forms. The bush is regarded as deciduous, although this habit is not very distinct in warm climates, and India has what might be termed evergreen selected cultivars. The plant may be trained with single or multiple trunks; the shoots are mostly spiny, the leaves narrow and the plant is prone to basal suckering. The flowers are usually a decorative bright scarlet colour. Ornamental cultivars may have double flowers and variegated flowers. Flowers are normally found on short spurs or laterally on shoots older than 1 year.

Flowers may be found singly or in clusters of up to five flowers. Flowering normally takes place in late spring. Many selections are self-fruitful, although cross-pollination can improve fruit set. Pomegranate flowers are insect pollinated. Plants should come into flower within 3 years of planting and possibly in the first year. Good commercial production should be achieved after 5–6 years.

Most commercial fruits range in width from 60 to 100 cm. The fruit is nearly round in shape and crowned by the remnants of the flower calyx. The impression is that of a round ball with a crown of points on top. The leathery skin (the rind or husk) may be yellow or red in colour. The fruit interior has membranous walls, creating compartments with the edible part of the fruit – a sweet, juicy, pink pulp – surrounding each seed. This pulp-like bag is termed an aril. The fruit should mature about 7 months after bloom. In Israel the cultivar Wonderful is harvested when soluble solids reach 15%, whereas in California maturity has been equated with 1.8% titratable acidity (TA) and soluble solid compounds of 17% or more. The fruits are cut rather than pulled from the plant. The fruit is not climacteric and will not ripen after harvest, other than soften with gradual moisture loss. Fruits may be used for fresh fruit consumption or juice extraction. Cultivars with soft seeds are best for fresh fruit consumption. Storage in modified atmosphere packaging (microperforated bags) at 6°C can give fruits a postharvest life of up to 16 weeks. Alternatively fruits may be stored between 0 and 2°C, at 80–85% RH for up to 7 months.

The pomegranate produces best in climates with cool winters and long, warm, dry summers. It can withstand low temperatures of up to −12°C. The bush is tolerant of poor soils and dry conditions, although more amenable soil conditions will assure better production. The plant is also able to tolerate saline growing conditions. The use of drip irrigation has allowed planting in semi-arid regions and some control over growth and crop timing through manipulation of water stress and its alleviation. Drought stress should be avoided at flowering or fruit set will be affected. Plants need moist soils for their initial establishment.

The pomengranate roots readily from cuttings. Propagation using suckers and air layers is also possible. There are many cultivars worldwide. The cultivar Wonderful is used in the USA and Israel. Mollar de Elche is grown in Spain, Bagua in India and Hicanzar in Turkey.

Quinces (*Cydonia oblonga*)

The quince has a historical connection with Ancient Greece and Rome and there still is some commercial production in Eastern Europe and Asia Minor, especially Turkey. The fruit is used for jams, jellies and preserves. It can be fragrant and small pieces may be used to flavour other fruit preserves.

The quince is the only species in the genera *Cydonia*. It is closely related to apple and to pear and can be used as a rootstock for some pear cultivars. The quince does not grow as large as pear trees and therefore is used as a rootstock to try and contain the vigour of the grafted pear scions. In spring the showy pink or white flowers are produced at the ends of leafy shoots. Ideally the flowers need to be cross-pollinated, although some selections are self-fruitful. Bees are the main pollenating agent. The young immature fruit is green with dense, grey-white hair covering the fruit skin. The hair will reduce, or can be rubbed off, as the fruit matures in late autumn and begins to change skin colour to yellow. Fruit thinning will improve fruit size. Most fruit are pear shaped and of a similar, if not larger, size than the typical pear. The fruit is hard, even when fully coloured, and has a strongly perfumed flesh. The fruit is normally quite acidic and astringent and most uses involve cooking and processing.

The quince grows in similar environments and conditions to the pear and apple. The plant can be propagated by layers, hardwood cuttings, stools or suckers, or can be grafted on to quince seedlings or clonal rootstocks. There are named cultivar selections, including Angers, Orange, Pineapple, Champion, Smyrna, Meeches Prolific, Morava Portugal (syn. Lusitanica) and Vrajna (syn. Bereczcki).

In addition to the normal run of pip fruit diseases, quinces are subject to two leaf spots, *Fabraea maculata* and *Sclerotinia cydoniae*. In some regions quince plants can be an alternate host for the disease fireblight, *Erwinia*, and therefore their planting is discouraged.

Tomatillos (*Physalis ixocarpa*)

The plant may be known as husk tomato, jamberberry, ground cherry, tomate verde or tomatillo. The botanical naming of tomatillo has variously been listed as *Physalis ixocarpa* and *Physalis philadelphica*, and a recent proposal suggests the name should be amended to *Physalis philadelphica* subsp. *ixocarpa*.

The tomatillo is a relative of the Cape gooseberry (*P. peruviana*). It has a similar fruit habit, producing a fruit enclosed within a papery husk. The 25–50 mm berry (within the husk) is larger than the Cape gooseberry fruit. It can be eaten green or later, when it becomes purple and has more flavour and sweetness. The fruit is used in the making of chilli sauce and dressings for Mexican food, such as taco and enchilada. It is grown in Central and South America and particularly in Mexico and Guatemala. It has also been introduced into Spain.

Fig. 20.3. Quince flowers and fruit.

The plant is an erect, 1–1.5 m tall, branching, frost-sensitive herbaceous plant and it is best treated as an annual. The first flowers on the main branches produce the largest fruit. The lateral- and sublateral-borne flowers do not set as readily and produce smaller fruit. Plantings are normally established using transplanted seedlings. Plants can be provided with trellises to assist management and fruit harvesting. Field plant populations of 40,000–50,000 plants/ha will provide good coverage and crop yields.

White Sapote (*Casimiroa edulis*)

This plant is a member of the *Rutaceae* family, although its common name creates confusion, suggesting it is a member of the family *Sapotaceae*, whose members include the sapodilla (*Achras sapota*), mamey sapote (*Pouteria sapota*) and star apple (*Chrysophyllum cainito*).

The white sapote is native to the highlands of Central America and will grow in similar climates to citrus. It will withstand very occasional frosts of up to −2°C. It is an evergreen tree, but leaf shedding can occur, making the tree look open and ragged. The branches hang when heavy with fruit. The foliage matures from light green to dark green and flowering is possible in most months, but it is more likely with the spring growth flush. The flowers may be grouped in panicles of 15–20 or more on terminal growths or in the axils of mature leaves.

Fruits are spherical to slightly oval, 6–15 cm in diameter. The fruit skin is green with a yellow tinge or, alternatively, more yellow than green. The skin is thin and inedible. The flesh is yellowish, tender, buttery, sweet, and mild except for a slight resinous flavour. The fruit has two to five seeds, which may occupy a large volume of the fruit. The seeds are reportedly toxic, fatally, if eaten. Fruits can be harvested by assessing skin colour and dry matter content.

Cultivar selection should be concerned with reducing the seed to flesh ratio, as well as improving other characteristics, such as fruit yield, adaptation to local climate conditions and cultivar

suitability for handling and transport. The fruit will ripen if picked a little before full maturity but will fall if allowed to ripen on the tree (it bruises easily). The ripe fruit is rather soft, which means picking and transporting before full ripeness.

A number of cultivars have been selected in the USA (e.g. Wilson, Blumenthal, Pike) and elsewhere. Seedling plants can be budded with clonal selections. At planting, young trees are headed back to encourage branching at about 1 m in height. Subsequent pruning will be required to shorten lateral branches, which become too long and straggling, in a similar way to citrus.

Relatively little is known specifically about nutrition, irrigation and other cultural details.

Further Reading

Campos, F.A.P., Dubeux, J.C.B. Jr and de Melo Silva, J. (eds) (2009) VI International Congress on Cactus Pear and Cochineal. *Acta Horticulturae* 811.

Chomchalow, N. and Sukhvibul, N. (eds) (2005) II International Symposium on Lychee, Longan, Rambutan and other Sapindaceae Plants. *Acta Horticulturae* 665.

Drew, R. (ed.) (2002) International Symposium on Tropical and Subtropical Fruits. *Acta Horticulturae* 575.

Huang, H. and Menzel, C. (eds) (2001) I International Symposium on Litchi and Longan. *Acta Horticulturae* 558.

Huang, X.M. and Janick, J. (eds) (2007) II International Symposium on Loquat. *Acta Horticulturae* 750.

Liu, M.J. (ed.) (2009) I International Jujube Symposium. *Acta Horticulturae* 840.

Mondragon Jacobo, C., Aranda Osorio, G. and Phippen, W.B. (eds) (2006) V International Congress on Cactus Pear and Cochineal. *Acta Horticulturae* 728.

Nefzaoui, A. and Inglese, P. (eds) (2002) IV International Congress on Cactus Pear and Cochineal. *Acta Horticulturae* 581.

Özgüven, A.I., Fideghelli, C. and Yalcin Mendi, Y. (eds) (2009) I International Symposium on Pomegranate and Minor Mediterranean Fruits. *Acta Horticulturae* 818.

Rohde, W. and Fermin, G. (eds) (2010) II International Symposium on Guava and Other Myrtaceae. *Acta Horticulturae* 849.

21 Edible Nuts

DAVID MCNEIL, DAVID JACKSON AND MICHAEL MORLEY-BUNKER

Nuts are such a healthy food that in 2003 the Federal Department of Agriculture (FDA) approved the first qualified health claim for a conventional food, saying that 'Scientific evidence suggests, but does not prove, that eating 37.5 grams (1.5 ounces) per day of most nuts, as part of a diet low in saturated fat and cholesterol, may reduce the risk of heart disease.'

Almonds

The almond probably originated in the hot, dry regions of western Asia and was carried by early humans to the areas around the Mediterranean and thence to North America (Table 21.1). It is cultivated mainly in southern Europe (Spain and Italy) and California, with plantings in North Africa (Tunisia, Morocco), across the Middle East (Iran, Turkey) and the equivalent southern hemisphere zone (Australia, South Africa), but is much less common in milder, moister regions. The efficient North American production (>70% of world exports) is causing severe competition for other producing regions around the world. World production is about 2.07 Mt (2007), the USA (51% of world total), Spain (10%) and Italy (5%) being the main producers.

Almonds are gaining popularity in a healthy diet because of their low saturated fat level (8%), high level of monounsaturates (70–75%) and moderate polyunsaturate levels (15–18%). They also have high levels of α-tocopherol (vitamin E), ranging from 40 mg/100 g oil up to 400 mg/100 g oil in the cultivar Supernova.

Key points

Cultivars

Nonpareil, grown in California, represents 40–45% of the Californian production. It has several advan-

tages: a mild flavour, good blanching and the 'paper shell' characteristic, which gives it over 70% crackout (ratio of kernel weight to total weight of kernel plus shell). It does show biennial bearing, and as a result production in California from an area of 190,000 ha (1998) to 248,000 ha (2007) has varied between 2.0 and 4.2 t/ha of in-shell nuts. Another main cultivar in California is Mission (Texas), with 50–60% crackout and a stronger flavour. Both these cultivars have been grown since 1900. Europe has numerous regional and local cultivars. All, however, have poor crackouts (e.g. Marcona (27%) and Tuono (35%)) and tend to be grown in dryland, extensive situations. The result is that, with 650,000 ha, Spain produces about 0.3 t in-shell nuts per hectare at about 30% kernel, around one-fifth of the Californian production. One result of this is that an additional industry has developed in Spain producing imitation timbers from almond shells.

Pest management

Due to the use of insect-susceptible paper shell varieties, the almond industry in California suffered severe damage by navel orange worm (*Amyelois transitella*) and peach twig borer (*Anarsia lineatella*, also a problem in Europe). Damage caused by these allowed *Aspergillus* spp. fungi to infect nuts, resulting in aflatoxin contamination of many damaged nuts. This mycotoxin is both poisonous and carcinogenic to humans. Chemicals used to control these pests were also disrupting natural predation of mites and other insects, causing secondary problems. As a result, there is now considerable use of integrated pest management (IPM) in Californian orchards, using 'soft pesticides', introduced predatory wasps, *Bacillus*

Table 21.1. Almonds.

Botanical, anatomical and physiological aspects	Climatic, geographic, soil and water requirements	Cultural aspects
Common name Almond	*Temperature needs* Mostly grown in areas with dry, warm to hot climates	*Propagation* Budded, can also grow from cuttings
Botanical name *Prunus amygdalus*	*Frost tolerance* A serious problem, due to very early blossom. Buds showing pink can tolerate −4 to −6°C. At full blossom, −2°C will cause damage	*Rootstocks* Selected peach (Nemaguard, Lovell), plum (Marianna 2624, Myr 029C) and hybrid almond/peach (Titan) rootstocks are used. Peach is slightly more disease resistant but somewhat more susceptible to drought than almond. Plum is less vigorous and can be incompatible and produce suckers but resists waterlogging, phytopthora, oak root, crown rot, gall and verticillium. All have some nematode resistance
Type of plant and size Trees 4–6 m high, 3–4 m at base. Round and bushy	*Water needs* Often grown in dry areas but crop much better with irrigation, especially in areas below 500 mm rain per annum. More tolerant of drought than other stone fruit	*Spacing* Formerly planted on square at 6–7.5 m. Closer spacing at 2 × 4.5 m is now recommended since, like peaches, almonds have limited life span
Sexuality Hermaphrodite	*Water tolerance* Dislikes wet climates, especially at blossoming and close to harvest	*Training and pruning* Grown as vase-shaped and more recently as central-leader trees (Chapter 5). After establishment of framework, pruning is restricted to maintaining a reasonably open tree, accessible for spraying. Prune if possible before leaf fall
Pollination Flowers are usually self-incompatible. Careful selection of pollinizers is essential. Insect pollinated. Considerable research on self-fertile cultivars, with varieties (Independence) available for release in California	*Humidity tolerance* Moderate to good	*Thinning* Not required
	Wind tolerance Moderate	*Tillage* Grass/legume or grass between rows but sprayed under the trees and along rows with herbicides. May use complete weed control between and along the rows. Barn owls for gopher control in USA
Flower buds Fruiting habit is similar to peaches – flower buds surround a vegetative bud on previous year's shoots. Flowers also form on spurs produced on old wood. Flowers have five sepals, five petals, a single carpel and numerous stamens; they are sessile	*Edaphic features* In cold districts, slopes can be beneficial for frost control	*Time to first harvest* 3–4 years – less at closer spacings *Time to full production* 7–9 years – less at closer spacings

Continued

Table 21.1. Continued.

Botanical, anatomical and physiological aspects	Climatic, geographic, soil and water requirements	Cultural aspects
Growth and type of fruit Unlike other stone fruit, the flesh (pericarp) does not grow in stage III but dries and splits on ripening to expose the endocarp. The endocarp, or shell, may be: (i) crumbling and soft, 'paper shell'; (ii) moderately soft, 'soft shell'; or (iii) 'hard shelled'. The latter cannot be broken by hand *Time of bud burst* Very early, ranging from late winter to early spring, depending on cultivar *Time of flowering* See above; flower buds open before leaf buds *Time of fruit maturity* Nuts mature in autumn	*Soil needs* Tolerant of poor soils (particularly on plum rootstock) but better in lighter, deep, fertile soils *Nutrient requirements* Like most stone fruit they respond to nitrogen. Suggest fertilizer regime provided for stone fruit in Chapter 13	*Expected yield* Yield of unshelled nuts at 7 × 7 m. 3 years: 1 kg/tree 5 years: 2 kg/tree 8 years: 9 kg/tree. This equals 1.8 t/ha at 8 years. Narrow-spaced trees may produce 2 t/ha in year 4 *Normal productive life* 15–25 years *Method of harvest* Hand or machine (shake then sweep) *Storage* Remove nuts from pericarp as soon as possible and store in a cool, dry place *Main pathogens* Silver leaf can be serious, although not quite as serious as in peaches since less detailed pruning is used. Blast and brown rot can cause problems. Rootstocks are used for protection against crown gall and rot, phytopthora, verticillium, nematodes and oak root fungus (*Armillaria mellea*), shot hole, brown rot, blossom blight, scab, anthracnose and mildew. Non-infectious bud failure (crazy top) is a genetic physiological disorder that increases with time and causes lack of fruiting. A local problem that builds up in some clones or sites, in California

Fig. 21.1. Almond kernels.

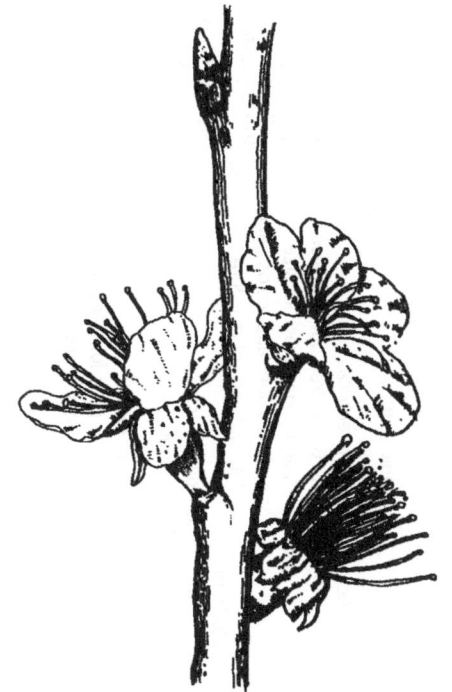

Fig. 21.2. Almond flowers.

thuringiensis, predatory mites, natural parasites, rapid harvest, cleanliness and other management methods. This has also been supplemented by BIOS (biologically integrated orchard systems) programmes in California, which integrate all elements of the farm system (fertilizer, cover crop, etc.). Other pests include ants, borers and leafroll caterpillars. However, fungicidal sprays around bloom

may still be needed for fungal diseases, which include shot hole, brown rot, blossom blight, scab, anthracnose and mildew. Non-infectious bud failure (crazy top) is a genetic physiological disorder that increases with time and causes lack of fruiting – a local problem that builds up in some clones or sites in California.

Chestnuts

About 12 species of chestnut originated over a wide area of the northern temperate zone (Table 21.2). The American, European, Japanese and Chinese chestnuts have all been cultivated for food since ancient times and, in certain areas, have been a staple item of diet. In contrast with most other nuts, which have high protein and low carbohydrate, the high moisture (45%), high carbohydrate (50%), low fat (1%) and moderate protein (4%) levels of chestnuts make them more suited to this purpose. In southern Europe they are often made into a gluten-free flour; in Japan they are much used in cooking and baking. Elsewhere, they are roasted or used for cordon bleu cooking. World production of chestnuts in 2007 was 1.26 Mt from 366,000 ha. Of this, 82% was grown in Asia and 7% in Europe. China, Korea, Japan (*Castanea mollissima*, *Castanea crenata*) and Turkey, Italy, Bolivia, Spain and Portugal (*Castanea sativa*) are the main producers. Limited production, aimed at the fresh, out of season market, has commenced in the southern hemisphere, in Peru, Bolivia, Australia and New Zealand.

Key points

Propagation

Graft incompatibility has proved to be a major problem in many areas where interspecific hybrids are being used to create improved chestnuts. This incompatibility results in the death of the scions or whole trees, often as late as 6–10 years after planting. There appear to be specific rootstock–scion combinations that lead to a high proportion of tree deaths, and careful matching of the scion to a seedling rootstock source (knowing the maternal and paternal parents) is needed to avoid this problem. While rooting of cuttings and tissue culture of chestnuts are possible, their cost and unreliability have generally precluded their use.

Table 21.2. Chestnuts.

Botanical, anatomical and physiological aspects	Climatic, geographic, soil and water requirements	Cultural aspects
Common name Chestnut	*Temperature needs* Will grow in a wide range of temperate climates. Dry heat at flowering may reduce pollination and yields	*Propagation* Not too difficult to propagate and may be whip-and-tongue grafted on to seedlings in spring, or summer budded using dormant scion wood. Nurse seed grafting can be used. Will propagate by cuttings but with difficulty
Botanical name *Castanea sativa* (variously termed 'European', 'Spanish' or 'Italian' chestnut)	*Frost tolerance* Can be damaged by frost of −4 to −6°C in spring	*Rootstocks* Use seedlings of the species to be propagated. See Key points below
Castanea crenata (Japanese chestnut) *Castanea mollissima* (Chinese chestnut) *Castanea dentata* (American chestnut) All species interbreed freely and many commercial plantings use hybrids	*Water needs* Needs plentiful water supply to produce good crops	*Spacing* Normally planted at 12 × 12 m. Planting much closer, down to 3 m, followed by later thinning or tree moving, will increase early yields
Type of plant and size Tall, spreading, 20–30 m (*C. sativa, C. dentata*) or medium height, 10–20 m (*C. crenata, C. mollissima*), spreading, deciduous tree, 20–25 m diameter	*Tolerance of wet soils* Moderate but very susceptible to *Phytophthora cinnamomi* (ink disease) in waterlogged soils. Some biocontrol may be possible using *Trichoderma* and management of compaction and soil organic matter	*Training and pruning* Establish a central-leader shape in early years, otherwise little pruning required
		Thinning Not required, but excess crop load may reduce nut size
Sexuality Monoecious	*Drought tolerance* Poor as young tree but improves at maturity. However, nut size (therefore value) and yields decline rapidly with lack of moisture	*Tillage* Would probably do better in clean-cultivated, weed-sprayed or close-cut grass sward
Pollination Most seem to be self-sterile, although some self-compatibility exists. Both wind and insect pollinated	*Humidity tolerance* Fair to good	*Time to first harvest* 3–5 years with improved cultivars
	Wind tolerance Moderate to poor for trees with upright, lanky branches	*Time to full production* 10–15 years, depending on planting spacing
Flower buds Buds are mixed and produced laterally on previous season's growth. These open to form shoots with male catkins, which open first, at the base and mixed male and female catkins, produced slightly further up the shoot	*Edaphic features* Use flat or gently sloping land but not frost pockets	*Expected yields* Well-managed, intensive orchards may achieve 8 t/ha. 5 t/ha would seem reasonable
		Normal productive life Probably 100 years or more if not diseased
		Method of harvest Hand picked from the ground after falling naturally or after shaking. Can be swept or vacuumed up by machine

Growth of fruit
C. sativa has two forms: 'Marrons' normally have a single chestnut in a spiky husk, which is larger, more aromatic and sweeter than 'Chataignes', which are the more common among seedlings. These have two or three smaller nuts in the husk. Relative size and flavour of other species: C. crenata – large with poor flavour; C. molissima – moderate size, easily removed pellicle and sweet; C. dentata – small, sweetest of the species

Time of bud burst
Spring

Time of flowering
Late spring

Time of fruit maturity
Autumn

Soil needs
Because it is a deep-rooted tree, it grows best in deep, well-drained soils. Tolerant of poor soils but crops poorly. Life of trees is short when grown in heavy clay or on soils with impermeable subsoils

Nutrient requirements
Crops better in soils which are fertile. Does not like calcareous soils. Responds well to added fertility. See notes under Walnuts

Storage
Fresh chestnuts can be stored for 12 or more weeks at 0–1°C. One hour in a water bath at 68°C after harvest may suppress mould, not injure nuts and improve such storage. Dry thoroughly before storing. Can be kept for 12 months at 4–5°C if dried to 10% moisture

Main pests and diseases
Main disease is chestnut blight, see Key points. Phytopthora can be a problem, as can phomopsis nut rots. Other problems are gall wasp, chestnut weevil, sunburn and various stem borers

Fig. 21.3. Chestnuts from *Castanea crenata* × *Castanea sativa* hybrid trees.

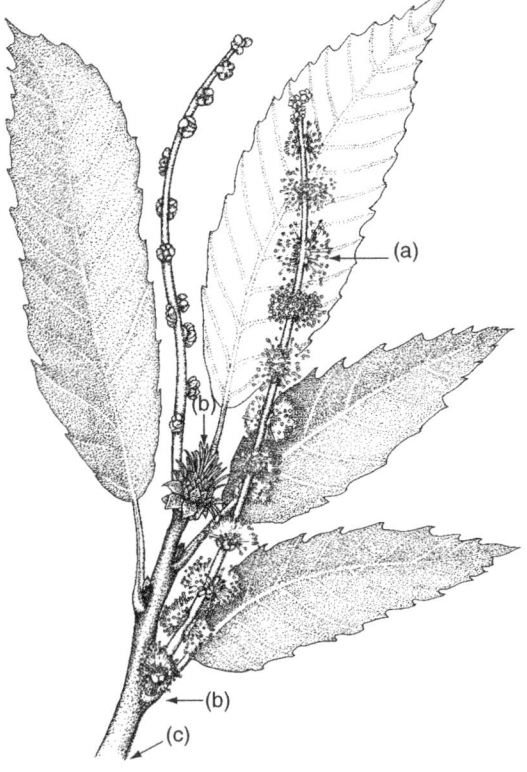

Fig. 21.4. Chestnuts: spring growth, which has originated from a lateral bud on last season's shoot. Some lateral buds on this shoot produce male catkins (a) which have female flowers at the base (b). Catkins lower down the shoot (c) have male flowers only.

Chestnut blight

Trees of *Castanea dentata* and *C. sativa* can be killed by this disease; *C. crenata* is moderately susceptible and *C. mollissima* is resistant. Chestnut blight (*Endothia parasitica* or *Cryphonectria parasitica*) has caused widespread destruction of chestnuts in the USA (where it eliminated one of the most common trees in US forests). European and, to some extent, Japanese production, has fallen but the decline in Europe has been halted through the use of numerous resistant lines and improved management methods. New hypervirulent strains have also been introduced that prevent destructive infections spreading in *C. sativa*. These hypervirulent fungal strains cause such a strong protective response from the tree that they do not themselves kill the tree, while other, more deadly, strains cannot infect the tree in the presence of the hypervirulent strain. In the USA, the American Chestnut Foundation is attempting to reforest the country by breeding and distributing American chestnuts containing one-sixteenth Chinese chestnut carrying resistance to blight. In 2007 the first 500 of these nuts became available for planting back into American forests. The USA also has a small industry based on Asian and European chestnuts.

The other main disease problem with chestnuts comes from internal rotting of nuts in storage. A variety of pathogens cause problems, including *Phomopsis*, *Penicillium*, *Fusarium* and *Botrytis*. All are reduced by low-temperature storage.

Cultivars

Chestnuts are an excellent timber as well as nut tree. Thus there are still many seedling plantings in northern China, as well as top-worked 'natural' forests in Europe. Varieties tend to be very localized in their use and often produce well in a limited area only. For example, none of the elite Asian lines imported into Australia or New Zealand, where chestnut blight has not been imported thus far, have done better than local selections. Chestnuts exhibit 'xenia', i.e. the male-pollen parent strongly affects nut quality and thus good combinations of trees need to be selected. Breeding efforts worldwide are including blight resistance into new releases as well as high-yielding, sweet, easily peeled 'Marron' nuts.

D. McNeil *et al.*

Harvesting and processing

Mature nuts abscise with or without their prickly receptacle (burr). While more difficult to hand harvest, nuts in the burr are more easily machine harvested and have a reduced rate of quality deterioration on the ground prior to pick up. *C. mollissima* tends to be more easily peeled than other species. Flame, steam and hand peeling are used worldwide to produce nuts with the skin and the, generally bitter, pellicle removed. The ability to easily peel the nuts substantially enhances their marketability.

Hazelnuts or Filberts

There are about 15 species of *Corylus* scattered through the temperate parts of the northern hemisphere (Table 21.3). Most commercial strains are probably hybrids of the European cobnut and the filbert (*Corylus avellana* and *Corylus maxima*).

Of the world's crop (816,000 t in 2007), approximately 70% is grown in Turkey. Production has grown slowly over the last 12 years, from 310,000 ha to 434,000 ha, although yields were relatively low, between 0.9 and 1.8 t/ha average over that period. Italy (15% of the world's supply) comes next, with Oregon, in the USA, producing about 5% of the world's supply. The Oregon growers use much more advanced production methods and achieve average regional yields of 1.2–3.8 t/ha. Production is declining in Spain (3% of world's supply) due to low yields (0.6–1.2 t/ha) and poor returns. Other producers include Iran, Georgia, China and Azerbaijan. Biennial bearing is a major problem, with annual yields in the major producing areas having varied by a factor of three between seasons.

Hazelnuts are rich in monounsaturated fats (about 80% of oil) and low in saturated fats (5%) and polyunsaturated fats (15%). They are high in vitamin E (56 mg/100 g in the New Zealand cultivar Whiteheart), protein (16%) and fibre (12%).

Key points

The hazelnut is not a difficult crop to grow but needs adequate inputs to achieve high yields. The Oregon Department of Agriculture and Oregon State University have developed an integrated pest management strategy for hazelnuts to reduce external inputs. Growers in several regions of the world are successfully growing the crop organically. Growers are advised to train the plant as a tree, rather than allow multiple stems to develop. Once established as a vase-shaped tree, little further pruning is required, except to prevent the trees becoming too dense for sprays or for light to penetrate the canopy. One tree in nine or one row in four should be of the pollinizer. Such rows should be at right angles to the prevailing wind in mid- to late winter.

Eastern filbert blight (EFB)

This is a devastating native American disease of hazelnuts (*Anisogramma anomola*) that has recently spread to Oregon from the east and has completely altered breeding practices and crop management. Daviana, a pollinizer for the most common cultivar Barcelona and its descendants (e.g. Ennis), is particularly sensitive. Some cultivars show tolerance (e.g. Barcelona) and these include some new releases (Willamette, Lewis, Clark), while others show very high tolerance (e.g. Tondo di Giffoni). Others in the breeding programme have complete resistance (Jefferson and Yamhill) due to the inclusion of the Gasaway (named after its source cultivator) resistance gene. So far this disease has still (2010) not been observed outside of North America.

Cultivars

Turkish cultivars are little used elsewhere, due to the adherence of the husk and suitability only for hand harvest. Some European cultivars for the small processing nut trade are widely grown (Negret, Tondo di Giffoni, Casina), as are some American cultivars for the large unshelled trade (Barcelona, Ennis). Other widely grown lines are the US (Butler) and European (Hall's Giant) pollinizers, but there are numerous local cultivars and breeding programmes selecting for yield, disease resistance, size, crackout and blanching ability. US cultivars are being replaced by EFB-resistant varieties.

Macadamias

The macadamia, or Queensland nut, originates from an area stretching along the eastern coastal

Table 21.3. Hazelnuts or filberts.

Botanical, anatomical and physiological aspects	Climatic, geographic, soil and water requirements	Cultural aspects
Common name Hazelnuts, cobnuts, filberts, Barcelona nuts	*Temperature needs* Grows in a wide range of temperate climates	*Propagation* Can be grown in stoolbeds or layered. Cuttings are not always easy to grow but root better if taken from suckers. For rapid propagation, grafting on to seedling or clonal, non-suckering stock is used (see below). Better results are obtained if graft is kept warm (21°C). Micro-propagation becoming common in USA
Botanical name *Corylus avellana* (European cobnut), *Corylus maxima* (European filbert). Many cultivars are hybrids between these and other species	*Frost tolerance* Good. Flowers will tolerate –5°C	
	Water needs Hardy but requires plentiful water supply for good yields	*Rootstocks* Generally own rooted but some use of seedlings of *C. avellana* or *C. maxima* or non-suckering 'Dundee' or 'Newburg'. May be compatibility problems if *C. colurna* seedlings are used
Botanical name of related and useful species *Corylus americana* (American filbert) *Corylus colurna* (Turkish filbert)	*Tolerance of wet soils* Moderate to good	
	Drought tolerance Poor for growth but good survival	*Spacing* Traditionally grown at 4.5 × 4.5 m. Close planting (3.6 × 2.4 m) followed by later tree removal is now used as it gives earlier returns
Type of plant and size Deciduous tree, 4–5 m tall, 4–5 m diameter; all except *C. colurna*, which is much taller, have a tendency to sucker	*Humidity tolerance* Good	*Training and pruning* Can be allowed to sucker and grow as a large bush, but it is better to restrict suckering and grow on a single trunk as an open-centre, vase-shaped tree pruned moderately to allow reasonable light penetration, air circulation and encourage growth of new flowering wood
	Wind tolerance Sensitive to wind; grows and crops better under sheltered conditions	
Sexuality Monoecious and dichogamous; the male flowers usually opening first	*Edaphic features* Use flat land or gentle slopes	*Thinning* Not required
Pollination Wind pollinated. Male catkins shed pollen early winter, female flowers are open from midwinter to spring. Females remain receptive for several weeks but male pollen has a short life. Pollinizer cultivars must be selected to overlap with female flowers of desired cultivars. Regionally, flower timing across varieties may differ	*Soil needs* The shallow rooting habit means it will grow satisfactorily in shallow soils. Also tolerant of clay, which it prefers to sandy soils	*Tillage* Use herbicides under tree and grass sward between rows. Total weed suppression with herbicides can also be considered, especially where mechanical sweeping is contemplated. Flailing to remove debris is used before and after harvest
		Time to first harvest 4–5 years, longer with seedlings
		Time to full production 10–12 years, longer with seedlings
		Expected yields (in shells) 5 years: 2 kg/tree 8 years: 4 kg/tree 3.5 t/ha is achievable

Flower buds
Buds with male and female flowers are separate, simple and borne laterally on previous season's growth

Growth and type of fruit
After pollination, late winter fruit development is very slow. More rapid development occurs in mid- to late spring. *C. avellana* has roundish nuts, not completely enveloped by husk; nuts may drop free at maturity. *C. maxima* has long nuts enveloped by the husk, which adheres to the nut

Time of bud burst
Male and female flower buds open early (see above). Leafing occurs in early spring

Time of fruit maturity
Early to mid-autumn

Nutrient requirements
Needs fertile soils with pH approximately 6. Regular applications of N, P, K fertilizers are required. Boron deficiency is often a problem on lower fertility soils. Regular (every 2–3 years) leaf analysis for boron is often recommended. Fertilizer recommendations made in Chapter 9 for deciduous fruit could be used as a guide

Normal productive life
40 or more years

Method of harvest
Pick by hand from tree (Turkey) or the ground, or sweep (USA) or vacuum up nuts (Italy)

Storage
Separate from husks and store in cool, dry shed with good air circulation

Main pests and diseases
Big bud mite, which causes swelling and dropping of buds in spring, seems to be the main problem. Bacterial blight, *Xanthomonas arboricola*, is a major problem in Australia. Whitefly and aphids can also occasionally cause problems. In Europe hazelnut weevil can be a problem. Filbert worm is a serious pest in Oregon, as is eastern filbert blight

Fig. 21.5. Hazelnuts showing large cultivar (Ennis) and smaller processing cultivar (Whiteheart), both in shell and after cracking.

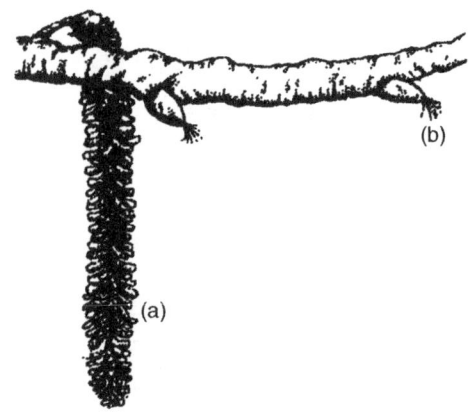

Fig. 21.7. Hazelnut. Buds open in winter or early spring to produce catkins (a) or female flowers (b).

zone of Australia (latitude 25–32 °S). The tree is a member of the *Proteaceae* family and related to the ornamental *Protea*, *Grevillea* and *Banksia* plants (Table 21.4). There are two important nut-bearing species, *Macadamia integrifolia* and *Macadamia*

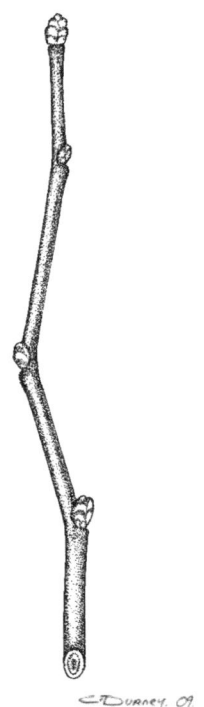

Fig. 21.6. Hazelnut winter shoot with apical and lateral buds.

tetraphylla; the latter has a more southerly distribution and is found in New South Wales, indicating greater cold tolerance. There has been some hybridization between the two species. Other macadamia species exist but have not been exploited commercially.

World macadamia production is estimated at 26,123 t of macadamia kernels for 2009, and world production is slowly increasing. Hawaii was where the commercialization of macadamia production first began, with the first seeds of *M. integrifolia* planted in Hawaii in 1878. Production in Hawaii dates from the 1920s, and large-scale production took place from 1950 onwards. However, Hawaiian production is now projected to decrease as existing trees age and the expectation of few new plantings. Australia is the largest world producer of macadamia and has about 16,000 ha planted, with most of the production area located in New South Wales and Queensland, and South Africa is also a significant producer, with about 12,000 ha. South Africa, Hawaii and Australia account for roughly 75–80% of the world total production. The South African industry is still increasing relative to Australia and may become the largest world producer in the near future. Other countries producing macadamia include Kenya, Malawi and Guatemala.

Clonal selections of *M. tetraphylla* are under test in the USA, Australia, South Africa and New Zealand in areas where *M. integrifolia* is considered too tender. It is considered important that the nuts of *M. tetraphylla* match the quality of the *M.*

D. McNeil *et al.*

Fig. 21.8. Hazelnut fruit with husks. Barcelona type on left, Turkish type on right.

integrifolia cultivars that were first selected for commercial production in Hawaii. This means having an oil content of 72% or more and good shape and size (preferably 6–8 g nut weight).

Key points

Cultivar selection

Cultivars were originally developed in Hawaii for Hawaiian conditions. However, many are now selected for local conditions elsewhere, particularly for local temperature regimes. *M. integrifolia* is used extensively in Hawaii and northern Australia; however, there is interest in *M. tetraphylla* and hybrids between *M. tetraphylla* and *M. integrifola* in cooler areas (Table 21.5). Recently a series of licensed cultivars have been released in Australia. Trials in south Queensland suggest that the cultivars Daddow and Heilscher may have potential.

Propagation

Macadamias are commonly grafted on to seedlings of *M. tetraphylla*. Fresh seeds germinate readily; the nut should be de-husked but not cracked. Seed viability decreases from 100% at harvest to 50% or less 6 months later. This nut should be placed in a moist, light, germinating medium, e.g. vermiculite. Macadamia roots are sensitive to transplanting stress; thus care must be taken when removing seedlings from the germination bed and again when planting in the field.

Grafting on to the seedling rootstock is done by a cleft or similar graft. The grafts appear to be very susceptible to dehydration and excessive reduction of carbohydrate reserves. Slow cambium healing and poor cross-diffusion between stock and scion is the likely cause. Various techniques are employed to overcome dehydration and carbohydrate draining. For example, to enhance carbohydrate reserves, scion wood can be taken from branches that were selected and ring-barked 2 or 3 weeks earlier.

A technique called 'seed grafting' has been used successfully. This involves cutting off at its base a seedling shoot that has reached 45–60 cm, leaving its base and cotyledons exposed. A scion piece (approximately 5 mm in diameter) is cut to form a wedge-shaped end, which is then inserted into a cut in the shoot base of the germinated seed, and both parts together are buried in a suitable medium until the graft has taken. Cuttings of terminal twigs, three or four nodes long, can be rooted. Cuttings are not usually favoured, however, as they tend to have shallow root systems. In California top-working and field grafting have been used. Graft incompatibility has been observed in California across species if *M. integrifolia* rootstocks are used with *M. tetraphylla* scions; hence the common use of *M. tetraphylla* as preferred rootstocks.

Table 21.4. Macadamias.

Botanical, anatomical and physiological aspects	Climatic, geographic, soil and water requirements	Cultural aspects
Common name Macadamia, Queensland nut	*Temperature needs* A subtropical plant, *M. tetraphylla* originated in northern NSW (Australia) and is more cold tolerant than *M. integrifolia*. High temperatures (>35°C) at flowering reduce flower set and temperatures >35°C during nut filling reduce yields	*Propagation* Seedling rootstocks are produced from selected nuts and parents. Nuts may be germinated in a warm, moist, well-drained medium. Leafy cuttings and marcottage techniques have been reported. Grafting and budding are difficult. Approach grafting is dependable but time consuming
Botanical name Macadamia integrifolia Macadamia tetraphylla		
Botanical name of related and useful species *Gevuina avellana*, which produces an edible nut. Family *Proteaceae* is also notable for many ornamental plants such as *Leucadendron, Hakea, Grevillea, Banksia, Leucospermum* and *Protea*	*Frost tolerance* Older trees will tolerate –6°C in winter but young trees, or old trees not properly hardened by the cool weather of autumn, will be damaged at –3°C	*Rootstocks* Seedlings of either species can be used. Seedlings of *M. tetraphylla* preferred. See Key points
	Water needs Well-distributed rainfall over 1000 mm required; native habitat receives 1500–2250 mm, mostly in summer	*Spacing* Double planted and thinned out 10–15 years after establishment; initial spacing: 4.5 m between rows and 2.25 m between trees
Type of plant and size Evergreen tree, 10–15 m high, tending to thick, upright growth. Branching with narrow angles and consequently splitting is common; three or four leaves are present at a node, in a whorl; each leaf has serial axillary buds. Flower spike 10–20 cm long; 75–400 flowers per spike. Fruit has fleshy husk, very hard brown shell. Seed 1.5–2.5 cm diameter, with high oil content required for processing (over 72%)	*Tolerance of wet soils* High rainfall is tolerated provided drainage is adequate	*Training and pruning* Prune to form a strong, self-supporting branch framework. Early pruning aims to form a single-leader tree with near-horizontal side branches. Lateral branches tend to occur in whorls. For further details see Key points
		Thinning Not used
	Drought tolerance Drought will cause small nuts, poor kernel development and reduced yield. However, tree survival may be fair to good	*Tillage* As nuts are often harvested after falling, clean cultivation and rolling, herbicides or close-cut mowing are desirable. Rows may be kept clean with herbicides
	Humidity tolerance Indications are that it tolerates a wide range of humidity and high rainfall conditions	*Time to first harvest* 5–12 years
Sexuality Flowers hermaphrodite	*Wind tolerance* Weak anchorage and poor branch angles make macadamias very susceptible to damage and breaking with wind. Shelter is essential in windy areas	*Time to full production* 10–15 years
Pollination Allow for cross-pollination. Self-pollination occurs but evidence of some partial self-incompatibility. Bees may be introduced into orchards		*Expected yields (in shell)* 10–12 years: 30–70 kg/tree
		Normal productive life 25–35 years

Growth of fruit
In the early phase of nut development, the endosperm is absorbed by the embryo and cotyledons until shell hardening begins about 3 months after flowering. Thereafter, the oil content rises until maturity

Time of bud burst
A number of growth flushes may be expected, particularly a spring flush

Time of flowering
Mostly in midsummer

Time of fruit maturity
Main crop is borne in winter

Edaphic features
Warm, sheltered conditions and/or sun-facing slopes with good air drainage needed in marginal areas

Soil needs
Macadamias grow on a wide range of soils, but a well-drained soil of pH 6–7 and 1 m depth is ideal. Do not tolerate saline soils

Nutrient requirements
Three to four balanced fertilizer applications are needed in one season, amounting to 5 kg per annum for a 10-year-old tree. Macadamias appear to be sensitive to a number of trace element deficiencies. Excess nitrogen may reduce yields and give excess, weak vegetative growth or chlorosis

Method of harvest
See under Key points

Storage
Air-dried nuts may be stored in the shell for several months in a dry, ventilated place

Main pests and diseases
Leafroller caterpillar, mealybug and thrips. Rats may damage fallen nuts. Phytophthora (rarely), anthracnose and botrytis sometimes cause problems

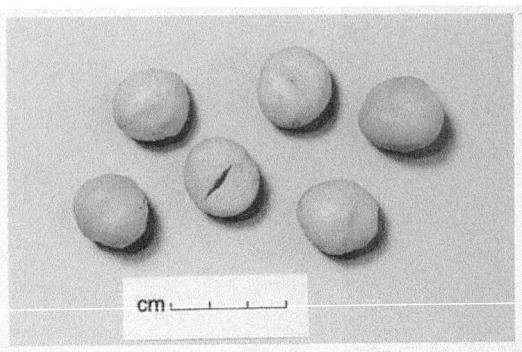

Fig. 21.9. Macadamia kernels.

Branching and primary and secondary buds

The aim in pruning and training the macadamia is to produce a strong structure with a main stem and wide-spreading branches. However, the growth habit of the plant favours narrow-angled, upright branches.

The plant has serial buds in the axil of each leaf. The primary and secondary buds are usually visible. The primary bud is bigger and borne above the secondary. When the primary bud commences growth, it tends to grow upright at a narrow angle from the main stem. Secondary buds will grow at a wider angle, especially if the primary bud or primary shoot still exists.

The common method favours the production of a central-leader tree with wide-angled side branches originating from secondary buds, beginning at a height of about 1 m from the ground. To produce a central leader, cut back the young tree to 1 m. This will induce shoots to form from the whorl of buds; when the resulting regrowth has hardened, the shoots arising from the primary buds can be reduced to one. If the secondary buds have not been stimulated to break, head back the unwanted primary shoots to a 1 cm stub to try to force the secondary buds to grow out more horizontally. This process of heading back the main leader is repeated at each level, usually at 40–50 cm intervals, where a tier of horizontal branches is required.

Harvesting and processing

The nuts are picked up from under the tree. Most authorities recommend that this occurs within 2 or 3 days after nut fall, otherwise oil content and flavour are reduced. Machines to sweep the orchard floor and nets to catch the nuts have been used. Finger reel pick-up machines are becoming common in Australia for smaller orchards.

The nuts should be de-husked and then cracked. The kernel should have a minimum weight of 3 g at 3% moisture and be 35–45% of the total nut weight. Kernels may be graded by specific gravity, which indicates oil content. Grade 1 nuts have kernels that float in water (i.e. a specific gravity of 1 or less), indicating an oil content of 72% or greater. Grade 2 kernels have a specific gravity of 1–1.025 and sink; they have approximately 66–72% oil. The kernels are then processed by drying and roasting.

Table 21.5. Properties of macadamia cultivars.

Cultivar	Species	Origin	% kernel	Nut size	Other
Keaau	I	Hawaii	45	m	V
Keauhou	I	Hawaii	39	m to l	S, V, AR
Waimanalo	I	Hawaii	38	l	(M)H, AR, P
Ikaika	I	Hawaii	39	m	V
Cate	T	California	40	m to l	P, NB, (M)H, L
Beaumont	I×T	Australia	40	m to l	L
Elimbah	T	Australia	47	m	L
Vista	I×T	California	46	s to m	P, M

Species: I, *Macadamia integrifolia*; T, *Macadamia tetraphylla*.

Nut size: s, m, l, small, medium or large.

Other: S, L, short or long harvest season; P, precocious; (M)H, (moderately) hardy; NB, not biennial; M, V, medium or vigorous tree growth; AR, anthracnose resistant.

D. McNeil *et al.*

Pecans

Pecans and hickories consist of about 20 species belonging to the genus *Carya*, which is native to eastern North America. The edible nuts are often collected in the wild, but it is only the pecan which has been extensively bred and cultivated (Table 21.6). Pecans are larger and more suited to commerce. Hickories are hardier and will survive in cooler climates, and some have particularly fine flavours. Some crosses between pecan and hickories, called 'hickans', have been produced in an attempt (not yet very successful) to gain the better of the two groups.

The USA, and particularly the southern states of Georgia, New Mexico, Texas, Arizona and Oklahoma, produces most of the world's pecans – about 75% of world production. USA yields tend to vary year to year, with a biennial pattern. Nearly 135,000 t (an 'on' year) was produced in 2009. Mexico produces about 20% of the world's pecans and Australia produces about 2% of the world's supply. Other countries with pecan production include South Africa, Peru and Israel. In the past, much of the supply of pecans came from seedling plantations or natural stands. This has declined and most of the production is now coming from improved, grafted material.

Key points

The native habitat of pecans is the more southerly eastern states of the USA. Consequently, they would be expected to do best in the warmer temperate areas of the world. A large plantation has successfully been established in the north of NSW in Australia, a warm, semi-arid region with available irrigation.

Pecans are much easier to propagate than walnuts, but, because they have long tap roots, nurseries find that using a long, thin pot or bag makes transplanting easier and more successful. The seeds have a vernalization (chilling) requirement before germination.

There is a wide range of varieties available in the USA, which enables growers to spread out flowering times for pollination, obtain pest and disease resistance, vary leafing out and harvest dates for different regions, avoid freezing injury, and maximize regional productivity and quality. A new release by the USDA, Kanza, is resistant to scab and leaf scorch. Cheyenne is a preferred variety, due to a high kernel quality, and Mohawk produces very large nuts. Wichita is a high producer but susceptible to freezing and scab injury.

Walnuts

Production of walnuts is widespread in the northern hemisphere. China is the largest producer, increasing production from 246,000 t in 1998 to 503,000 t in 2008 (nearly 49% increase). Another major producer is the USA, with 290,000 t in 2008, with California producing about 98% of the USA crop. Although total USA production is lower than China, a significant proportion of the USA crop is exported. Turkey, Iran and the Ukraine are also large producers of walnuts. Chile (23,000 t in 2008) and Argentina (10,000 t in 2008) are the largest producers in the southern hemisphere.

There has been considerable interest in the health benefits of walnuts, as their consumption in moderate amounts has been shown to reduce low-density lipoprotein (LDL) cholesterol levels by about 16% in men. This level of reduction has been shown to be sufficient to reduce the risk of stroke and heart attacks. It is believed the effect may be due to the kernel's combination of fibre (10%), good-quality protein (14%) and particularly the high level of polyunsaturated fats (70%), including both omega-3 and omega-6 fatty acids, which lower blood triglycerides and LDL cholesterol, respectively.

Key points

Cultivars

Most new cultivars are the highly productive lateral bearers, which produce more nuts on younger trees. In California, Franquette and Hartley are the most common producing trees, with Sunland and Chandler the most common in new plantings. Tulare is a recently released cultivar in the USA, with a slightly darker pellicle, higher yield and stronger walnut flavour than Chandler. Some earlier lines, e.g. Serr, were found to suffer from premature nut drop and have fallen out of favour, even though capable of producing high yields of quality nuts in some years. Other specialty cultivars exist that

Table 21.6. Pecans.

Botanical, anatomical and physiological aspects	Climatic, geographic, soil and water requirements	Cultural aspects
Common name Pecan	*Temperature needs* Reported to need between 1200 and 1700°C days (base 10°C) to mature. May also have a chilling requirement for good flower bud development, and seeds need chilling for germination	*Propagation* Grafted or budded trees using the whip-and-tongue graft, skin or patch budding. Long tap root of seedling causes problems in transplanting. Use long planter bags
Botanical name Carya illinoensis		*Rootstocks* Seedling pecans
Botanical name of related and useful species Carya ovata (shagbark hickory) Carya laciniosa (shellbark hickory) Carya tomentosa (mockernut hickory)	*Frost tolerance* Freezing injury has been severe in New Mexico, with damage from spring, autumn and winter freezes. It occurs mostly in young trees but also in old trees. Appears to be more sensitive than walnut, but less sensitive cultivars have been selected	*Spacing* Have been planted on square at 10–20 m, depending on fertility and climate. More recent plantings have been established at 4.5 × 9.0, 7.5 × 7.5 or 4.5 × 4.5 m, but trees may need thinning later
Type of plant and size Large, deciduous tree up to 50 m in height, 35 m in diameter. Long tap root makes transplanting difficult, though large trees can be successfully moved with care		*Training and pruning* Initial training as central leader, after which little pruning is needed
Sexuality Monoecious, dichogamous. Use several cultivars in small plantations to ensure overlapping of pollen shed and pistil receptivity. There may be enough overlap in large plantations of one cultivar	*Water needs* Although growing naturally in both arid and humid areas, their water requirements in the summer are reported to be quite high	*Thinning* Not practical. Natural thinning is achieved in high crop years in the southern USA by failing to control the case bearer moth
	Tolerance of wet soils Moderate but will develop chlorosis, root disease and dieback in waterlogged soils	*Tillage* Usually grown with clean cultivation or closely cut sward
Pollination Wind pollinated		*Time to first harvest* 4–5 years
	Humidity tolerance Rain at blossoming restricts pollination. High humidity causes disease and insect problems	*Time to full production* 15–20 years
Flower buds Male catkins develop in simple buds borne laterally; female flower buds are mixed and produced terminally – both on the previous season's growth. Flowers are positioned terminally on the shoot that grows from the terminal bud	*Wind tolerance* Poor to moderate, with leaf damage and limb breakage	*Expected yields* For good cultivars on good soils, planted at relatively high density: 5 years: 1 t/ha 10 years: 3.5 t/ha 15 years: 4.0 t/ha
	Edaphic features Flat land or gentle slopes preferred	*Normal productive life* 50–100 years or longer
		Method of harvest Hand harvest as nuts fall; remove hull and dry. Commercially, nuts are shaken off the tree and picked off the ground by mechanical sweepers. Ethephon may be used to assist hull dehiscence

D. McNeil *et al.*

Growth of fruit
Hull and nut grow with a sigmoid growth curve. Has a tendency to biennial bearing

Time of bud burst
About the same time as apples

Time of flowering
2–3 weeks after bud burst. Male catkins open before the female flower

Time of fruit maturity
In autumn, 180–220 days after anthesis

Soil needs
Deep alluvial soils considered to be the best

Nutrient requirements
pH 6–6.5 optimum. In Texas 120–200 kg N/ha recommended for mature orchards. Sensitive to zinc deficiencies (particularly in alkaline soils), which results in small crinkled leaves. Nitrogen required

Storage
Store the dry nuts in cool, dry, vermin-proof conditions

Main pests and diseases
Few serious pests and diseases. Rats and leaf-eating insects may be a problem. *Phytophthora* and *Phymatotrichum* may cause root rot. Scab is a major problem unless resistant cultivars are used. Hickory weevil and shuck-worm are the worst pests in the USA. Susceptible to phylloxera, thrips and aphids, crown gall and mildew

Fig. 21.10. Pecan kernels.

produce the following: very large nuts but will have less than 20% kernel (Wilson's Wonder); purple nuts (G1239); or darker, stronger-tasting nuts (Vina). Europe, the former USSR and other parts of Asia have all produced their own favourite cultivars.

Propagation

Grafting offers the best method but is not always successful. In California, English walnuts are grafted outdoors on to Paradox seedling rootstocks. Under Californian conditions, these seedlings will make 3–4 m of growth in one season. This is cut back to 45 cm in spring, left for 1–2 weeks and then recut and grafted with strong scion wood using the whip-and-tongue method. Unless conditions which promote similar growth can be provided, grafting in the field is difficult.

Paradox seedlings are hybrids between *J. hindsii* and *J. regia*. Seeds are selected from *J. hindsii* trees growing in an area with *J. regia* trees in the vicinity. The seeds produce two types of seedlings: the hybrid, called Paradox, which is more vigorous and has larger leaflets, and *J. hindsii*, which is weaker and has smaller leaflets. Both can be used but Paradox is preferred.

With weaker stocks and scions, bench grafting is the norm, and temperature following grafting becomes critical. The optimum is 26–28°C. Seeds may be grown in pots or bags. In the winter after one season's growth, they are cut, whip-and-tongue grafted and placed into a hot box or glasshouse at the correct temperature for 2–3 weeks.

When the callus has developed, the plants are removed from the warm temperature and placed in a cool glasshouse. They are planted out in spring.

The graft and the tip of the scion must be carefully sealed with grafting wax. A humid atmosphere must be maintained during callusing, for example by placing a bag or sphagnum moss over the graft.

Other ingenious techniques are being investigated, but the strength and health of scion and stock and/or the correct temperatures seem to be the important factors promoting success. Graft wood should be collected in early winter and stored outdoors in a cool position, buried in moist sand or placed in a cold store. Grafting has been done successfully from early winter to early spring.

Pests and diseases

Bacterial blight (*Xanthomonas juglandis*) is a problem, especially in wetter districts, and may limit commercial production to drier climates. Bordeaux (containing copper) mixture or various proprietary mixes containing copper (e.g. 'Kocide') can reduce infection and should be applied at the first signs of bud burst and 7–10 days thereafter in wet weather. In California, strains resistant to copper are emerging, and new materials are under test. Organic walnut production remains possible as copper is an acceptable organic application. Strains of walnut that may be resistant to anthracnose and blight have been imported to the USA from Eastern Europe and are being evaluated. Root and crown rots (*Phytopthora* spp., *Armillaria mellea*) and deep bark canker (*Erwinia rubrifaciens*) can also cause problems at times. Codling moth will sometimes infest nuts, but it is not normally serious enough to require spraying.

Blackline disease (cherry leafroll virus) can cause death of cells between the scion and the rootstock and often appears 10 or more years after planting trees in an orchard. This is caused by a virus which is not present in all countries. Overzealous importation of material from overseas is not to be encouraged. Blackline disease is more prevalent on trees with black walnut rootstocks, including Paradox. It does not develop when *J. regia* is used as rootstock.

Table 21.7. Walnuts.

Botanical, anatomical and physiological aspects	Climatic, geographic, soil and water requirements	Cultural aspects
Common name English or Persian walnut *Botanical name* *Juglans regia* *Botanical name of related and useful species* *Juglans nigra* (eastern American black walnut) *Juglans hindsii* (western American black walnut) *Juglans allantifolia* var. *cordiformis* (heartnut) *Juglans cinerea* (butternut) *Type of plant and size* Large, deciduous tree, 20–30 m tall, spreading, not especially dense *Sexuality* Monoecious and displaying dichogamy. The pollen of catkins is normally shed before the female flowers open, although in some the reverse occurs. To ensure good pollination, it is best to select cultivars where male and female flowers overlap *Pollination* Wind pollinated	*Temperature needs* Will grow in cool to warm temperate climates. Cooler climates need early-maturing cultivars. Very high temperatures (>38°C) can damage bark and hulls, and lead to poor-quality kernels *Frost tolerance* Late spring frosts, below −1°C, can damage flowers as well as new shoots and young trees *Water needs* At least 760 mm of rain required, and irrigation will benefit trees in the drier areas. Deficit irrigation (where a mild water stress is applied mid-season after flowering but before the plant has a strong carbohydrate demand for nut-fill) may improve flowering and nut production and reduce vegetative growth *Tolerance of wet soils* Wet weather at harvest causes shell and kernel stain. Trees can be damaged by a high water table and need drainage to at least 3 m. Susceptible to fungal root rots under prolonged exposure to wet soils	*Propagation* Normally grafted, occasionally patch budded. May be done outdoors in spring using dormant scion wood in warm climates. With poorer scion and rootstock material, bench grafting and temperatures of 26–28°C for 2–3 weeks need to be used. Recent work has shown rooting potential can be promoted using *Agrobacterium*, which induces tumours on branch cuttings, from which roots may then arise more easily *Rootstocks* Seedlings of *J. nigra, J. hindsii* or *J. regia* may be used. Hybrid between *J. regia* and *J. hindsii* is called Paradox (tolerant of lesion nematodes but more susceptible to crown gall) and is considered a superior rootstock. Blackline disease may be a problem with black walnut stock (see Key points) *Spacing* Formerly planted on the square at 12–18 m. Newer Californian cultivars are more compact and 9 m square is recommended. Plantations may be double or quadruple planted and subsequently thinned *Training and pruning* Initially pruned and trained as a central leader but subsequently little pruning required *Thinning* Not required *Tillage* In California, most are grown under clean cultivated conditions where mechanical harvesting is used. The ground must be level and either free of herbage or cut very closely *Time to first harvest* 4–5 years with good trees, 8–10 years with poor cultivars or seedlings *Time to full production* 10–20 years, depending on cultivars and density of planting

Continued

Table 21.7. Continued.

Botanical, anatomical and physiological aspects	Climatic, geographic, soil and water requirements	Cultural aspects
Flower buds Male catkins develop in simple buds, laterally, on the previous season's growth. Female flower buds are mixed and produced terminally and, in some new cultivars, laterally, also on the previous year's growth. Flowers are formed terminally on the spring shoot that originates from the bud	*Humidity tolerance* Dislikes cool, wet weather in spring. This predisposes the trees to bacterial blight, which can be a serious disease in wetter climates	*Expected yields* Based on good cultivars and good soils, initially planted at double spacing: 4 years: 1 t/ha 8 years: 4 t/ha 12 years: 5 t/ha
Growth of fruit Both the outer husk or hull and the nut grow with a sigmoid growth curve	*Wind tolerance* Will grow and crop much more slowly if exposed to persistent winds. Susceptible to limb breakage also. Thus, shelter essential to promote good growth during establishment years	*Normal productive life* 50–100 years *Method of harvest* Pick up by hand soon after nuts fall from tree. Remove husks as soon as possible. Mechanically, trees may be shaken and nuts swept up into appropriate containers
Time of bud burst Spring. This is one of the latest deciduous trees to come into leaf	*Edaphic features* Flat land preferred *Soil needs* In California it is considered that walnuts are only economic if grown on deep, fertile soils	*Storage* When husks are removed and fruit is dry, keep in dry, airy, vermin-proof shed in boxes or bins
Time of flowering Pollen is normally released 10–12 days after bud burst; full bloom of female flowers is usually 15–18 days later, although this varies between cultivars	*Nutrient requirements* Considered to need moderate amounts of nitrogen and potassium and small amounts of phosphorus. Walnuts respond to calcium, and liming is beneficial to maintain pH above 5.8. Suggest the fertilizer regime given for deciduous fruits in Chapter 9 be followed and regular leaf analysis be used to monitor nutrient levels	*Main pests and diseases* Bacterial blight, *Xanthomonas juglandis*, especially in moist conditions. Other less severe problems are codling moth, navel orange worm, aphids, scale and husk fly. Rats are a problem with nuts on the ground or in store
Time of nut maturity Autumn		

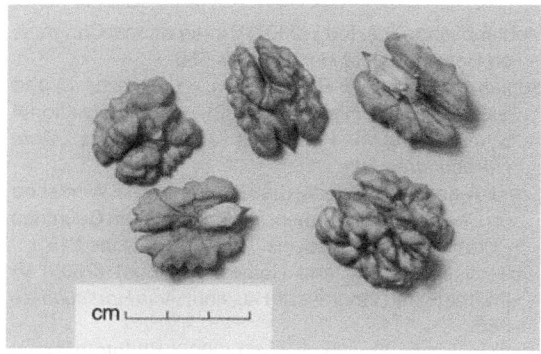

Fig. 21.11. Walnut kernels.

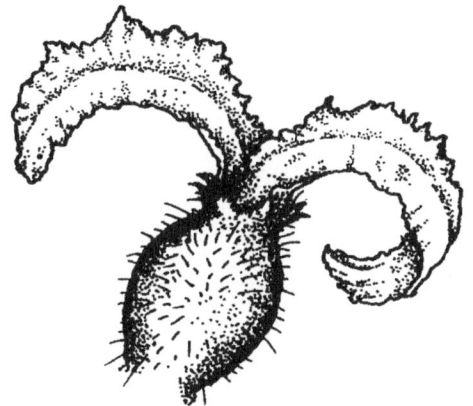

Fig. 21.12. Female flower.

Fig. 21.13. Male catkins.

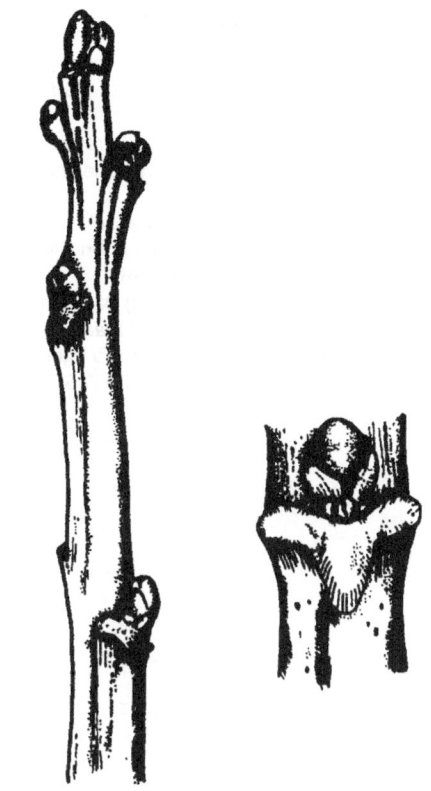

Fig. 21.14. Dormant shoot with buds.

Fig. 21.15. New growth of walnut shoot, showing female flower and buds for later growth.

Further Reading

Abreu, C.G., Rosa, E. and Monteiro, A.A. (eds) (2005) III International Chestnut Congress. *Acta Horticulturae* 693.

Abreu, C.G., Peixoto, F.P. and Gomes-Laranjo, J. (eds) (2008) II Iberian Congress on Chestnut. *Acta Horticulturae* 784.

Javanshah, A., Facelli, E. and Wirthensohn, M. (eds) (2006) IV International Symposium on Pistachios and Almonds. *Acta Horticulturae* 726.

Lin, Q. and Hong-Wen, H. (eds) (2009) IV International Chestnut Symposium. *Acta Horticulturae* 844.

Malvolti, M.E. and Avanzato, D. (eds) (2005) V International Walnut Symposium. *Acta Horticulturae* 705.

Mehlenbacher, S.A. (ed.) (2001) V International Congress on Hazelnut. *Acta Horticulturae* 556.

Socias i Company, R., Batlle, I., Hormaza, I. and Espiau, M.T. (eds) (2002) III International Symposium on Pistachios and Almonds. *Acta Horticulturae* 591.

Soylu, A. and Mert, C. (eds) (2009) International Workshop on Chestnut Management in Mediterranean Countries - Problems and Prospects. *Acta Horticulturae* 815.

Tous, J., Rovira, M. and Romero, A. (eds) (2005) VI International Congress on Hazelnut. *Acta Horticulturae* 686.

Varvaro, L. and Franco, S. (eds) (2009) VII International Congress on Hazelnut. *Acta Horticulturae* 845.

D. McNeil *et al.*

Index

Page numbers in **bold** type refer to figures and tables.